# Graduate Texts in Mathematics 91

# Graduate Texts in Mathematics

*(continued on back page)*

Alan F. Beardon

# The Geometry
# of Discrete Groups

With 93 Illustrations

Springer-Verlag
New York Heidelberg Berlin

Alan F. Beardon
University of Cambridge
Department of Pure Mathematics
  and Mathematical Statistics
16 Mill Lane
Cambridge CB2 1SB
England

AMS Subject Classifications (1980): 30-01, 30 CXX, 20F32, 30 FXX, 51 M10, 20 HXX

Library of Congress Cataloging in Publication Data
Beardon, Alan F.
   The geometry of discrete groups.

   (Graduate texts in mathematics; 91)
   Includes bibliographical references and index.
   1. Discrete groups.   2. Isometries (Mathematics)
3. Möbius transformations.   4. Geometry, Hyperbolic.
I. Title.   II. Series.
QA171.B364   1983      512′.2      82-19268

Typeset by Composition House Ltd., Salisbury, England.
Printed and bound by R. R. Donnelley & Sons, Harrisonburg, VA.
Printed in the United States of America.

9 8 7 6 5 4 3 2 1

ISBN 0-387-90788-2 Springer-Verlag New York Heidelberg Berlin
ISBN 3-540-90788-2 Springer-Verlag Berlin Heidelberg New York

*To Toni*

# Preface

This text is intended to serve as an introduction to the geometry of the action of discrete groups of Möbius transformations. The subject matter has now been studied with changing points of emphasis for over a hundred years, the most recent developments being connected with the theory of 3-manifolds: see, for example, the papers of Poincaré [77] and Thurston [101]. About 1940, the now well-known (but virtually unobtainable) Fenchel–Nielsen manuscript appeared. Sadly, the manuscript never appeared in print, and this more modest text attempts to display at least some of the beautiful geometrical ideas to be found in that manuscript, as well as some more recent material.

The text has been written with the conviction that geometrical explanations are essential for a full understanding of the material and that however simple a matrix proof might seem, a geometric proof is almost certainly more profitable. Further, wherever possible, results should be stated in a form that is invariant under conjugation, thus making the intrinsic nature of the result more apparent. Despite the fact that the subject matter is concerned with groups of isometries of hyperbolic geometry, many publications rely on Euclidean estimates and geometry. However, the recent developments have again emphasized the need for hyperbolic geometry, and I have included a comprehensive chapter on analytical (not axiomatic) hyperbolic geometry. It is hoped that this chapter will serve as a "dictionary" of formulae in plane hyperbolic geometry and as such will be of interest and use in its own right. Because of this, the format is different from the other chapters: here, there is a larger number of shorter sections, each devoted to a particular result or theme.

The text is intended to be of an introductory nature, and I make no apologies for giving detailed (and sometimes elementary) proofs. Indeed, many geometric errors occur in the literature and this is perhaps due, to some extent, to an omission of the details. I have kept the prerequisites to a minimum and, where it seems worthwhile, I have considered the same topic from different points of view. In part, this is in recognition of the fact that readers do not always read the pages sequentially. The list of references is not comprehensive and I have not always given the original source of a result. For ease of reference, Theorems, Definitions, etc., are numbered collectively in each section (2.4.1, 2.4.2, ...).

I owe much to many colleagues and friends with whom I have discussed the subject matter over the years. Special mention should be made, however, of P. J. Nicholls and P. Waterman who read an earlier version of the manuscript, Professor F. W. Gehring who encouraged me to write the text and conducted a series of seminars on parts of the manuscript, and the notes and lectures of L. V. Ahlfors. The errors that remain are mine.

*Cambridge, 1982*                                         ALAN F. BEARDON

# Contents

# CHAPTER 1
# Preliminary Material

## §1.1. Notation

We use the following notation. First, $\mathbb{Z}$, $\mathbb{Q}$, $\mathbb{R}$ and $\mathbb{C}$ denote the integers, the rationals, the real and complex numbers respectively: $\mathbb{H}$ denotes the set of quaternions (Section 2.4).

As usual, $\mathbb{R}^n$ denotes Euclidean $n$-space, a typical point in this being $x = (x_1, \ldots, x_n)$ with

$$|x| = (x_1^2 + \cdots + x_n^2)^{1/2}.$$

Note that if $y > 0$, then $y^{1/2}$ denotes the positive square root of $y$. The standard basis of $\mathbb{R}^n$ is $e_1, \ldots, e_n$ where, for example, $e_1 = (1, 0, \ldots, 0)$. Certain subsets of $\mathbb{R}^n$ warrant special mention, namely

$$B^n = \{x \in \mathbb{R}^n : |x| < 1\},$$
$$H^n = \{x \in \mathbb{R}^n : x_n > 0\},$$

and

$$S^{n-1} = \{x \in \mathbb{R}^n : |x| = 1\}.$$

In the case of $\mathbb{C}$ (identified with $\mathbb{R}^2$) we shall use $\Delta$ and $\partial\Delta$ for the unit disc and unit circle respectively.

The notation $x \mapsto x^2$ (for example) denotes the function mapping $x$ to $x^2$: the domain will be clear from the context. Functions (maps or transformations) act on the *left*: for brevity, the image $f(x)$ is often written as $fx$ (omitting brackets). The composition of functions is written as $fg$: this is the map $x \mapsto f(g(x))$.

Two sets $A$ and $B$ *meet* (or $A$ meets $B$) if $A \cap B \neq \emptyset$. Finally, a property $P(n)$ holds *for almost all n* (or all sufficiently large $n$) if it fails to hold for only a finite set of $n$.

## §1.2. Inequalities

All the inequalities that we need are derivable from Jensen's inequality: for a proof of this, see [90], Chapter 3.

**Jensen's Inequality.** *Let $\mu$ be a positive measure on a set $X$ with $\mu(X) = 1$, let $f : X \to (a, b)$ be $\mu$-integrable and let $\phi : (a, b) \to \mathbb{R}$ be any convex function. Then*

$$\phi\left(\int_X f \, d\mu\right) \leq \int_X (\phi f) \, d\mu. \tag{1.2.1}$$

Jensen's inequality includes Hölder's inequality

$$\int_X fg \, d\mu \leq \left(\int_X f^2 \, d\mu\right)^{1/2} \left(\int_X g^2 \, d\mu\right)^{1/2}$$

as a special case: the discrete form of this is the Cauchy–Schwarz inequality

$$\left|\sum a_i b_i\right| \leq \left(\sum |a_i|^2\right)^{1/2} \left(\sum |b_i|^2\right)^{1/2}$$

for real $a_i$ and $b_i$. The complex case follows from the real case and this can, of course, be proved by elementary means.

Taking $X = \{x_1, \ldots, x_n\}$ and $\phi(x) = e^x$, we find that (1.2.1) yields the general Arithmetic–Geometric mean inequality

$$y_1^{\mu_1} \cdots y_n^{\mu_n} \leq \mu_1 y_1 + \cdots + \mu_n y_n,$$

where $\mu$ has mass $\mu_j$ at $x_j$ and $y_j = \phi f(x_j)$.

In order to apply (1.2.1) we need a supply of convex functions: a sufficient condition for $\phi$ to be convex is that $\phi^{(2)} \geq 0$ on $(a, b)$. Thus, for example, the functions cot, tan and $\cot^2$ are all convex on $(0, \pi/2)$. This shows, for instance, that if $\theta_1, \ldots, \theta_n$ are all in $(0, \pi/2)$ then

$$\cot\left(\frac{\theta_1 + \cdots + \theta_n}{n}\right) \leq \frac{\cot \theta_1 + \cdots + \cot \theta_n}{n}. \tag{1.2.2}$$

As another application, we prove that if $x$ and $y$ are in $(0, \pi/2)$ and $x + y < \pi/2$ then

$$\tan x \tan y \leq \tan^2\left(\frac{x + y}{2}\right). \tag{1.2.3}$$

Writing $w = (x + y)/2$, we have

$$\frac{\tan x + \tan y}{1 - \tan x \tan y} = \tan(x + y)$$

$$= \frac{2 \tan w}{1 - \tan^2 w}.$$

As tan is convex, (1.2.1) yields

$$\tan x + \tan y \geq 2 \tan w$$

and the desired inequality follows immediately (noting that $\tan^2 w < 1$ so both denominators are positive).

## §1.3. Algebra

We shall assume familiarity with the basic ideas concerning groups and (to a lesser extent) vector spaces. For example, we shall use elementary facts about the group $S_n$ of permutations of $\{1, 2, \ldots, n\}$: in particular, $S_n$ is generated by transpositions. As another example, we mention that *if $\theta: G \to H$ is a homomorphism of the group $G$ onto the group $H$, then the kernel $K$ of $\theta$ is a normal subgroup of $G$ and the quotient group $G/K$ is isomorphic to $H$.*

Let $g$ be an element in the group $G$. The elements *conjugate* to $g$ are the elements $hgh^{-1}$ in $G$ ($h \in G$) and the conjugacy classes $\{hgh^{-1} : h \in G\}$ partition $G$. In passing, we mention that the maps $x \mapsto xgx^{-1}$ and $x \mapsto gxg^{-1}$ (both of $G$ onto itself) play a special role in the later work. The *commutator* of $g$ and $h$ is

$$[g, h] = ghg^{-1}h^{-1}:$$

for our purposes this should be viewed as the composition of $g$ and a conjugate of $g^{-1}$.

Let $G$ be a group with subgroups $G_i$ ($i$ belonging to some indexing set). We assume that the union of the $G_i$ generate $G$ and that different $G_i$ have only the identity in common. Then $G$ is the *free product* of the $G_i$ if and only if each $g$ in $G$ has a unique expression as $g_1 \cdots g_n$ where no two consecutive $g_i$ belong to the same $G_j$. Examples of this will occur later in the text.

## §1.4. Topology

We shall assume a knowledge of topology sufficient, for example, to discuss Hausdorff spaces, connected spaces, compact spaces, product spaces and homeomorphisms. In particular, *if $f$ is a 1–1 continuous map of a compact* *space X onto a Hausdorff space Y, then f is a homeomorphism.* As special examples of topologies we mention the discrete topology (in which every subset is open) and the topology derived from a metric $\rho$ on a set $X$. An isometry $f$ of one metric space $(X, \rho)$ onto another, say $(Y, \sigma)$, satisfies

$$\sigma(fx, fy) = \rho(x, y)$$

and is necessarily a homeomorphism.

Briefly, we discuss the construction of the quotient topology induced by a given function. Let $X$ be any topological space, let $Y$ be any non-empty set and let $f: X \to Y$ be any function. A subset $V$ of $Y$ is *open* if and only if $f^{-1}(V)$ is an open subset of $X$: the class of open subsets of $Y$ is indeed a topology $\mathcal{T}_f$ on $Y$ and is called the *quotient topology induced by $f$*. With this topology, $f$ is automatically continuous. The following two results on the quotient topology are useful.

**Proposition 1.4.1.** *Let $X$ be a topological space and suppose that $f$ maps $X$ onto $Y$. Let $\mathcal{T}$ be any topology on $Y$ and let $\mathcal{T}_f$ be the quotient topology on $Y$ induced by $f$.*

(1) *If $f: X \to (Y, \mathcal{T})$ is continuous, then $\mathcal{T} \subset \mathcal{T}_f$.*
(2) *If $f: X \to (Y, \mathcal{T})$ is continuous and open, then $\mathcal{T} = \mathcal{T}_f$.*

PROOF. Suppose that $f: X \to (Y, \mathcal{T})$ is continuous. If $V$ is in $\mathcal{T}$, then $f^{-1}(V)$ is in open in $X$ and so $V$ is in $\mathcal{T}_f$. If, in addition, $f: X \to (Y, \mathcal{T})$ is an open map then $V$ in $\mathcal{T}_f$ implies that $f^{-1}(V)$ is open in $X$ and so $f(f^{-1}V)$ is in $\mathcal{T}$. As $f$ is surjective, $f(f^{-1}V) = V$ so $\mathcal{T}_f \subset \mathcal{T}$. $\qquad\square$

**Proposition 1.4.2.** *Suppose that $f$ maps $X$ into $Y$ where $X$ and $Y$ are topological spaces, $Y$ having the quotient topology $\mathcal{T}_f$. For each map $g: Y \to Z$ define $g_1: X \to Z$ by $g_1 = gf$. Then $g$ is continuous if and only if $g_1$ is continuous.*

PROOF. As $f$ is continuous, the continuity of $g$ implies that of $g_1$. Now suppose that $g_1$ is continuous. For an open subset $V$ of $Z$ (we assume, of course, that $Z$ is a topological space) we have

$$(g_1)^{-1}(V) = f^{-1}(g^{-1}V)$$

and this is open in $X$. By the definition of the quotient topology, $g^{-1}(V)$ is open in $Y$ so $g$ is continuous. $\qquad\square$

An alternative approach to the quotient topology is by equivalence relations. If $X$ carries an equivalence relation $R$ with equivalence classes $[x]$, then $X/R$ (the space of equivalence classes) inherits the quotient topology induced by the map $x \mapsto [x]$. Equally, any surjective function $f: X \to Y$ induces an equivalence relation $R$ on $X$ by $xRy$ if and only if $f(x) = f(y)$ and $Y$ can be identified with $X/R$. As an example, let $G$ be a group of homeomorphisms of a topological space $X$ onto itself and let $f$ map each $x$ in $X$

to its $G$-orbit $[x]$ in $X/G$. If $X/G$ is given the induced quotient topology, then $f: X \to X/G$ is continuous. In this case, $f$ is also an open map because if $V$ is open in $X$ then so is

$$f^{-1}(fV) = \bigcup_{g \in G} g(V).$$

Finally, the reader will benefit from an understanding of covering spaces and Riemann surfaces although most of the material in this book can be read independently of these ideas. Some of this is discussed briefly in Chapter 6: for further information, the reader is referred to (for example) [4], [6], [28], [50], [63] and [100].

## §1.5. Topological Groups

A *topological group* $G$ is both a group and a topological space, the two structures being related by the requirement that the maps $x \mapsto x^{-1}$ (of $G$ onto $G$) and $(x, y) \mapsto xy$ (of $G \times G$ onto $G$) are continuous: obviously, $G \times G$ is given the product topology. Two topological groups are *isomorphic* when there is a bijection of one onto the other which is both a group iso-morphism and a homeomorphism: this is the natural identification of topological groups.

For any $y$ in $G$, the space $G \times \{y\}$ has a natural topology with open sets $A \times \{y\}$ where $A$ is open in $G$. The map $x \mapsto (x, y)$ is a homeomorphism of $G$ onto $G \times \{y\}$ and the map $(x, y) \mapsto xy$ is a continuous map of $G \times \{y\}$ onto $G$. It follows that $x \mapsto xy$ is a continuous map of $G$ onto itself with continuous inverse $x \mapsto xy^{-1}$ and so we have the following elementary but useful result.

**Proposition 1.5.1.** *For each $y$ in $G$, the map $x \mapsto xy$ is a homeomorphism of $G$ onto itself: the same is true of the map $x \mapsto yx$.*

A topological group $G$ is *discrete* if the topology on $G$ is the discrete topology: thus we have the following Corollary of Proposition 1.5.1.

**Corollary 1.5.2.** *Let $G$ be a topological group such that for some $g$ in $G$, the set $\{g\}$ is open. Then each set $\{y\}$ ($y \in G$) is open and $G$ is discrete.*

Given a topological group $G$, define the maps

$$\phi(x) = xax^{-1}$$

and

$$\psi(x) = xax^{-1}a^{-1} = [x, a],$$

where $a$ is some element of $G$. We shall be interested in the iterates $\phi^n$ and $\psi^n$ of these maps and with this in mind, observe that $\phi$ has a unique fixed point, namely $a$. The iterates are related by the equation

$$\phi^n(x) = \psi^n(x)a,$$

because (by induction)

$$\begin{aligned}
\phi^{n+1}(x) &= [\psi^n(x)a]a[\psi^n(x)a]^{-1} \\
&= \psi^n(x)a[\psi^n(x)]^{-1} \\
&= \psi^{n+1}(x)a.
\end{aligned}$$

In certain circumstances, the iterated commutator $\psi^n(x)$ converges to the identity (equivalently, the iterates $\phi^n(x)$ converge to the unique fixed point $a$ of $\phi$) and if the group in question is discrete, then we must have $\phi^n(x) = a$ for some $n$. For examples of this, see [106], [111: Lemma 3.2.5] and Chapter 5 of this text.

Finally, let $G$ be a topological group and $H$ a normal subgroup of $G$. Then $G/H$ carries both the usual structures of a quotient group and the quotient topology.

**Theorem 1.5.3.** *If $H$ is a normal subgroup of a topological group $G$, then $G/H$ with the usual structures is a topological group.*

For a proof and for further information, see [20], [23], [39], [67], [69] and [94].

# §1.6. Analysis

We assume a basic knowledge of analytic functions between subsets of the complex plane and, in particular, the fact that these functions map open sets of open sets. As specific examples, we mention Möbius transformations and hyperbolic functions (both of which form a major theme in this book).

A map $f$ from an open subset of $\mathbb{R}^n$ to $\mathbb{R}^n$ is *differentiable* at $x$ if

$$f(y) = f(x) + (y - x)A + |y - x|\varepsilon(y),$$

where $A$ is an $n \times n$ matrix and where $\varepsilon(y) \to 0$ as $y \to x$. We say that a differentiable $f$ is *conformal* at $x$ if $A$ is a positive scalar multiple $\mu(x)$ of an orthogonal matrix $B$. More generally, $f$ is directly or indirectly conformal according as $\det B$ is positive or negative. If $f$ is an analytic map between plane domains, then the Cauchy–Riemann equations show that $f$ is directly conformal except at those $z$ where $f^{(1)}(z) = 0$.

If $D$ is a subdomain of $\mathbb{R}^n$ and if $\lambda$ is a density (that is, a positive continuous function) on $D$ we define

$$\rho(x, y) = \inf \int_\gamma \lambda(\gamma(t))|\dot{\gamma}(t)|\, dt,$$

the infimum being over all (smooth) curves $\gamma$ (with derivative $\dot{\gamma}$) joining $x$ to $y$ in $D$. It is easy to see that $\rho$ is a metric on $D$; indeed, $\rho$ is obviously symmetric, non-negative and satisfies the Triangle inequality. As $\rho(x, x) = 0$, we need only prove that $\rho(x, y) > 0$ when $x$ and $y$ are distinct. Choosing a suitably small open ball $N$ with centre $x$ and radius $r$, we may assume (by continuity) that $\lambda$ has a positive lower bound $\lambda_0$ on $N$ and that $y \notin N$. Thus $\lambda$ is at least $\lambda_0$ on a section of $\gamma$ of length at least $r$ so $\rho(x, y) > 0$.

More generally, let $\gamma = (\gamma_1, \ldots, \gamma_n)$ be any differentiable curve in $D$ and suppose that

$$q(t) = \sum_{i,j} a_{ij}(\gamma t)\dot{\gamma}_i(t)\dot{\gamma}_j(t)$$

is positive on $D$ (except when $\dot{\gamma} = 0$). Then we can define a metric as above by integrating $[q(t)]^{1/2}$ and the metric topology is the Euclidean topology.

If $f$ is a conformal bijection of $D$ onto the domain $D_1$, then

$$\lim_{y \to x} \frac{|f(y) - f(x)|}{|y - x|} = \mu(x).$$

and $D_1$ inherits the density $\sigma$ where

$$\sigma(fx) = \lambda(x)/\mu(x)$$

and hence a metric $\rho_1$. In fact, $f$ is then a isometry of $(D, \rho)$ onto $(D_1, \rho_1)$. If, in addition, $D = D_1$ and

$$\lambda(fx)\mu(x) = \lambda(x),$$

then $f$ is an isometry of $(D, \rho)$ onto itself: in terms of differentials, this condition can be expressed as

$$\lambda(y)|dy| = \lambda(x)|dx|, \quad y = f(x).$$

As an example, let $D = H^2$, $\lambda(z) = 1/\mathrm{Im}[z]$ and

$$f(z) = \frac{az + b}{cz + d},$$

where $a, b, c$ and $d$ are real and $ad - bc > 0$. Then $f$ maps $H^2$ onto itself and as

$$\mathrm{Im}[fz] = \mathrm{Im}[z]|f^{(1)}(z)|,$$

we see that $f$ is an isometry of $(H^2, \rho)$ onto itself: this is the *hyperbolic metric* on $H^2$.

We shall need the *Poisson kernel* for the unit disc $\Delta$ and the upper half-plane $H^2$. For each $z$ in $\Delta$ and each $\zeta$ in $\partial\Delta$, the Poisson kernel is

$$P_\Delta(z, \zeta) = \frac{1 - |z|^2}{|z - \zeta|^2}.$$

Obviously, $P_\Delta$ is positive on $\Delta$ and zero on $\partial\Delta$ except at the point $\zeta$. Because

$$P_\Delta(z, \zeta) = \mathrm{Re}\left[\frac{\zeta + z}{\zeta - z}\right],$$

we see immediately that $P_\Delta$ is (for each $\zeta$) a harmonic function of $z$ with a pole at $\zeta$.

The map

$$f(z) = \frac{\zeta + z}{\zeta - z}$$

maps $\Delta$ onto $\{z : x > 0\}$ and $\zeta$ to $\infty$ with

$$\mathrm{Re}[f(z)] = P_\Delta(z, \zeta).$$

It follows immediately that the level curves of $P_\Delta(z, \zeta)$ (for a fixed $\zeta$) are the images under $f^{-1}$ of the vertical lines in $H^2$ and these are circles in $\Delta$ which are tangent to $\partial\Delta$ at $\zeta$.

The most general Möbius transformation preserving $\Delta$ is of the form

$$g(z) = \frac{az + \bar{c}}{cz + \bar{a}}, \qquad |a|^2 - |c|^2 = 1,$$

and a computation shows that

$$1 - |g(z)|^2 = |g^{(1)}(z)|(1 - |z|^2).$$

As $g$ is a Möbius transformation, we also have

$$|g(z) - g(\zeta)|^2 = |z - \zeta|^2 |g^{(1)}(z)| \, |g^{(1)}(\zeta)|$$

and so we obtain the relation

$$P_\Delta(gz, g\zeta)|g^{(1)}(\zeta)| = P_\Delta(z, \zeta).$$

The Poisson kernel for the half-plane $H^2$ is

$$P(z, \zeta) = \begin{cases} y & \text{if } \zeta = \infty, \\ y/|z - \zeta|^2 & \text{if } \zeta \neq \infty, \end{cases}$$

and the reader is invited to explore its properties.

# CHAPTER 2
# Matrices

## §2.1. Non-singular Matrices

If $ad - bc \neq 0$, the $2 \times 2$ complex matrix

$$A = \begin{pmatrix} a & b \\ c & d \end{pmatrix} \tag{2.1.1}$$

induces the Möbius transformation

$$g(z) = \frac{az + b}{cz + d}$$

of the extended complex plane onto itself. As these transformations are our primary concern, it is worthwhile to study the class of $2 \times 2$ complex matrices.

Given $A$ as in (2.1.1), the *determinant* $\det(A)$ of $A$ is given by

$$\det(A) = ad - bc$$

and $A$ is *non-singular* if and only if $\det(A) \neq 0$. If $A$ is non-singular then the inverse

$$A^{-1} = \begin{pmatrix} \lambda d & -\lambda b \\ -\lambda c & \lambda a \end{pmatrix}, \qquad \lambda = (ad - bc)^{-1}$$

exists and is also non-singular.

For any matrices $A$ and $B$ we have

$$\det(AB) = \det(A)\det(B) \qquad (2.1.2)$$
$$= \det(BA),$$

and so

$$\det(BAB^{-1}) = \det(AB^{-1}B)$$
$$= \det(A). \qquad (2.1.3)$$

The class of non-singular $2 \times 2$ complex matrices is a group with respect to the usual matrix multiplication: it is the *General Linear Group* and is denoted by $GL(2, \mathbb{C})$. We shall be more concerned with the subgroup $SL(2, \mathbb{C})$, the *Special Linear Group*, which consists of those matrices with $\det(A) = 1$. We denote the identity matrix (of any size) by $I$ although sometimes, for emphasis, we use $I_n$ for the $n \times n$ identity matrix.

Much of the material in this chapter can be written in terms of $n \times n$ complex matrices. The determinant can be defined (by induction on $n$) and a matrix $A$ is non-singular with inverse $A^{-1}$ if and only if $\det(A) \neq 0$. The identities (2.1.2) and (2.1.3) remain valid.

The $n \times n$ real matrix $A$ is *orthogonal* if and only if

$$|x| = |xA|$$

for every $x$ in $\mathbb{R}^n$: this is equivalent to the condition $A^{-1} = A^t$ where $A^t$ denotes the transpose of $A$. Observe that if $A$ is orthogonal then, because $\det(A) = \det(A^t)$, we have $\det(A)$ is $1$ or $-1$. The class of orthogonal $n \times n$ matrices is denoted by $O(n)$.

For $z_1, \ldots, z_n$ in $\mathbb{C}^n$, we write

$$|z| = [|z_1|^2 + \cdots + |z_n|^2]^{1/2}.$$

A complex $n \times n$ matrix is *unitary* if and only if

$$|z| = |zA|$$

for every $z$ in $\mathbb{C}^n$: this is equivalent to the condition $A^{-1} = \bar{A}^t$ where $\bar{A}$ is obtained in the obvious way by taking the complex conjugate of each element of $A$.

From a geometric point of view, the following result is of interest.

**Selberg's Lemma.** *Let $G$ be a finitely generated group of $n \times n$ complex matrices. Then $G$ contains a normal subgroup of finite index which contains no non-trivial element of finite order.*

This result is used only once in this text and we omit the proof which can be found in [92] and [17], [18]: see also [16], [27], [31], [35], [85] and [104] where it is discussed in the context of discrete groups.

EXERCISE 2.1

1. Show that the matrices

$$\begin{pmatrix} 1 & 1 \\ 0 & 1 \end{pmatrix}, \quad \begin{pmatrix} 1 & -1 \\ 0 & 1 \end{pmatrix}$$

   are conjugate in $SL(2, \mathbb{C})$ but not in $SL(2, \mathbb{R})$ (the real matrices in $SL(2, \mathbb{C})$).

2. Show that $A \mapsto \det(A)$ is a homomorphism of $GL(2, \mathbb{C})$ onto the multiplicative group of non-zero complex numbers and identify the kernel.

3. The centre of a group is the set of elements that commute with every element of the group. Show that the centres of $GL(2, \mathbb{C})$ and $SL(2, \mathbb{C})$ are

$$H = \{tI : t \neq 0\}, \quad K = \{I, -I\}$$

   respectively. Prove that the groups

$$GL(2, \mathbb{C})/H, \quad SL(2, \mathbb{C})/K$$

   are isomorphic.

4. Find the centres $H_1$ and $K_1$ of $GL(2, \mathbb{R})$ and $SL(2, \mathbb{R})$ respectively. Are

$$GL(2, \mathbb{R})/H_1, \quad SL(2, \mathbb{R})/K_1$$

   isomorphic?

## §2.2. The Metric Structure

The *trace* $\mathrm{tr}(A)$ of the matrix $A$ in (2.1.1) is defined by

$$\mathrm{tr}(A) = a + d.$$

A simple computation shows that

$$\mathrm{tr}(AB) = \mathrm{tr}(BA)$$

and we deduce that

$$\mathrm{tr}(BAB^{-1}) = \mathrm{tr}(AB^{-1}B) = \mathrm{tr}(A):$$

thus tr *is invariant under conjugation*. Other obvious facts are

$$\mathrm{tr}(\lambda A) = \lambda\,\mathrm{tr}(A) \quad (\lambda \in \mathbb{C})$$

and

$$\mathrm{tr}(A^t) = \mathrm{tr}(A),$$

where $A^t$ denotes the transpose of $A$.

The trace function also acts in an important way on pairs of matrices. First, we recall that the class of $2 \times 2$ matrices is a vector space over the complex numbers and the Hermitian transpose $A^*$ of $A$ is defined by

$$A^* = (\bar{A})^t = \begin{pmatrix} \bar{a} & \bar{c} \\ \bar{b} & \bar{d} \end{pmatrix}. \tag{2.2.1}$$

Given any matrices

$$A = \begin{pmatrix} a & b \\ c & d \end{pmatrix}, \qquad B = \begin{pmatrix} \alpha & \beta \\ \gamma & \delta \end{pmatrix},$$

we define $[A, B]$ by

$$[A, B] = \text{tr}(AB^*)$$
$$= a\bar{\alpha} + b\bar{\beta} + c\bar{\gamma} + d\bar{\delta}.$$

This is a scalar product on the vector space of all $2 \times 2$ matrices: explicitly

(i) $[A, A] \geq 0$ with equality if and only if $A = 0$;

(ii) $[\lambda_1 A_1 + \lambda_2 A_2, B] = \lambda_1 [A_1, B] + \lambda_2 [A_2, B]$;

and

(iii) $[B, A] = \overline{[A, B]}$.

Any scalar product, say $[x, y]$, induces a norm $[x, x]^{1/2}$ and hence a metric $[x - y, x - y]^{1/2}$. In our case the norm $\|A\|$ is given explicitly by

$$\|A\| = [A, A]^{1/2}$$
$$= (|a|^2 + |b|^2 + |c|^2 + |d|^2)^{1/2}$$

and for completeness, we shall show that this satisfies the defining properties of a norm, namely

(iv) $\|A\| \geq 0$ with equality if and only if $A = 0$;

(v) $\|\lambda A\| = |\lambda| . \|A\| \qquad (\lambda \in \mathbb{C})$

and

(vi) $\|A + B\| \leq \|A\| + \|B\|$.

Of these, (iv) and (v) are trivial: (vi) will be proved shortly. We also have the additional relations

(vii) $|\det(A)| . \|A^{-1}\| = \|A\|$;

(viii) $|[A, B]| \leq \|A\| . \|B\|$;

(ix) $\|AB\| \leq \|A\| . \|B\|$

and

(x) $2|\det(A)| \leq \|A\|^2$.

Of these, (vii) is immediate. To prove (viii) let

$$C = \lambda A - \mu B,$$

where $\lambda = [B, A]$ and $\mu = \|A\|^2$. By (iv), $\|C\|^2 \geq 0$ and this simplifies to give (viii). As

$$\|A + B\|^2 = \|A\|^2 + [A, B] + [B, A] + \|B\|^2,$$

(vi) follows directly from (viii) and (iii).

To prove (ix), note that if

$$AB = \begin{pmatrix} p & q \\ r & s \end{pmatrix},$$

then, for example,

$$|p|^2 = |a\alpha + b\gamma|^2$$
$$\leq (|a|^2 + |b|^2)(|\alpha|^2 + |\gamma|^2),$$

(the last line by the Cauchy–Schwarz inequality). A similar inequality holds for $q$, $r$ and $s$ and (ix) follows.

Finally, (x) holds as

$$\|A\|^2 - 2|\det(A)| \geq |a|^2 + |b|^2 + |c|^2 + |d|^2 - 2(|ad| + |bc|)$$
$$= (|a| - |d|)^2 + (|b| - |c|)^2$$
$$\geq 0.$$

Next, the norm $\|A\|$ induces a metric $\|A - B\|$ for

$$\|A - B\| = 0 \quad \text{if and only if } A = B;$$

$$\|B - A\| = \|(-1)(A - B)\| = \|A - B\|$$

and

$$\|A - B\| = \|(A - C) + (C - B)\|$$
$$\leq \|A - C\| + \|C - B\|.$$

The metric is given explicitly by

$$\|A - B\| = [|a - \alpha|^2 + \cdots + |d - \delta|^2]^{1/2}$$

and we see that

$$\begin{pmatrix} a_n & b_n \\ c_n & d_n \end{pmatrix} \rightarrow \begin{pmatrix} a & b \\ c & d \end{pmatrix}$$

in this metric if and only if $a_n \rightarrow a$, $b_n \rightarrow b$, $c_n \rightarrow c$ and $d_n \rightarrow d$. Note that this is a metric on the vector space of all $2 \times 2$ matrices.

Observe that the norm, the determinant and the trace function are all continuous functions. The map $A \mapsto A^{-1}$ is also continuous (on $GL(2, \mathbb{C})$) and if $A_n \rightarrow A$ and $B_n \rightarrow B$ then $A_n B_n \rightarrow AB$. These facts show that $GL(2, \mathbb{C})$ is a topological group with respect to the metric $\|A - B\|$.

EXERCISE 2.2

1. Show that if A and B are in SL(2, $\mathbb{C}$) then

   (i) $\text{tr}(AB) + \text{tr}(A^{-1}B) = \text{tr}(A)\,\text{tr}(B)$;
   (ii) $\text{tr}(BAB) + \text{tr}(A) = \text{tr}(B)\,\text{tr}(AB)$;
   (iii) $\text{tr}^2(A) + \text{tr}^2(B) + \text{tr}^2(AB) = \text{tr}(A)\,\text{tr}(B)\,\text{tr}(AB) + 2 + \text{tr}(ABA^{-1}B^{-1})$.

   Replace $B$ by $A^n B$ in (i) and hence obtain $\text{tr}(A^n B)$ as a function of $\text{tr}(A)$, $\text{tr}(B)$, $\text{tr}(AB)$ and $n$.

2. Find subgroups $G_1$ and $G_2$ of GL(2, $\mathbb{C}$) and a map $f$ of $G_1$ onto $G_2$ which is an isomorphism but not a homeomorphism.

3. Let $V$ be the metric space of all $2 \times 2$ complex matrices with metric $\|A - B\|$. Prove that as subsets of $V$,

   (i) GL(2, $\mathbb{C}$) is open but not closed;
   (ii) SL(2, $\mathbb{C}$) is closed but not open;
   (iii) GL(2, $\mathbb{R}$) is disconnected;
   (iv) GL(2, $\mathbb{C}$) is connected;
   (v) $\{A : \text{tr}(A) = 1\}$ is closed but not compact.

   [In (iv), show that every matrix in GL(2, $\mathbb{C}$) is conjugate to an upper triangular matrix $T$ and that $T$ can be joined to $I$ by a curve in GL(2, $\mathbb{C}$).]

4. For an $n \times n$ complex matrix $A = (a_{ij})$, define

   $$\text{tr}(A) = a_{11} + \cdots + a_{nn}.$$

   Prove that

   $$\text{tr}(BAB^{-1}) = \text{tr}(A)$$

   and that $\text{tr}(AB^*)$ is a metric on the space of all such matrices.

## §2.3. Discrete Groups

In this section we shall confine our attention to subgroups of the topological group GL(2, $\mathbb{C}$). We recall that a subgroup $G$ of GL(2, $\mathbb{C}$) is *discrete* if and only if the subspace topology on $G$ is the discrete topology. It follows that if $G$ is discrete and if $X, A_1, A_2, \ldots$ are in $G$ with $A_n \to X$ then $A_n = X$ for all sufficiently large $n$. It is not necessary to assume that $X \in G$ here but only that $X$ is in GL(2, $\mathbb{C}$). Indeed, in this case,

$$A_n(A_{n+1})^{-1} \to XX^{-1} = I$$

and so for almost all $n$, we have $A_n = A_{n+1}$ and hence $A_n = X$.

In order to prove that $G$ is discrete, it is only necessary to prove that one point of $G$ is isolated: for example, it is sufficient to prove that

$$\inf\{\|X - I\| : X \in G,\ X \neq I\} > 0,$$

so that $\{I\}$ is open in $G$ (Corollary 1.5.2). In terms of sequences, $G$ is discrete if and only if $A_n \to I$ and $A_n \in G$ implies that $A_n = I$ for almost all $n$.

We shall mainly be concerned with SL(2, $\mathbb{C}$) and in this case an alternative formulation of discreteness can be given directly in terms of the norm. The subgroup $G$ of SL(2, $\mathbb{C}$) is discrete if and only if for each positive $k$, the set

$$\{A \in G: \|A\| \leq k\} \tag{2.3.1}$$

is finite. If this set is finite for each $k$, then $G$ clearly cannot have any limit points (the norm function is continuous) and so $G$ is discrete. On the other hand, if this set is infinite then there are distinct elements $A_n$ in $G$ with $\|A_n\| \leq k, n = 1, 2, \ldots$ . If $A_n$ has coefficients $a_n, b_n, c_n$ and $d_n$ then $|a_n| \leq k$ and so the sequence $a_n$ has a convergent subsequence. The same is true of the other coefficients and using the familiar "diagonal process" we see that there is a subsequence on which each of the coefficients converge. On this subsequence, $A_n \to B$ say, for some $B$ and as det is continuous, $B \in$ SL(2, $\mathbb{C}$): thus $G$ is not discrete.

The criterion (2.3.1) shows that *a discrete subgroup $G$ of* SL(2, $\mathbb{C}$) *is countable*. In fact,

$$G = \bigcup_{n=1}^{\infty} G_n,$$

where $G_n$ is the finite set of $A$ in $G$ with $\|A\| \leq n$. *Any subgroup of a discrete group is also discrete*: this is obvious. Finally, *if $G$ is discrete then so is any conjugate group $BGB^{-1}$*, because $X \mapsto BXB^{-1}$ is a homeomorphism of GL(2, $\mathbb{C}$) onto itself.

There are other more delicate consequences of and criteria for discreteness but these are best considered in conjunction with Möbius transformations (which we shall consider in later chapters). For a stronger version of discreteness, see [11]. We end with an important example.

**Example 2.3.1.** The *Modular group* is the subgroup of SL(2, $\mathbb{R}$) consisting of all matrices $A$ with $a, b, c$ and $d$ integers. This group is obviously discrete. More generally, *Picard's group* consisting of all matrices $A$ in SL(2, $\mathbb{C}$) with $a, b, c$ and $d$ Gaussian integers (that is, $m + in$ where $m$ and $n$ are integers) is discrete.

EXERCISE 2.3

1. Show that $\{2^n I : n \in \mathbb{Z}\}$ is a discrete subgroup of GL(2, $\mathbb{C}$) and that in this case, (2.3.1) is infinite.

2. Find all discrete subgroups of GL(2, $\mathbb{C}$) which contain only diagonal matrices.

3. Prove that a discrete subgroup of GL(2, $\mathbb{C}$) is countable.

4. Suppose that a subgroup $G$ of GL(2, $\mathbb{R}$) contains a discrete subgroup of finite index. Show that $G$ is also discrete.

## §2.4. Quaternions

A *quaternion* is a $2 \times 2$ complex matrix of the form

$$q = \begin{pmatrix} z & w \\ -\bar{w} & \bar{z} \end{pmatrix}: \tag{2.4.1}$$

the set of quaternions is denoted by $\mathbb{H}$ (after Hamilton). The addition and multiplication of quaternions is as for matrices and the following facts are easily verified:

(i) $\mathbb{H}$ is an abelian group with respect to addition;
(ii) the non-zero quaternions form a non-abelian group with respect to multiplication;
(iii) $\mathbb{H}$ is a four-dimensional real vector space with basis

$$\mathbf{1} = \begin{pmatrix} 1 & 0 \\ 0 & 1 \end{pmatrix}, \qquad \mathbf{i} = \begin{pmatrix} i & 0 \\ 0 & -i \end{pmatrix},$$

$$\mathbf{j} = \begin{pmatrix} 0 & 1 \\ -1 & 0 \end{pmatrix}, \qquad \mathbf{k} = \begin{pmatrix} 0 & i \\ i & 0 \end{pmatrix},$$

(note that $\mathbf{1}$ is not the same as 1, likewise $\mathbf{i} \neq i$).

As multiplication of matrices is distributive, the multiplication of quaternions is determined by the products of the four elements $\mathbf{1}$, $\mathbf{i}$, $\mathbf{j}$ and $\mathbf{k}$. In fact, these elements generate a multiplicative group of order 8 and

$$\mathbf{i}^2 = \mathbf{j}^2 = \mathbf{k}^2 = -1;$$
$$\mathbf{ij} = \mathbf{k}, \quad \mathbf{jk} = \mathbf{i}, \quad \mathbf{ki} = \mathbf{j};$$
$$\mathbf{ji} = -\mathbf{k}, \quad \mathbf{kj} = -\mathbf{i}, \quad \mathbf{ik} = -\mathbf{j}.$$

The quaternions contain a copy of $\mathbb{C}$ for the map

$$x + iy \mapsto x\mathbf{1} + y\mathbf{i}$$

of $\mathbb{C}$ into $\mathbb{H}$ clearly preserves both addition and multiplication. Returning to (2.4.1) we write $x + iy = z$ and $u + iv = w$ so that

$$q = (x\mathbf{1} + y\mathbf{i}) + (u\mathbf{j} + v\mathbf{k})$$
$$= (x\mathbf{1} + y\mathbf{i}) + (u\mathbf{1} + v\mathbf{i})\mathbf{j}. \tag{2.4.2}$$

In view of this, it is convenient to change our notation and rewrite (2.4.2) in the form

$$q = z + wj,$$

where such expressions are to be multiplied by the rule

$$(z_1 + w_1 j)(z_2 + w_2 j) = (z_1 z_2 - w_1 \bar{w}_2) + (z_1 w_2 + w_1 \bar{z}_2)j.$$

In particular, if $z$ and $w$ are in $\mathbb{C}$, then

$$jz = \bar{z}j$$

and

$$(z + wj)(\bar{z} - wj) = |z|^2 + |w|^2.$$

This last identity gives the form of the multiplicative inverse, namely

$$(z + wj)^{-1} = (\bar{z} - wj)/(|z|^2 + |w|^2)$$

where, of course,

$$\det(z + wj) = |z|^2 + |w|^2.$$

EXERCISE 2.4

1. Show that the non-zero quaternions form a multiplicative group with centre $\{tI : t \text{ real and non-zero}\}$.

2. Show that $SL(2, \mathbb{C})$ is not compact whereas

$$\{q \in \mathbb{H} : \det(q) = 1\}$$

is compact.

3. Let $S$ be the set of quaternions of the form $z + tj$ where $t$ is real. Show that $S$ is invariant under the map $q \mapsto jqj^{-1}$. By identifying $z + tj$ with $(x, y, t)$ in $\mathbb{R}^3$, give a geometric description of this map.

4. As in Question 3, show that the map $q \mapsto kqk^{-1}$ also leaves $S$ invariant and give a geometric description of this map.

## §2.5. Unitary Matrices

The matrix $A$ is said to be *unitary* if and only if

$$AA^* = I,$$

where $A^*$ is given by (2.2.1). Any unitary matrix satisfies

$$1 = \det(A)\det(A^*) = |\det(A)|^2$$

and we shall focus our attention on the class $SU(2, \mathbb{C})$ of unitary matrices with determinant one.

**Theorem 2.5.1.** *Let $A$ be in $SL(2, \mathbb{C})$. The following statements are equivalent and characterize elements of $SU(2, \mathbb{C})$.*

(i) *$A$ is unitary;*
(ii) *$\|A\|^2 = 2$;*
(iii) *$A$ is a quaternion.*

*In particular*

$$SU(2, \mathbb{C}) = SL(2, \mathbb{C}) \cap \mathbb{H}.$$

PROOF. Suppose that

$$A = \begin{pmatrix} a & b \\ c & d \end{pmatrix}, \qquad ad - bc = 1,$$

then

$$AA^* = \begin{pmatrix} |a|^2 + |b|^2 & a\bar{c} + b\bar{d} \\ \bar{a}c + \bar{b}d & |c|^2 + |d|^2 \end{pmatrix} \tag{2.5.1}$$

and

$$|a - \bar{d}|^2 + |b + \bar{c}|^2 = \|A\|^2 - 2. \tag{2.5.2}$$

First, (2.5.1) shows that if $A$ is unitary then $\|A\|^2 = 2$. Next, if $\|A\|^2 = 2$ we deduce from (2.5.2) that $a = \bar{d}$ and $b = -\bar{c}$ so $A$ is a quaternion. Finally, if $A$ is a quaternion, then $a = \bar{d}$, $b = -\bar{c}$ and recalling that $ad - bc = 1$, we find from (2.5.1) that $A$ is unitary. $\qquad\qquad\square$

A simple computation shows that each $A$ in $SU(2, \mathbb{C})$ preserves the quadratic form $|z|^2 + |w|^2$: explicitly, if

$$(z, w)A = (z', w'),$$

then

$$|z'|^2 + |w'|^2 = |z|^2 + |w|^2.$$

A similar result holds for column vectors and so for any matrix $X$,

$$\|AX\| = \|XA\| = \|X\|.$$

This shows that

$$\|AXA^{-1} - AYA^{-1}\| = \|A(X - Y)A^{-1}\| = \|X - Y\|$$

and so we have the following result.

**Theorem 2.5.2.** *Suppose that $A$ is in $SU(2, \mathbb{C})$. Then the map $X \mapsto AXA^{-1}$ is an isometry of the space of matrices onto itself.*

*Remark.* Theorems 2.5.1 and 2.5.2 will appear later in a geometric form.

EXERCISE 2.5

1. Show that $SU(2, \mathbb{C})$ is compact and deduce that any discrete subgroup of $SU(2, \mathbb{C})$ is finite.

2. Is $SU(2, \mathbb{C})$ connected?

3. The group of real orthogonal matrices $A(AA^t = I)$ in SL(2, $\mathbb{R}$) is denoted by SO(2). Show that there is a map of SO(2) onto the unit circle in the complex plane which is both an isomorphism and a homeomorphism.

4. Show that every matrix in SU(2, $\mathbb{C}$) can be expressed in the form

$$\begin{pmatrix} e^{i\theta} & 0 \\ 0 & e^{-i\theta} \end{pmatrix} \begin{pmatrix} \cos\phi & -\sin\phi \\ \sin\phi & \cos\phi \end{pmatrix} \begin{pmatrix} e^{i\psi} & 0 \\ 0 & e^{-i\psi} \end{pmatrix}$$

for some real $\theta$, $\phi$ and $\psi$.

# CHAPTER 3
# Möbius Transformations on $\mathbb{R}^n$

## §3.1. The Möbius Group on $\mathbb{R}^n$

The sphere $S(a, r)$ in $\mathbb{R}^n$ is given by

$$S(a, r) = \{x \in \mathbb{R}^n : |x - a| = r\}$$

where $a \in \mathbb{R}^n$ and $r > 0$. The *reflection* (or *inversion*) in $S(a, r)$ is the function $\phi$ defined by

$$\phi(x) = a + \left(\frac{r}{|x - a|}\right)^2 (x - a). \tag{3.1.1}$$

In the special case of $S(0, 1)$ ($= S^{n-1}$), this reduces to

$$\phi(x) = x/|x|^2$$

and it is convenient to denote this by $x \mapsto x^*$ where $x^* = x/|x|^2$. The general reflection (3.1.1) may now be rewritten as

$$\phi(x) = a + r^2(x - a)^*.$$

The reflection in $S(a, r)$ is not defined when $x = a$ and this is overcome by adjoining an extra point to $\mathbb{R}^n$. We select any point not in $\mathbb{R}^n$ (for any $n$), label it $\infty$ and form the union

$$\hat{\mathbb{R}}^n = \mathbb{R}^n \cup \{\infty\}.$$

As $|\phi(x)| \to +\infty$ when $x \to a$ it is natural to define $\phi(a) = \infty$: likewise, we define $\phi(\infty) = a$. The reflection $\phi$ now acts on $\hat{\mathbb{R}}^n$ and, as is easily verified, $\phi^2(x) = x$ for all $x$ in $\hat{\mathbb{R}}^n$. Clearly $\phi$ is a 1–1 map of $\hat{\mathbb{R}}^n$ onto itself: also, $\phi(x) = x$ if and only if $x \in S(a, r)$.

We shall call a set $P(a, t)$ a *plane* in $\hat{\mathbb{R}}^n$ if it is of the form

$$P(a, t) = \{x \in \mathbb{R}^n : (x \cdot a) = t\} \cup \{\infty\},$$

where $a \in \mathbb{R}^n$, $a \neq 0$, $(x \cdot a)$ is the usual scalar product $\sum x_j a_j$ and $t$ is real. Note that by definition, $\infty$ *lies in every plane*. The *reflection* $\phi$ in $P(a, t)$ (or, as we sometimes say, in $(x \cdot a) = t$) is defined in the usual way; that is

$$\phi(x) = x + \lambda a,$$

where the real parameter $\lambda$ is chosen so that $\frac{1}{2}(x + \phi(x))$ is on $P(a, t)$. This gives the explicit formula

$$\phi(x) = x - 2[(x \cdot a) - t]a^*, \tag{3.1.2}$$

when $x \in \mathbb{R}^n$ and, of course, $\phi(\infty) = \infty$. Again, $\phi$ acts on $\hat{\mathbb{R}}^n$, $\phi^2(x) = x$ for all $x$ in $\hat{\mathbb{R}}^n$ and so $\phi$ is a 1–1 map of $\hat{\mathbb{R}}^n$ onto itself. Also, $\phi(x) = x$ if and only if $x \in P(a, t)$.

It is clear that any reflection $\phi$ (in a sphere or a plane) is continuous in $\mathbb{R}^n$ except at the points $\infty$ and $\phi^{-1}(\infty)$ where continuity is not yet defined. We shall now construct a metric on $\hat{\mathbb{R}}^n$ and shall show that $\phi$ is actually continuous (with respect to this metric) throughout $\hat{\mathbb{R}}^n$.

We first embed $\hat{\mathbb{R}}^n$ in $\hat{\mathbb{R}}^{n+1}$ in the natural way by making the points $(x_1, \ldots, x_n)$ and $(x_1, \ldots, x_n, 0)$ correspond. Specifically, we let $x \mapsto \tilde{x}$ be the map defined by

$$\tilde{x} = (x_1, \ldots, x_n, 0), \qquad x = (x_1, \ldots, x_n),$$

and, of course, $\tilde{\infty} = \infty$. Thus $x \mapsto \tilde{x}$ is a 1–1 map of $\hat{\mathbb{R}}^n$ onto the plane $x_{n+1} = 0$ in $\hat{\mathbb{R}}^{n+1}$. The plane $x_{n+1} = 0$ in $\hat{\mathbb{R}}^{n+1}$ can be mapped in a 1–1 manner onto the sphere

$$S^n = \{y \in \mathbb{R}^{n+1} : |y| = 1\}$$

by projecting $\tilde{x}$ towards (or away from) $e_{n+1}$ until it meets the sphere $S^n$ in the unique point $\pi(\tilde{x})$ other than $e_{n+1}$. This map $\pi$ is known as the *stereographic projection* of $\hat{\mathbb{R}}^n$ onto $S^n$.

It is easy to describe $\pi$ analytically. Given $x$ in $\mathbb{R}^n$, then

$$\pi(\tilde{x}) = \tilde{x} + t(e_{n+1} - \tilde{x}),$$

where $t$ is chosen so that $|\pi(\tilde{x})| = 1$. The condition $|\pi(\tilde{x})|^2 = 1$ gives rise to a quadratic equation in $t$ which has the two solutions $t = 1$ and (as $|\tilde{x}| = |x|$)

$$t = \frac{|x|^2 - 1}{|x|^2 + 1}.$$

We conclude that

$$\pi(\tilde{x}) = \left( \frac{2x_1}{|x|^2 + 1}, \ldots, \frac{2x_n}{|x|^2 + 1}, \frac{|x|^2 - 1}{|x|^2 + 1} \right), \qquad x \in \mathbb{R}^n,$$

and, by definition, $\pi(\infty) = e_{n+1}$.

As $x \mapsto \pi(\tilde{x})$ is a 1–1 map of $\hat{\mathbb{R}}^n$ onto $S^n$ we can transfer the Euclidean metric from $S^n$ to a metric $d$ on $\hat{\mathbb{R}}^n$. This is the *chordal metric* $d$ and is defined on $\hat{\mathbb{R}}^n$ by

$$d(x, y) = |\pi(\tilde{x}) - \pi(\tilde{y})|, \qquad x, y \in \hat{\mathbb{R}}^n.$$

A tedious (but elementary) computation now yields an explicit expression for $d$, namely

$$d(x, y) = \begin{cases} \dfrac{2|x - y|}{(1 + |x|^2)^{1/2}(1 + |y|^2)^{1/2}} & \text{if } x, y \neq \infty; \\[4mm] \dfrac{2}{(1 + |x|^2)^{1/2}} & \text{if } y = \infty. \end{cases} \tag{3.1.3}$$

A shorter proof of this will be given in Section 3.4.

This formula shows that the metric $d$ restricted to $\mathbb{R}^n$ induces the same topology as does the Euclidean metric; thus a function from a subset of $\mathbb{R}^n$ to $\mathbb{R}^n$ is continuous with respect to both or to neither of these two metrics. It is now easy to see that each reflection $\phi$ is a homeomorphism (with respect to $d$) of $\hat{\mathbb{R}}^n$ onto itself. Indeed, as $\phi = \phi^{-1}$ we need only show that $\phi$ is continuous at each point $x$ in $\hat{\mathbb{R}}^n$ and this is known to be so whenever $x$ is distinct from $\infty$ and $\phi(\infty) (= \phi^{-1}(\infty))$. If $\phi$ denotes reflection in $S(a, r)$ then, for example,

$$d(\phi(x), \phi(a)) = d(\phi(x), \infty)$$

$$= \frac{2}{(1 + |\phi(x)|^2)^{1/2}},$$

$$\to 0$$

as $x \to a$. Thus $\phi$ is continuous at $x = a$: a similar argument shows $\phi$ to be continuous at $\infty$ also. If $\psi$ is the reflection in the plane $P(a, t)$ then (as is easily seen)

$$|\psi(x)|^2 = |x|^2 + O(|x|)$$

as $|x| \to \infty$ and so $|\psi(x)| \to +\infty$. This shows that $\psi$ is continuous at $\infty$ and so is also a homeomorphism of $\hat{\mathbb{R}}^n$ onto itself.

**Definition 3.1.1.** A *Möbius transformation* acting in $\hat{\mathbb{R}}^n$ is a finite composition of reflections (in spheres or planes).

Clearly, each Möbius transformation is a homeomorphism of $\hat{\mathbb{R}}^n$ onto itself. The composition of two Möbius transformations is again a Möbius transformation and so also is the inverse of a Möbius transformation for if $\phi = \phi_1 \cdots \phi_m$ (where the $\phi_j$ are reflections) then $\phi^{-1} = \phi_m \cdots \phi_1$. Finally, for any reflection $\phi$ say, $\phi^2(x) = x$ and so the identity map is a Möbius transformation.

**Definition 3.1.2.** The group of Möbius transformations acting in $\hat{\mathbb{R}}^n$ is called the *General Möbius group* and is denoted by $GM(\hat{\mathbb{R}}^n)$.

Let us now consider examples of Möbius transformations. First, the translation $x \mapsto x + a$, $a \in \mathbb{R}^n$, is a Möbius transformation for it is the reflection in $(x . a) = 0$ followed by the reflection in $(x . a) = \frac{1}{2}|a|^2$. Next, the magnification $x \mapsto kx$, $k > 0$, is also a Möbius transformation for it is the reflection in $S(0, 1)$ followed by the reflection in $S(0, \sqrt{k})$.

If $\phi$ and $\phi^*$ denote reflections in $S(a, r)$ and $S(0, 1)$ respectively and if $\psi(x) = rx + a$, then (by computation)

$$\phi = \psi \phi^* \psi^{-1}. \tag{3.1.4}$$

As $\psi$ is a Möbius transformation, we see that any two reflections in spheres are conjugate in the group $GM(\hat{\mathbb{R}}^n)$.

As further examples of Möbius transformations we have the entire class of Euclidean isometries. Note that each isometry $\phi$ of $\mathbb{R}^n$ is regarded as acting on $\hat{\mathbb{R}}^n$ with $\phi(\infty) = \infty$.

**Theorem 3.1.3.** *Each Euclidean isometry of $\mathbb{R}^n$ is a composition of at most $n + 1$ reflections in planes. In particular each isometry is a Möbius transformation.*

PROOF. As each reflection in a plane is an isometry, it is sufficient to consider only those isometries $\phi$ which satisfy $\phi(0) = 0$. Such isometries preserve the lengths of vectors because

$$|\phi(x)| = |\phi(x) - \phi(0)| = |x - 0| = |x|$$

and also scalar products because

$$\begin{aligned} 2(\phi(x) . \phi(y)) &= |\phi(x)|^2 + |\phi(y)|^2 - |\phi(x) - \phi(y)|^2 \\ &= |x|^2 + |y|^2 - |x - y|^2 \\ &= 2(x . y). \end{aligned}$$

This means that the vectors $\phi(e_1), \ldots, \phi(e_n)$ are mutually orthogonal and so are linearly independent. As there are $n$ of them, they are a basis of the vector space $\mathbb{R}^n$ and so for each $x$ in $\mathbb{R}^n$ there is some $\mu$ in $\mathbb{R}^n$ with

$$\phi(x) = \sum_{j=1}^{n} \mu_j \phi(e_j).$$

But as the $\phi(e_j)$ are mutually orthogonal,

$$\begin{aligned} \mu_j &= (\phi(x) . \phi(e_j)) \\ &= (x . e_j) \\ &= x_j. \end{aligned}$$

Thus

$$\phi\left(\sum_{j=1}^{n} x_j e_j\right) = \sum_{j=1}^{n} x_j \phi(e_j)$$

and this shows that $\phi$ is a linear transformation of $\mathbb{R}^n$ into itself. As any isometry is 1–1, the kernel of $\phi$ has dimension zero: thus $\phi(\mathbb{R}^n) = \mathbb{R}^n$.

If $A$ is the matrix of $\phi$ with respect to the basis $e_1, \ldots, e_n$ then $\phi(x) = xA$ and $A$ has rows $\phi(e_1), \ldots, \phi(e_n)$. This shows that the $(i, j)$th element of the matrix $AA^t$ is $(\phi(e_i) . \phi(e_j))$ and as this is $(e_i . e_j)$, it is 1 if $i = j$ and is zero otherwise. We conclude that $A$ is an orthogonal matrix.

We shall now show that $\phi$ is a composition of at most $n$ reflections in planes. First, put

$$a_1 = \phi(e_1) - e_1.$$

If $a_1 \neq 0$, we let $\psi_1$ be the reflection in the plane $P(a_1, 0)$ and a direct computation using (3.1.2) shows that $\psi_1$ maps $\phi(e_1)$ to $e_1$. If $a_1 = 0$ we let $\psi_1$ be the identity so that in all cases, $\psi_1$ maps $\phi(e_1)$ to $e_1$. Now put $\phi_1 = \psi_1\phi$: thus $\phi_1$ is an isometry which fixes 0 and $e_1$.

In general, suppose that $\phi_k$ is an isometry which fixes each of $0, e_1, \ldots, e_k$ and let

$$a_{k+1} = \phi_k(e_{k+1}) - e_{k+1}.$$

Again, we let $\psi_{k+1}$ be the identity (if $a_{k+1} = 0$) or the reflection in $P(a_{k+1}, 0)$ (if $a_{k+1} \neq 0$) and exactly as above, $\psi_{k+1}\phi_k$ fixes 0 and $e_{k+1}$. In addition, if $1 \leq j \leq k$ then

$$\begin{aligned}
(e_j . a_{k+1}) &= (e_j . \phi_k(e_{k+1})) - (e_j . e_{k+1}) \\
&= (\phi_k(e_j) . \phi_k(e_{k+1})) - 0 \\
&= (e_j . e_{k+1}) \\
&= 0
\end{aligned}$$

and so by (3.1.2),

$$\psi_{k+1}(e_j) = e_j.$$

As $\phi_k$ also fixes $0, e_1, \ldots, e_k$ we deduce that $\psi_{k+1}\phi_k$ fixes each of $0, e_1, \ldots, e_{k+1}$. In conclusion, then, there are maps $\psi_j$ (each the identity or a reflection in a plane) so that the isometry $\psi_n \cdots \psi_1 \phi$ fixes each of $0, e_1, \ldots, e_n$. By our earlier remarks, such a map is necessarily a linear transformation and so is the identity: thus $\phi = \psi_1 \cdots \psi_n$. This completes the proof of Theorem 3.1.3 as any isometry composed with a suitable reflection is of the form $\phi$. $\square$

There is an alternative formulation available.

**Theorem 3.1.4.** *A function $\phi$ is a Euclidean isometry if and only if it is of the form*

$$\phi(x) = xA + x_0,$$

*where $A$ is an orthogonal matrix and $x_0 \in \mathbb{R}^n$.*

PROOF. As an orthogonal matrix preserves lengths, it is clear that any $\phi$ of the given form is an isometry. Conversely, if $\phi$ is an isometry, then $\phi(x) - \phi(0)$ is an isometry which fixes the origin and so is given by an orthogonal matrix (as in the proof of Theorem 3.1.3). $\qquad\square$

More detailed information on Euclidean isometries is available: for example, we have the following result.

**Theorem 3.1.5.** *Given any real orthogonal matrix $A$ there is a real orthogonal matrix $Q$ such that*

$$
QAQ^{-1} = \begin{pmatrix} A_1 & & & & & & 0 \\ & \ddots & & & & & \\ & & A_r & & & & \\ & & & I_s & & \\ 0 & & & & -I_t \end{pmatrix},
$$

*where $r$, $s$, $t$ are non-negative integers and*

$$
A_k = \begin{pmatrix} \cos\theta_k & -\sin\theta_k \\ \sin\theta_k & \cos\theta_k \end{pmatrix}.
$$

Any Euclidean isometry which fixes the origin can therefore be represented (with a suitable choice of an orthonormal basis) by such a matrix and this explicitly displays all possible types of isometries.

We now return to discuss again the general reflection $\phi$. It seems clear that $\phi$ is orientation-reversing and we shall now prove that this is so.

**Theorem 3.1.6.** *Every reflection is orientation-reversing and conformal.*

PROOF. Let $\phi$ be the reflection in $P(a, t)$. Then we can see directly from (3.1.2) that $\phi$ is differentiable and that $\phi^{(1)}(x)$ is the constant symmetric matrix $(\phi_{ij})$ where

$$
\phi_{ij} = \delta_{ij} - \frac{2a_i a_j}{|a|^2},
$$

($\delta_{ij}$ is the Kronecker delta and is 1 if $i = j$ and is zero otherwise). We prefer to write this in the form

$$
\phi'(x) = I - 2Q_a,
$$

where $Q_a$ has elements $a_i a_j / |a|^2$. Now $Q_a$ is symmetric and $Q_a^2 = Q_a$, so

$$
\phi'(x) \cdot \phi'(x)^t = (I - 2Q_a)^2 = I.
$$

This shows that $\phi'(x)$ is an orthogonal matrix and so establishes the conformality of $\phi$.

Now let $D = \det \phi'(x)$. As $\phi'(x)$ is orthogonal, $D \neq 0$ (in fact, $D = \pm 1$). Moreover, $D$ is a continuous function of the vector $a$ in $\mathbb{R}^n - \{0\}$ and so is a continuous map of $\mathbb{R}^n - \{0\}$ into $\mathbb{R}^1 - \{0\}$. As $\mathbb{R}^n - \{0\}$ is connected (we assume that $n \geq 2$), $D$ is either positive for all non-zero $a$ or is negative for all non-zero $a$. If $a = e_1$, then $\phi$ becomes

$$\phi(x_1, \ldots, x_n) = (-x_1 + 2t, x_2, \ldots, x_n)$$

and in this case, $D = -1$. We conclude that for all non-zero $a$, $D < 0$ and so every reflection in a plane is orientation reversing.

A similar argument holds for reflections in spheres. First, let $\phi$ be the reflection in $S(0, 1)$. Then for $x \neq 0$, the general element of $\phi'(x)$ is

$$\frac{\delta_{ij}}{|x|^2} - \frac{2x_i x_j}{|x|^4},$$

so

$$\phi'(x) = |x|^{-2}(I - 2Q_x).$$

This shows (as above) that $\phi$ is conformal at each non-zero $x$.

Now let $D(x)$ be $\det \phi'(x)$. As $\phi(\phi(x)) = x$, the Chain Rule yields

$$D(\phi(x))D(x) = 1$$

and so exactly as above, $D$ is either positive throughout $\mathbb{R}^n - \{0\}$ or negative throughout $\mathbb{R}^n - \{0\}$. Taking $x = e_1$, a simple computation yields $D(e_1) = -1$ and so $D(x) < 0$ for all non-zero $x$.

The proof for the general reflection is now a simple application of (3.1.4): the details are omitted. □

The argument given above shows that the composition of an even number of reflections is orientation-preserving and that the composition of an odd number is orientation-reversing.

**Definition 3.1.7.** The *Möbius group* $M(\hat{\mathbb{R}}^n)$ acting in $\hat{\mathbb{R}}^n$ is the subgroup of $GM(\hat{\mathbb{R}}^n)$ consisting of all orientation-preserving Möbius transformations in $GM(\hat{\mathbb{R}}^n)$.

We end this section with a simple but useful formula. If $\sigma$ is the reflection in the Euclidean sphere $S(a, r)$ then

$$|\sigma(y) - \sigma(x)| = r^2|(y - a)^* - (x - a)^*|$$

$$= r^2\left[\frac{1}{|y - a|^2} - \frac{2(x - a) \cdot (y - a)}{|x - a|^2|y - a|^2} + \frac{1}{|x - a|^2}\right]^{1/2}$$

$$= \frac{r^2|y - x|}{|x - a||y - a|}. \tag{3.1.5}$$

This shows that

$$\lim_{h \to 0} \frac{|\sigma(x+h) - \sigma(x)|}{|h|} = \frac{r^2}{|x-a|^2}$$

and this measures the local magnification of $\sigma$ at $x$.

### EXERCISE 3.1

1. Show that the reflections in the planes $x \cdot a = 0$ and $x \cdot b = 0$ commute if and only if $a$ and $b$ are orthogonal.

2. Show that if $\phi$ is the reflection in $x \cdot a = t$, then

$$|\phi(x)|^2 = |x|^2 + O(x)$$

as $|x| \to +\infty$.

3. Let $\phi$ be the reflection in $S(a, r)$. Prove analytically that

   (i) $\phi(x) = x$ if and only if $x \in S(a, r)$;
   (ii) $\phi^2(x) = x$;
   (iii) $|x - a| \cdot |\phi(x) - a| = r^2$.

   Repeat (with a modified (iii)) for the reflection in $P(a, t)$.

4. Prove (analytically and geometrically) that for all non-zero $x$ and $y$ in $\mathbb{R}^n$,

$$|x| \cdot |y - x^*| = |y| \cdot |x - y^*|.$$

5. Show that if $\phi_t$ denotes reflection in $S(ta, t \, |a|)$ then

$$x \mapsto \phi(x) = \lim_{t \to +\infty} \phi_t(x)$$

denotes reflection in the plane $x \cdot a = 0$.

6. Verify the formula (3.1.3).

7. Let $\pi$ be the stereographic projection of $x_{n+1} = 0$ onto $S^n$. Show that if $y \in S^n$ then

$$\pi^{-1}(y) = \frac{1}{(1 - y_{n+1})}(y_1, \ldots, y_n, 0).$$

8. Let $\phi$ denote reflection in $S(e_{n+1}, \sqrt{2})$. Show that $\phi = \pi$ on the plane $x_{n+1} = 0$ and find $\phi(H^{n+1})$.

9. Show that the map $z \mapsto 1 + \bar{z}$ in $\mathbb{C}$ is a composition of three (and no fewer) reflections. (Thus $n + 1$ in Theorem 3.1.3 can be attained.)

10. Use Theorem 3.1.5 and Definition 3.1.7 to show that if $n$ is odd and if $\phi \in M(\hat{\mathbb{R}}^n)$ then $\phi$ has an axis (a line of fixed points).

## §3.2. Properties of Möbius Transformations

We shall show that a Möbius transformation maps each sphere and plane onto some sphere or plane and because of this, it is convenient to modify our earlier terminology. *Henceforth we shall use "sphere" to denote either a sphere of the form $S(a, r)$ or a plane.* A sphere $S(a, r)$ will be called a *Euclidean sphere* or will simply be said to be *of the form $S(a, r)$*.

**Theorem 3.2.1.** *Let $\phi$ be any Möbius transformation and $\Sigma$ any sphere. Then $\phi(\Sigma)$ is also a sphere.*

PROOF. It is easy to see that $\phi(\Sigma)$ is a sphere whenever $\phi$ is a Euclidean isometry: in particular, this holds when $\phi$ is the reflection in a plane. It is equally easy to see that $\phi(\Sigma)$ is a sphere when $\phi(x) = kx$, $k > 0$.

Each sphere $\Sigma$ is the set of points $x$ in $\hat{\mathbb{R}}^n$ which satisfy some equation

$$\varepsilon |x|^2 - 2(x \cdot a) + t = 0,$$

where $\varepsilon$ and $t$ are real, $a \in \mathbb{R}^n$ and where, by convention, $\infty$ satisfies this equation if and only if $\varepsilon = 0$.

If $x \in \Sigma$, then writing $y = x^*$ we have

$$\varepsilon - 2(y \cdot a) + t|y|^2 = 0$$

and this is the equation of another sphere $\Sigma_1$. Thus if $\phi^*$ is the map $x \mapsto x^*$ then $\phi^*(\Sigma) \subset \Sigma_1$. The same argument shows that $\phi^*(\Sigma_1) \subset \Sigma$ and so $\phi^*(\Sigma) = \Sigma_1$.

By virtue of (3.1.4) and the above remarks, $\phi(\Sigma)$ is a sphere whenever $\phi$ is the reflection in any Euclidean sphere. As each Möbius transformation is a composition of reflections the result now follows.  $\square$

Any detailed discussion of the geometry of Möbius transformations depends essentially on Theorem 3.2.1 and the fact that Möbius transformations are conformal. A useful substitute for conformality is the elegant concept of the *inversive product* $(\Sigma, \Sigma')$ of two spheres $\Sigma$ and $\Sigma'$. This is an explicit real expression which depends only on $\Sigma$ and $\Sigma'$ and which is invariant under all Möbius transformations. When $\Sigma$ and $\Sigma'$ intersect it is a function of their angle of intersection: when $\Sigma$ and $\Sigma'$ are disjoint it is a function of the hyperbolic distance between them (this will be explained later). Without doubt, it is the invariance and explicit nature of $(\Sigma, \Sigma')$ which makes it a powerful and elegant tool.

The equation defining a sphere $\Sigma$, say $S(a, r)$ or $P(a, t)$, is

$$|x|^2 - 2(x \cdot a) + |a|^2 - r^2 = 0,$$

or

$$-2(x \cdot a) + 2t = 0,$$

respectively, and these can be written in the common form

$$a_0|x|^2 - 2(x \cdot a) + a_{n+1} = 0,$$

where $a = (a_1, \ldots, a_n)$. The *coefficient vector* of $\Sigma$, namely $(a_0, a_1, \ldots, a_n, a_{n+1})$ is not uniquely determined by $\Sigma$ but it is determined to within a real non-zero multiple. Moreover if $(a_0, \ldots, a_{n+1})$ is any coefficient vector of $\Sigma$ then (as is easily checked in the two cases)

$$|a|^2 > a_0 a_{n+1}.$$

**Definition 3.2.2.** Let $\Sigma$ and $\Sigma'$ have coefficient vectors $(a_0, \ldots, a_{n+1})$ and $(b_0, \ldots, b_{n+1})$ respectively. The *inversive product* $(\Sigma, \Sigma')$ of $\Sigma$ and $\Sigma'$ is

$$(\Sigma, \Sigma') = \frac{|2(a \cdot b) - a_0 b_{n+1} - a_{n+1} b_0|}{2(|a|^2 - a_0 a_{n+1})^{1/2}(|b|^2 - b_0 b_{n+1})^{1/2}}. \tag{3.2.1}$$

Note that this is uniquely determined by $\Sigma$ and $\Sigma'$: the bracketed terms in the denominator are positive and we take positive square roots. If we define a bilinear form $q$ on $\mathbb{R}^{n+2}$ by

$$q(x, y) = 2(x_1 y_1 + \cdots + x_n y_n) - (x_0 y_{n+1} + x_{n+1} y_0),$$

then we can write the inversive product more concisely as

$$(\Sigma, \Sigma') = \frac{|q(a', b')|}{|q(a', a')|^{1/2}|q(b', b')|^{1/2}},$$

where $a' = (a_0, a_1, \ldots, a_n, a_{n+1})$ and similarly for $b'$.

It is helpful to obtain explicit expressions for $(\Sigma, \Sigma')$ in the following three cases.

*Case I.* If $\Sigma = S(a, r)$ and $\Sigma' = S(b, t)$ then

$$(\Sigma, \Sigma') = \left| \frac{r^2 + t^2 - |a - b|^2}{2rt} \right|. \tag{3.2.2}$$

*Case II.* If $\Sigma = S(a, r)$ and $\Sigma' = P(b, t)$ then

$$(\Sigma, \Sigma') = \frac{|(a \cdot b) - t|}{r|b|}. \tag{3.2.3}$$

*Case III.* If $\Sigma = P(a, r)$ and $\Sigma' = P(b, t)$ then

$$(\Sigma, \Sigma') = \frac{|(a \cdot b)|}{|a||b|}. \tag{3.2.4}$$

These formulae are easily verified. Note that in all cases, if $\Sigma$ and $\Sigma'$ intersect then $(\Sigma, \Sigma') = \cos \theta$ where $\theta$ is one of the angles of intersection. In particular, $(\Sigma, \Sigma') = 0$ if and only if $\Sigma$ and $\Sigma'$ are *orthogonal*. Observe also that in Case II,

$$(\Sigma, \Sigma') = \delta/r,$$

where $\delta$ is the distance of the centre of $S(a, r)$ from the plane $P(b, t)$: thus $(\Sigma, \Sigma') = 0$ if and only if $a \in P(b, t)$.

We shall now establish the invariance of $(\Sigma, \Sigma')$.

**Theorem 3.2.3.** *For any Möbius transformation $\phi$ and any spheres $\Sigma$ and $\Sigma'$,*

$$(\phi(\Sigma), \phi(\Sigma')) = (\Sigma, \Sigma').$$

PROOF. A Möbius transformation maps a sphere $\Sigma$ to a sphere $\Sigma'$ and so induces a map

$$(a_0, a_1, \ldots, a_n, a_{n+1}) \mapsto (a'_0, a'_1, \ldots, a'_n, a'_{n+1})$$

between the coefficient vectors (to within a scalar multiple) of $\Sigma$ and $\Sigma'$. For example, an orthogonal transformation $x \mapsto xA = y$ or $\mathbb{R}^n$ (and this includes all reflections in planes through the origin) satisfies

$$|x|^2 = |y|^2, \qquad (x \cdot a) = (xA \cdot aA) = (y \cdot aA)$$

and so maps the sphere

$$a_0|x|^2 - 2(x \cdot a) + a_{n+1} = 0$$

to the sphere

$$a_0|y|^2 - 2(y \cdot aA) + a_{n+1} = 0.$$

The induced map between the coefficients is thus

$$a_0 \mapsto a_0, \qquad a \mapsto aA, \qquad a_{n+1} \mapsto a_{n+1}$$

and it is clear that (3.2.1) is invariant if both coefficient vectors are subjected to this transformation. We deduce that $(\Sigma, \Sigma')$ is invariant under the map $x \mapsto xA$.

In a similar way, the maps (i) $x \mapsto kx$ $(k > 0)$; (ii) $x \mapsto x^*$; (iii) $x \mapsto x + u$ induce the maps:

(i) $(a_0, a_1, \ldots, a_n, a_{n+1}) \mapsto (a_0, ka_1, \ldots, ka_n, k^2 a_{n+1})$;

(ii) $(a_0, a_1, \ldots, a_n, a_{n+1}) \mapsto (a_{n+1}, a_1, \ldots, a_n, a_0)$;

(iii) $(a_0, a_1, \ldots, a_n, a_{n+1}) \mapsto (a_0, a_1 + a_0 u_1, \ldots, a_n + a_0 u_n, a_{n+1}$
$\qquad\qquad\qquad\qquad\qquad\qquad + 2(a \cdot u) + a_0|u|^2)$.

It is easy to check that (3.2.1) remains invariant under all of these transformations and, as the corresponding Möbius transformations generate the Möbius group, the proof is complete. Algebraically, one is simply observing that a Möbius transformation induces a linear transformation with matrix $A$ on the coefficient vectors and that $A$ leaves the quadratic form $q$ invariant. $\qquad\square$

The proof of the next result illustrates the use of the inversive product in place of conformality.

**Theorem 3.2.4.** *Let* $\Sigma$ *be any sphere,* $\sigma$ *the reflection in* $\Sigma$ *and* $I$ *the identity map. If* $\phi$ *is any Möbius transformation which fixes each* $x$ *in* $\Sigma$, *then either* $\phi = I$ *or* $\phi = \sigma$.

PROOF. First, we consider the case when $\Sigma$ is the plane $x_n = 0$ in $\hat{\mathbb{R}}^n$. Let $\Sigma' = S(a, r)$ where $a \in \Sigma$ and $r > 0$. As $\infty \in \Sigma$, $\phi$ fixes $\infty$: thus $\phi$ maps $\Sigma'$ to a Euclidean sphere, say $\Sigma'' = S(b, t)$. As $a \in \Sigma$ we have $(\Sigma, \Sigma') = 0$. The invariance described by Theorem 3.2.3 yields $(\Sigma, \Sigma'') = 0$ and so $b \in \Sigma$: thus $a_n = b_n = 0$. Each point of $\Sigma \cap \Sigma'$ is fixed by $\phi$, thus

$$(x_1 - a_1)^2 + \cdots + (x_{n-1} - a_{n-1})^2 = r^2,$$

if and only if

$$(x_1 - b_1)^2 + \cdots + (x_{n-1} - b_{n-1})^2 = t^2.$$

We conclude that $a = b$ and $t = r$: hence $\phi$ maps $\Sigma'$ onto itself.

Next, we select any $x$ not in $\Sigma$ and let $y = \phi(x)$. Now select any $a$ in $\Sigma$ and let $r = |x - a|$ so $x \in S(a, r)$. As $\phi$ preserves $S(a, r)$, $y$ is on $S(a, r)$ and so

$$|x|^2 - 2(x \cdot a) + |a|^2 = |y|^2 - 2(y \cdot a) + |a|^2:$$

note that this holds for all $a$ in $\Sigma$. Taking $a = 0$ we find that $|x| = |y|$. As a consequence of this we find that for all $a$ in $\Sigma$,

$$(x \cdot a) = (y \cdot a)$$

and taking $a$ to be $e_1, \ldots, e_{n-1}$ we find that $x_j = y_j$, $j = 1, \ldots, n - 1$. As $|x| = |y|$ we now see that $y_n = \pm x_n$: thus $\phi(x) \, (= y)$ is either $x$ or $\sigma(x)$. As $\phi$ leaves $\Sigma$ invariant, it permutes the components of $\hat{\mathbb{R}}^n - \Sigma$ and so

$$\phi = I \quad \text{or} \quad \phi = \sigma.$$

We can now complete the proof in the general case. First, given any sphere $\Sigma$ there exists a Möbius transformation $\psi$ which maps $\Sigma$ onto the plane $x_n = 0$: we omit the details of this. Now let $\sigma$ be the reflection in $\Sigma$ and $\eta$ the reflection in plane $x_n = 0$. The transformation $\psi \sigma \psi^{-1}$ fixes each point of the plane $x_n = 0$ and is not the identity: thus by the first part of the proof, $\psi \sigma \psi^{-1} = \eta$.

If $\phi$ is now any Möbius transformation which fixes each point of $\Sigma$, then $\psi \phi \psi^{-1}$ is either $I$ or $\eta$: thus $\phi$ is either $I$ or $\sigma$.  $\square$

This proof also shows that any reflection $\sigma$ is conjugate to the fixed reflection $\eta$. Thus we have obtained the following generalization of (3.1.4).

**Corollary.** *Any two reflections are conjugate in* GM($\hat{\mathbb{R}}^n$).

There is an alternative formulation of Theorem 3.2.4 in terms of inverse points. Let $\sigma$ denote reflection in the sphere $\Sigma$: then $x$ and $y$ *are inverse points with respect to* $\Sigma$ if and only if $y = \sigma(x)$ (and, of course, $x = \sigma(y)$).

Now let $x$ and $y$ be inverse points with respect to $\Sigma$, let $\phi$ be any Möbius transformation and let $\sigma_1$ be the reflection in the sphere $\phi(\Sigma)$. According to Theorem 3.2.4, $\phi^{-1}\sigma_1\phi = \sigma$ or equivalently, $\sigma_1\phi = \phi\sigma$. This is the same as saying that for all $x$, $\sigma_1$ maps $\phi(x)$ to $\phi(y)$: thus $\phi(x)$ and $\phi(y)$ are inverse points with respect to $\phi(\Sigma)$. We state this as a second formulation of Theorem 3.2.4.

**Theorem 3.2.5.** *Let $x$ and $y$ be inverse points with respect to the sphere $\Sigma$ and let $\phi$ be any Möbius transformation. Then $\phi(x)$ and $\phi(y)$ are inverse points with respect to the sphere $\phi(\Sigma)$.*

**Theorem 3.2.6.** *The points $x$ and $y$ are inverse points with respect to the sphere $\Sigma$ if and only if every sphere through $x$ and $y$ is orthogonal to $\Sigma$.*

PROOF. This is clearly true when $\Sigma$ is a plane: it is true in general by the invariance of both inverse points and orthogonality.                                 $\square$

We end this section with a brief discussion of cross-ratios. Given four distinct points $x$, $y$, $u$, $v$ in $\mathbb{R}^n$, the *cross-ratio* of these points is

$$[x, y, u, v] = \frac{d(x, u)\, d(y, v)}{d(x, y)\, d(u, v)}. \tag{3.2.5}$$

By virtue of (3.1.3) (the expression for the chordal distance $d$) we also have

$$[x, y, u, v] = \frac{|x - u| \cdot |y - v|}{|x - y| \cdot |u - v|}, \tag{3.2.6}$$

with appropriate interpretations (which are completely justified by (3.2.5)) when one of the variables is $\infty$.

**Theorem 3.2.7.** *A map $\phi: \hat{\mathbb{R}}^n \to \hat{\mathbb{R}}^n$ is a Möbius transformation if and only if it preserves cross-ratios.*

PROOF. As each Möbius map that changes Euclidean distance by a constant factor leaves the expression (3.2.6) invariant, it is only necessary to consider the map $x \mapsto x^*$. As (see (3.1.5))

$$|x^* - y^*| = \frac{|x - y|}{|x||y|},$$

cross-ratios are also invariant under $x \mapsto x^*$. It follows that all Möbius maps preserve cross-ratios.

Suppose now that $\phi: \hat{\mathbb{R}}^n \to \hat{\mathbb{R}}^n$ preserves cross-ratios. By composing $\phi$ with a Möbius transformation, we see that it is sufficient to consider only the case when $\phi(\infty) = \infty$. Take four distinct points $x$, $y$, $u$, $v$ in $\mathbb{R}^n$: as

$$[\infty, y, u, v]/[x, y, \infty, v]$$

is invariant under $\phi$, we obtain

$$\frac{|\phi(x) - \phi(y)|}{|x - y|} = \frac{|\phi(u) - \phi(v)|}{|u - v|}.$$

The restriction that $\{x, y\} \cap \{u, v\} = \emptyset$ is unnecessary (compare each side with a similar expression for two points $a$ and $b$ chosen to be distinct from all of $x, y, u, v$) so $\phi$ is a Euclidean similarity and so is a Möbius map. $\square$

EXERCISE 3.2

1. Verify (3.2.2), (3.2.3) and (3.2.4).

2. Verify the details in the proof of Theorem 3.2.3.

3. Let $d$ be the chordal metric in $\hat{\mathbb{R}}^n$. Show that

$$d(x^*, y^*) = d(x, y).$$

## §3.3. The Poincaré Extension

Poincaré observed that each Möbius transformation $\phi$ acting in $\hat{\mathbb{R}}^n$ has a natural extension to a Möbius transformation $\tilde{\phi}$ acting in $\hat{\mathbb{R}}^{n+1}$ and that in this way, $GM(\hat{\mathbb{R}}^n)$ may be regarded as a subgroup of $GM(\hat{\mathbb{R}}^{n+1})$. This extension depends on the embedding

$$x \mapsto \tilde{x} = (x_1, \ldots, x_n, 0), \qquad x = (x_1, \ldots, x_n),$$

of $\hat{\mathbb{R}}^n$ into $\hat{\mathbb{R}}^{n+1}$.

For each reflection $\phi$ acting in $\hat{\mathbb{R}}^n$, we define a reflection $\tilde{\phi}$ acting in $\hat{\mathbb{R}}^{n+1}$ as follows. If $\phi$ is the reflection in $S(a, r)$, $a \in \mathbb{R}^n$, then $\tilde{\phi}$ is the reflection in $S(\tilde{a}, r)$: if $\phi$ is the reflection in $P(a, t)$ then $\tilde{\phi}$ is the reflection in $P(\tilde{a}, t)$. If $x \in \hat{\mathbb{R}}^n$ and $y = \phi(x)$, then from (3.1.1) and (3.1.2)

$$\tilde{\phi}(x_1, \ldots, x_n, 0) = (y_1, \ldots, y_n, 0) = \widetilde{\phi(x)}, \qquad (3.3.1)$$

and it is in this sense that $\tilde{\phi}$ is regarded as an extension of $\phi$. Alternatively, we can identify $\mathbb{R}^{n+1}$ with $\mathbb{R}^n \times \mathbb{R}^1$ and write (3.3.1) as

$$\tilde{\phi}(x, 0) = (\phi(x), 0).$$

Note that $\tilde{\phi}$ leaves invariant the plane $x_{n+1} = 0$ (this is $\hat{\mathbb{R}}^n$) and each of the half-spaces $x_{n+1} > 0$ and $x_{n+1} < 0$: these facts follow directly from (3.1.1) and (3.1.2).

As each Möbius transformation $\phi$ acting in $\hat{\mathbb{R}}^n$ is a finite composition of reflections $\phi_j$, say $\phi = \phi_1 \cdots \phi_m$, there is at least one Möbius transformation $\tilde{\phi}$, namely $\tilde{\phi}_1 \cdots \tilde{\phi}_m$, which extends the action of $\phi$ to $\hat{\mathbb{R}}^{n+1}$ in the sense of (3.3.1) and which preserves

$$H^{n+1} = \{(x_1, \ldots, x_{n+1}): x_{n+1} > 0\}.$$

In fact, there can be at most one extension for if $\psi_1$ and $\psi_2$ are two such extensions, then $\psi_2^{-1}\psi_1$ fixes each point of the plane $x_{n+1} = 0$ and preserves $H^{n+1}$. Thus by Theorem 3.2.4, $\psi_1 = \psi_2$.

**Definition 3.3.1.** The *Poincaré extension* of $\phi$ in $GM(\hat{\mathbb{R}}^n)$ is the transformation $\tilde{\phi}$ in $GM(\hat{\mathbb{R}}^{n+1})$ as defined above.

Observe that if $\phi$ and $\psi$ are in $GM(\hat{\mathbb{R}}^n)$ with say $\phi = \phi_1 \cdots \phi_m$ and $\psi = \psi_1 \cdots \psi_k$ then the Poincaré extension of $\phi\psi$ is given by

$$\begin{aligned}(\phi\psi)\tilde{} &= (\phi_1 \cdots \phi_m \psi_1 \cdots \psi_k)\tilde{} \\ &= \tilde{\phi}_1 \cdots \tilde{\phi}_m \tilde{\psi}_1 \cdots \tilde{\psi}_k \\ &= \tilde{\phi}\tilde{\psi},\end{aligned}$$

so the map $\phi \mapsto \tilde{\phi}$ is an injective homomorphism of $GM(\hat{\mathbb{R}}^n)$ into $GM(\hat{\mathbb{R}}^{n+1})$: this is a trivial but nonetheless important remark.

We shall now focus our attention on the action of the Poincaré extension $\tilde{\phi}$ in $H^{n+1}$. First, if $\tilde{\phi}$ is the reflection in the sphere $S(\tilde{a}, r)$, $a \in \mathbb{R}^n$, then by (3.1.5),

$$\frac{|\tilde{\phi}(y) - \tilde{\phi}(x)|}{|y - x|} = \frac{r^2}{|x - \tilde{a}||y - \tilde{a}|}.$$

For the moment, let $[\tilde{\phi}(x)]_j$ denote the $j$th component of $\tilde{\phi}(x)$. As

$$\tilde{\phi}(x) = \tilde{a} + r^2(x - \tilde{a})^*,$$

we find that

$$[\tilde{\phi}(x)]_{n+1} = 0 + \frac{r^2 x_{n+1}}{|x - \tilde{a}|^2} \tag{3.3.2}$$

and this shows that

$$\frac{|y - x|^2}{y_{n+1} x_{n+1}} \tag{3.3.3}$$

is invariant under $\tilde{\phi}$.

The reflection $\tilde{\phi}$ in the plane $P(\tilde{a}, t)$, $a \in \mathbb{R}^n$, is a Euclidean isometry and moreover,

$$[\tilde{\phi}(x)]_{n+1} = x_{n+1}:$$

thus (3.3.3) is also invariant under this reflection. We conclude that (3.3.3) *is invariant under all Poincaré extensions*. It is a direct consequence of this invariance that the Poincaré extension of any $\phi$ in $GM(\hat{\mathbb{R}}^n)$ is an isometry of the space $H^{n+1}$ endowed with the Riemannian metric $\rho$ given by

$$ds = \frac{|dx|}{x_{n+1}}.$$

This is our first model of *hyperbolic space* and $\rho$ is the *hyperbolic metric* in $H^{n+1}$. The rich structure of the hyperbolic geometry of $(H^{n+1}, \rho)$ is now available as an important tool for studying any subgroup $G$ of $GM(\hat{\mathbb{R}}^n)$ for we can form the Poincaré extension of each $\phi$ in $G$ and thereby study $G$ as a group of isometries of $H^{n+1}$.

We shall study the geometry of the hyperbolic plane $H^2$ in great detail in Chapter 7 and some of the results (and proofs) given there extend without difficulty to $H^{n+1}$. One such result is that if $x = se_{n+1}$ and $y = te_{n+1}$, then

$$\rho(x, y) = |\log(s/t)|,$$

so

$$\cosh \rho(x, y) = 1 + \frac{|x - y|^2}{2x_{n+1}y_{n+1}}. \tag{3.3.4}$$

As both sides of (3.3.4) are invariant under all $\tilde{\phi}$, we see that this is actually valid for all $x$ and $y$ in $H^{n+1}$.

In particular, the hyperbolic sphere

$$\{x \in H^{n+1} : \rho(x, y) = r\}$$

with hyperbolic centre $(y_1, \ldots, y_{n+1})$ and hyperbolic radius $r$ is precisely the Euclidean sphere

$$(x_1 - y_1)^2 + \cdots + (x_n - y_n)^2 + (x_{n+1} - y_{n+1} \cosh r)^2 = (y_{n+1} \sinh r)^2. \tag{3.3.5}$$

In addition to this, we mention that given two distinct points of $H^{n+1}$ there is a unique curve $\gamma$ joining them which minimizes the integral

$$\int_\gamma \frac{|dx|}{x_{n+1}} :$$

such a curve is an arc of a geodesic and the geodesics are the Euclidean semi-circles orthogonal to $\mathbb{R}^n$ together with the vertical Euclidean lines in $H^{n+1}$.

EXERCISE 3.3

1. Show that if $x$ and $y$ are in $H^{n+1}$ then

$$\sinh^2 \tfrac{1}{2}\rho(x, y) = \frac{|x - y|^2}{4x_{n+1}y_{n+1}}.$$

2. Show that if $x \in H^{n+1}$ then

$$\cosh \rho(x, |x|e_{n+1}) = |x|/x_{n+1}$$

and interpret this geometrically.

3. Let $S$ be the hyperbolic sphere in $H^{n+1}$ with hyperbolic centre $y$ and hyperbolic radius $r$. Let $\bar{y}$ denote the reflection of $y$ in the plane $x_{n+1} = 0$. Show that

$$S = \left\{ x : \frac{|x - y|}{|x - \bar{y}|} = \tanh(\tfrac{1}{2}r) \right\}.$$

4. Suppose that $\phi \in GM(\hat{\mathbb{R}}^{n+1})$ and that $\phi$ leaves $H^{n+1}$ invariant. Prove that $\phi$ is the Poincaré extension of some $\psi$ in $GM(\hat{\mathbb{R}}^n)$.

## §3.4. Self-mappings of the Unit Ball

We have seen that the elements of $GM(\hat{\mathbb{R}}^n)$ act as hyperbolic isometries of $H^{n+1}$ and we can obviously transform this situation to obtain other models of hyperbolic space. We shall now map $H^{n+1}$ onto $B^{n+1}$ and so obtain another (isomorphic) copy of $GM(\hat{\mathbb{R}}^n)$ in which the elements leave $B^{n+1}$ invariant. This new model has a greater symmetry and the point $\infty$ no longer plays a special role.

Let $\phi_0$ denote the reflection in $S(e_{n+1}, \sqrt{2})$ so that

$$\phi_0(x) = e_{n+1} + \frac{2(x - e_{n+1})}{|x - e_{n+1}|^2}.$$

If $x \in \mathbb{R}^n$, then

$$\phi_0(\tilde{x}) = e_{n+1} + \frac{2(x_1, \ldots, x_n, -1)}{1 + |x|^2}$$

$$= \left( \frac{2x_1}{1 + |x|^2}, \ldots, \frac{2x_n}{1 + |x|^2}, \frac{|x|^2 - 1}{|x|^2 + 1} \right)$$

and this is precisely the formula for the stereographic projection $\pi$ of $\hat{\mathbb{R}}^n$ onto $S^n$ in $\hat{\mathbb{R}}^{n+1}$ considered in Section 3.1.

This realization of stereographic projection as a reflection leads to an easy proof of the formula for the chordal distance given in (3.1.3). If $x \in \mathbb{R}^n$ then

$$|x - e_{n+1}|^2 = 1 + |x|^2$$

and this with (3.1.5) yields (as before)

$$d(x, y) = |\pi(\tilde{x}) - \pi(\tilde{y})|$$

$$= |\phi_0(\tilde{x}) - \phi_0(\tilde{y})|$$

$$= \frac{2|x - y|}{(1 + |x|^2)^{1/2}(1 + |y|^2)^{1/2}}.$$

Let us now return to the reflection $\phi_0$ defined above. If $x \in \hat{\mathbb{R}}^{n+1}$ then

$$|\phi_0(x)|^2 = 1 + \frac{4}{|x - e_{n+1}|^2} + \frac{4(e_{n+1} \cdot [x - e_{n+1}])}{|x - e_{n+1}|^2}$$

$$= 1 + \frac{4x_{n+1}}{|x - e_{n+1}|^2}: \qquad (3.4.1)$$

this shows that $\phi_0$ maps the lower half-space $x_{n+1} < 0$ into $B^{n+1}$.

Now let $\phi = \phi_0 \sigma$ where $\sigma$ is the reflection in the plane $x_{n+1} = 0$: this maps the plane $x_{n+1} = 0$ onto $S^n$ and $H^{n+1}$ onto $B^{n+1}$. Also, we find from (3.1.5) that

$$\lim_{y \to x} \frac{|\phi(y) - \phi(x)|}{|y - x|} = \lim_{y \to x} \frac{|\phi_0(\sigma(y)) - \phi_0(\sigma(x))|}{|y - x|}$$

$$= \lim_{y \to x} \frac{|\phi_0(\sigma(y)) - \phi_0(\sigma(x))|}{|\sigma(y) - \sigma(x)|}$$

$$= \frac{2}{|\sigma(x) - e_{n+1}|^2}.$$

Now (3.4.1) with $x$ replaced by $\sigma(x)$ gives

$$1 - |\phi(x)|^2 = 1 - |\phi_0(\sigma(x))|^2$$

$$= \frac{4x_{n+1}}{|\sigma(x) - e_{n+1}|^2}$$

and so we find that

$$\lim_{y \to x} \frac{|\phi(y) - \phi(x)|}{|y - x|} = \frac{1 - |\phi(x)|^2}{2x_{n+1}}.$$

It now follows from Section 1.6 that the hyperbolic metric $\rho$ in $H^{n+1}$ transforms to the metric

$$ds = \frac{2|dx|}{1 - |x|^2}$$

in $B^{n+1}$ and that the isometries $\psi$ of $H^{n+1}$ transform by $\psi \mapsto \phi\psi\phi^{-1}$ to isometries of $B^{n+1}$ with this metric. This shows that $GM(\hat{\mathbb{R}}^n)$ is conjugate in $GM(\hat{\mathbb{R}}^{n+1})$ to the subgroup of $GM(\hat{\mathbb{R}}^{n+1})$ consisting of those elements which leave $B^{n+1}$ invariant.

We shall now undertake a study of those Möbius transformations which leave the unit ball invariant. As there is no longer any need to consider $\hat{\mathbb{R}}^{n+1}$ we revert to a consideration of the space $\hat{\mathbb{R}}^n$: thus we shall study the elements $\phi$ in $GM(\hat{\mathbb{R}}^n)$ with $\phi(B^n) = B^n$.

Before proceeding further, we mention that we can derive a formula for $B^n$ analogous to (3.3.4): see Chapter 7. In fact we only need to know that if $x \in B^n$, then

$$\rho(0, x) = \log\left(\frac{1 + |x|}{1 - |x|}\right)$$

and we leave the details of this to the reader.

**Theorem 3.4.1.** *Let $\phi$ be a Möbius transformation with $\phi(0) = 0$ and $\phi(B^n) = B^n$. Then $\phi(x) = xA$ for some orthogonal matrix $A$.*

PROOF. By Theorem 3.2.5, $\phi$ fixes $\infty$ and, as in the proof of Theorem 3.2.7, we see that $\phi$ is a Euclidean similarity. Because $\phi$ fixes the origin and leaves $S^{n-1}$ invariant, it is actually a Euclidean isometry. The result now follows from Theorem 3.1.4.                                                      □

It is easy to see that the reflection in the plane $P(a, t)$ leaves $B^n$ invariant if and only if $t = 0$. Better still, this reflection leaves $B^n$ invariant if and only if $P(a, t)$ is orthogonal to $S^{n-1}$ and in this form the statement is true for all reflections.

**Theorem 3.4.2.** *Let $\phi$ be the reflection in $S(a, r)$. Then the following are equivalent:*

(i) *$S(a, r)$ and $S^{n-1}$ are orthogonal;*
(ii) *$\phi(a^*) = 0$ (equivalently, $\phi(0) = a^*$);*
(iii) *$\phi(B^n) = B^n$.*

PROOF. As

$$\begin{aligned} \phi(0) &= a - r^2 a^* \\ &= (|a|^2 - r^2)a^* \end{aligned}$$

we see that (i) and (ii) are equivalent. The assertion that (iii) implies (ii) is simply the fact that $a$ and $a^*$ map to inverse points with respect to $S^{n-1}$ (Theorem 3.2.5).

Finally, (i) and (ii) together with (3.1.5) imply that

$$\begin{aligned} |\phi(x)| &= |\phi(x) - \phi(a^*)| \\ &= \frac{r^2|x - a^*|}{|x - a|.|a^* - a|} \\ &= \frac{|a|.|x - a^*|}{|x - a|}, \end{aligned}$$

so

$$1 - |\phi(x)|^2 = \frac{(1 - |x|^2)r^2}{|x - a|^2} \qquad (3.4.2)$$

and this proves (iii). □

As another application of (3.4.2) we observe that if $\phi$ preserves $B^n$ then

$$\frac{|\phi(x) - \phi(y)|^2}{(1 - |\phi(x)|^2)(1 - |\phi(y)|^2)} = \frac{|x - y|^2}{(1 - |x|^2)(1 - |y|^2)}: \qquad (3.4.3)$$

this follows immediately from (3.1.5) and (3.4.2). In addition, (3.4.3) holds whenever $\phi$ is the reflection in a plane $P(a, 0)$ and hence for all Möbius $\phi$ which preserve $B^n$.

The invariance expressed by (3.4.3) also yields

$$\lim_{y \to x} \frac{|\phi(y) - \phi(x)|}{|y - x|} = \frac{1 - |\phi(x)|^2}{1 - |x|^2}$$

and this confirms once again the invariance of the hyperbolic metric in $B^n$.

In two dimensions the complex conjugate $\bar{z}$ of $z$ is available and in our notation this may be written as

$$z^* = 1/\bar{z}.$$

The familiar expression $|1 - \bar{z}w|$ (where $z$ and $w$ are complex numbers) satisfies

$$|1 - \bar{z}w| = |z| \, |z^* - w|$$

and this suggests the definition

$$[u, v] = |u| \, |u^* - v| \qquad (u, v \in R^n).$$

Observe that

$$\begin{aligned} [u, v]^2 &= |u|^2 |v|^2 + 2(u \cdot v) + 1 \\ &= |u - v|^2 + (|u|^2 - 1)(|v|^2 - 1) \end{aligned} \qquad (3.4.4)$$

and this shows that

$$[u, v] = [v, u].$$

The identity (3.4.4) also shows that if $|a| > 1$ then

$$\frac{|x - a^*|}{[x, a^*]} = 1$$

if and only if $|x| = 1$. Thus

$$S^{n-1} = \left\{ x \in R^n : \frac{|x - a^*|}{[x, a^*]} = 1 \right\}$$

and this is the $n$-dimensional version of the equation

$$\left| \frac{z - w}{1 - \bar{z}w} \right| = 1$$

of the unit circle in the complex plane.

Finally, we observe that (3.4.4) together with the invariance expressed by (3.4.3) yields the invariance

$$\frac{[\phi(x), \phi(y)]^2}{(1 - |\phi(x)|^2)(1 - |\phi(y)|^2)} = \frac{[x, y]^2}{(1 - |x|^2)(1 - |y|^2)}. \tag{3.4.5}$$

EXERCISE 3.4

1. Show that for $x$ in $B^n$,

$$\rho(0, x) = \log\left( \frac{1 + |x|}{1 - |x|} \right).$$

Deduce that if $x$ and $y$ are in $B^n$ then

$$\sinh^2 \tfrac{1}{2}\rho(x, y) = \frac{|x - y|^2}{(1 - |x|^2)(1 - |y|^2)}.$$

[Use (3.4.3).]

2. Let $\phi$ and $\psi$ be reflections in the spheres $S(a, r)$ and $S(b, t)$ respectively. Show that these spheres are orthogonal if and only if $\phi(b) = \psi(a)$.

3. Use Questions 1 and 2 to show that if $S(a, r)$ is orthogonal to $S(0, 1)$ and if $\phi$ denotes reflection in $S(a, r)$ then

$$\sinh \tfrac{1}{2}\rho(0, \phi 0) = 1/r$$

and, for all $x$,

$$|\phi(x) - a| \cdot |x - a| = 1/\sinh^2 \tfrac{1}{2}\rho(0, \phi 0).$$

# §3.5. The General Form of a Möbius Transformation

We shall establish the following characterization of Möbius transformations.

**Theorem 3.5.1.** *Let $\phi$ be a Möbius transformation.*

(i) *If $\phi(B^n) = B^n$ then*

$$\phi(x) = (\sigma x)A,$$

*where $\sigma$ is a reflection in some sphere orthogonal to $S^{n-1}$ and $A$ is an orthogonal matrix.*

(ii) *If $\phi(\infty) = \infty$ then*

$$\phi(x) = r(xA) + x_0,$$

*where $r > 0$, $x_0 \in \mathbb{R}^n$ and $A$ is orthogonal.*
(iii) *If $\phi(\infty) \neq \infty$ then*

$$\phi(x) = r(\sigma x)A + x_0$$

*for some $r$, $x_0$, $A$ and some reflection $\sigma$.*

*Remark.* $\sigma(x)A$ denotes $\sigma$ followed by $A$: the matrix $A$ appears on the right as we are using row vectors.

PROOF. If $\phi$ preserves $B^n$, let $\sigma$ be the reflection in the sphere $S(a, r)$ where $a = \phi^{-1}(\infty)$ and $|a|^2 = 1 + r^2$. By Theorem 3.4.2, $\sigma$ (and hence $\phi\sigma$) preserves $B^n$. By computation, $\sigma(0) = a^*$ so

$$\phi(\sigma(0)) = \phi(a^*) = 0,$$

(because $\phi$ preserves inverse points): thus $\phi(\sigma x) = xA$. Replacing $x$ by $\sigma x$, we obtain (i).

If $\phi$ fixes $\infty$ then, for a suitable

$$\psi : x \mapsto (x - x_0)/r,$$

the map $\psi\phi$ fixes $\infty$ and $B^n$ and hence also the origin. Now (ii) follows from Theorem 3.4.1. Finally, (iii) follows by applying (ii) to $\phi\sigma$ for a suitable reflection $\sigma$ mapping $\infty$ to $\phi^{-1}(\infty)$.          □

The characterization in (iii) leads to the notion of an isometric sphere. Suppose that $\phi(\infty) \neq \infty$ so that

$$\phi(x) = r(\sigma x)A + x_0,$$

where $\sigma$ is the reflection in some sphere $S(a, t)$ and (necessarily) $a = \phi^{-1}(\infty)$. By (3.1.5),

$$|\phi(x) - \phi(y)| = r|\sigma(x) - \sigma(y)|$$

$$= \frac{rt^2|x - y|}{|x - a|.|y - a|}$$

and so $\phi$ acts as a Euclidean isometry on the sphere with equation $|x - a| = t_1$ where $t_1 = t\sqrt{r}$. Indeed,

$$\lim_{y \to x} \frac{|\phi(y) - \phi(x)|}{|y - x|}$$

is greater than, equal to or less than one according as $x$ is inside, on or outside $S(a, t_1)$. For this reason, $S(a, t_1)$ is called the *isometric sphere* of $\phi$.

Note that if $\sigma$ denotes reflection in the isometric sphere of $\phi$ then $\phi\sigma$ fixes $\infty$ and also acts as a Euclidean isometry on the isometric sphere. It follows that the expression in Theorem 3.5.1(ii) must take the form

$$\phi\sigma(x) = xA + x_0,$$

so in general, we see that

$$\phi(x) = \psi\sigma(x),$$

where $\sigma$ is the reflection in the isometric sphere and $\psi$ is a Euclidean isometry.

In the special case when $\phi$ preserves $B^n$, the reflection $\sigma$ in Theorem 3.5.1(i) must be the reflection in the isometric sphere of $\phi$ as $\sigma$ and $A$ act as Euclidean isometries on this sphere. We deduce that in this case, *the isometric sphere is orthogonal to $S^{n-1}$*.

### EXERCISE 3.5

1. Show that if $\phi$ preserves $B^n$ then the Euclidean radius of the isometric sphere of $\phi$ is $1/(\sinh \frac{1}{2}\rho(0, \phi 0))$.

2. Show that if $\Sigma$ is the isometric sphere of $\phi$, then $\phi(\Sigma)$ is the isometric sphere of $\phi^{-1}$.

## §3.6. Distortion Theorems

We prove two sharp distortion theorems for Möbius transformations.

**Theorem 3.6.1.** *Let $\phi$ be a Möbius transformation acting in $\hat{\mathbb{R}}^n$ and let $\rho$ be the hyperbolic metric in $H^{n+1}$. Then*

$$\sup_{x, y \in \hat{\mathbb{R}}^n} \frac{d(\phi x, \phi y))}{d(x, y)} = \exp \rho(e_{n+1}, \phi e_{n+1}).$$

*Remark.* This shows that $\phi$ satisfies a Lipschitz condition on $\hat{\mathbb{R}}^n$ with respect to the chordal metric $d$ and actually exhibits the best Lipschitz constant in terms of $\phi$ acting on the hyperbolic space $(H^{n+1}, \rho)$.

The second result shows that if a family of Möbius transformations omits two values $\xi$ and $\zeta$ in a domain $D$, then the family is equicontinuous on compact subsets of $D$: this enables one to develop, for example, the theory of normal families for $GM(\hat{\mathbb{R}}^n)$.

**Theorem 3.6.2.** *Let $D$ be a subdomain of $\hat{\mathbb{R}}^n$ and suppose that $\xi$ and $\zeta$ are distinct points in $\hat{\mathbb{R}}^n$. If $\phi$ in $GM(\hat{\mathbb{R}}^n)$ does not assume the values $\xi$ and $\zeta$ in $D$, then for all $x$ and $y$ in $D$,*

$$d(\phi x, \phi y) \le \frac{8d(x, y)}{d(\xi, \zeta)d(x, \partial D)^{1/2}d(y, \partial D)^{1/2}}.$$

*The constant 8 is best possible.*

PROOF OF THEOREM 3.6.1. By reflecting in $x_{n+1} = 0$ and applying stereographic projection, we may assume that $\phi$ preserves $B^{n+1}$: now we need to show that

$$\sup_{x, y \in S^n} \frac{|\phi x - \phi y|}{|x - y|} = \exp \rho(0, \phi 0).$$

By Theorem 3.5.1(i), the Euclidean distortion under $\phi$ is the same as the distortion under the reflection $\sigma$ in the isometric sphere $S(a, r)$ of $\phi$. This is maximal (as a limiting value) at the point of $S^n$ closest to the centre $a$ of $S(a, r)$. Thus from (3.1.5),

$$\sup_{x, y \in S^n} \frac{|\phi(x) - \phi(y)|}{|x - y|} = \frac{r^2}{(|a| - 1)^2}$$

$$= \frac{|a| + 1}{|a| - 1},$$

because $S(a, r)$ is orthogonal to $S^n$ (Section 3.5). Now

$$|a| = |\phi^{-1}(\infty)| = 1/|\phi^{-1}(0)|$$

and so the supremum is

$$\frac{1 + |\phi^{-1}(0)|}{1 - |\phi^{-1}(0)|} = \exp \rho(0, \phi^{-1}(0))$$

$$= \exp \rho(\phi 0, 0). \qquad \square$$

PROOF OF THEOREM 3.6.2. Suppose that $x$ and $y$ are distinct points in $D$ and that $\alpha$ and $\beta$ are distinct points outside of $D$. By Theorem 3.2.7, the product

$$[x, \alpha, y, \beta] . [x, \beta, y, \alpha]$$

of cross-ratios is invariant under $\phi$. Thus

$$\left[\frac{d(\phi x, \phi y)}{d(x, y)}\right]^2 \leq \left[\frac{d(\alpha, \beta)}{d(\phi \alpha, \phi \beta)}\right]^2 \left[\frac{16}{d(x, \alpha)d(x, \beta)d(y, \alpha)d(y, \beta)}\right]$$

$$\leq \left[\frac{4}{d(\phi \alpha, \phi \beta)}\right]^2 \left[\frac{1}{d(x, \alpha)} + \frac{1}{d(x, \beta)}\right]\left[\frac{1}{d(y, \alpha)} + \frac{1}{d(y, \beta)}\right]$$

$$\leq \frac{64}{d(\phi \alpha, \phi \beta)^2 d(x, \partial D)d(y, \partial D)}.$$

The inequality follows by writing $\alpha = \phi^{-1}(\xi)$ and $\beta = \phi^{-1}(\zeta)$.

To show that the constant 8 cannot be improved, consider $\phi(z) = z + 2m$ acting on $\hat{C}$ with $D = \hat{C} - \{\infty, -m\}$. Clearly, $\phi$ omits the values $\infty$ and $m$ in $D$ and if $x = -2m$, we have

$$\lim_{y \to x} \frac{d(\phi x, \phi y)}{d(x, y)} \sim \frac{8}{d(\infty, m)d(x, \partial D)},$$

as $m \to +\infty$. $\qquad \square$

As an application of Theorem 3.6.2, we mention (briefly) the concept of a normal family. A family $\mathscr{F}$ of functions from one metric space $(X, d)$ to another, say to $(X', d')$, is *equicontinuous* on $X$ if and only if for every positive $\varepsilon$ there is a positive $\delta$ such that for all $x$ and $y$ in $X$ and all $f$ in $\mathscr{F}$,

$$d'(fx, fy) < \varepsilon \quad \text{whenever } d(x, y) < \delta.$$

Each function in an equicontinuous family is uniformly continuous on $X$ and the uniformity is with respect to $f$ as well as to the pair $(x, y)$.

A family $\mathscr{F}$ (as above) is said to be *normal* in $X$ if every sequence $f_1, f_2, \ldots$ chosen from $\mathscr{F}$ has a subsequence that converges uniformly on each compact subset of $X$. There is a general result (the Arzela–Ascoli Theorem) which relates the concepts of equicontinuity and normal families. In the context in which we are primarily interested, it is sufficient to obtain the following special case.

**Proposition 3.6.3.** *A family $\mathscr{F}$ of Möbius transformations of $(\hat{\mathbb{R}}^n, d)$ onto itself is normal in a subdomain $D$ of $\hat{\mathbb{R}}^n$ if it is equicontinuous on every compact subset of $D$.*

PROOF. We only sketch the proof as the interested reader can find a proof of the Arzela–Ascoli Theorem elsewhere in the literature. Find a sequence $x_1, x_2, \ldots$ which is dense in $D$. Given a sequence $\phi_1, \phi_2, \ldots$ in $\mathscr{F}$ we can find (because $\hat{\mathbb{R}}^n$ is compact) a subsequence which converges at $x_1$, then a subsequence of this which converges at $x_2$ and so on. By choosing a subsequence of the $\phi_n$ suitably, we can obtain a subsequence which is ultimately a subsequence of each of these chosen subsequences: thus we have constructed a subsequence which converges at each point $x_j$.

Now take any compact subset $K$ of $D$ and consider any positive $\varepsilon$. We can cover $K$ by a *finite* number of open balls (in the $d$-metric) of radius $\delta$ (corresponding to $\varepsilon$ in the definition of equicontinuity). Select one point $x_j$ in each: let the selected points be (after relabelling) $x_1, x_2, \ldots, x_s$. If $y$ is in $K$ then $d(y, x_j) < \delta$ for some $j$ and hence

$$d(\phi_n y, \phi_m y) \leq d(\phi_n y, \phi_n x_j) + d(\phi_n x_j, \phi_m x_j) + d(\phi_m x_j, \phi_m y)$$
$$\leq 2\varepsilon + d(\phi_n x_j, \phi_m x_j).$$

For $n, m \geq n_0$, say, the last term is at most $\varepsilon$ for all $x_1, \ldots, x_s$: hence $d(\phi_n y, \phi_m y) \leq 3\varepsilon$ on $K$.  $\square$

We can now combine Theorem 3.6.2 and Proposition 3.6.3.

**Theorem 3.6.4.** *Let $D$ be a subdomain of $\hat{\mathbb{R}}^n$ and let $\mathscr{F}$ be a family of Möbius transformations. Suppose that for every $\phi$ in $\mathscr{F}$, there are two points $\alpha_\phi, \beta_\phi$ in $\hat{\mathbb{R}}^n$ which are not taken as values of $\phi$ in $D$ and suppose that also,*

$$\inf_\phi d(\alpha_\phi, \beta_\phi) > 0.$$

*Then $\mathscr{F}$ is normal in $D$.*

*Remark.* We can rewrite the inequality in Theorem 3.6.4 as

$$\inf_{\phi} [\text{chordal diameter } \phi(\hat{\mathbb{R}}^n - D)] > 0.$$

PROOF. We simply apply Theorem 3.6.2 with $\zeta = \alpha_\phi, \zeta = \beta_\phi$ and we find that $\mathscr{F}$ is equicontinuous (in fact, it satisfies a uniform Lipschitz condition) on every compact subset of $D$. $\qquad\square$

Finally, this leads to the following result.

**Theorem 3.6.5.** *Let $\phi_1, \phi_2, \ldots$ be Möbius transformations and suppose that $\phi_n(x_j) \to y_j$ for three distinct points $x_1, x_2, x_3$ and three distinct points $y_1, y_2, y_3$. Then $\phi_1, \phi_2, \ldots$ contains a subsequence which converges uniformly on $\hat{\mathbb{R}}^n$ to a Möbius transformation.*

PROOF. By the deletion of a finite number of the $\phi_j$ (which clearly does not affect the result) we may assume that for each $n, i$ and $j$ ($i \neq j$) we have

$$d(\phi_n x_i, \phi_n x_j) \geq \tfrac{1}{2} d(y_i, y_j) > 0.$$

It follows that the family $\{\phi_1, \phi_2, \ldots\}$ is normal in each of the sets $\hat{\mathbb{R}}^n - \{x_i, x_j\}$ (Theorem 3.6.4) and hence in their union, namely $\hat{\mathbb{R}}^n$. Thus there is a subsequence of the $\phi_j$ converging uniformly to some $\phi$ in $\hat{\mathbb{R}}^n$ and by Theorem 3.2.7 (and its proof), $\phi$ is a Möbius transformation. $\qquad\square$

EXERCISE 3.6

1. Show that a family $F$ of Möbius transformations is normal in $\hat{\mathbb{R}}^n$ if and only if

$$\sup_{\phi \in F} \rho(e_{n+1}, \phi e_{n+1}) < +\infty$$

where $e_{n+1} = (0, \ldots, 0, 1)$ in $H^{n+1}$.

2. Prove that if two Möbius transformations are equal on an open subset $D$ of $\hat{\mathbb{R}}^n$ then they are the same transformation on $\hat{\mathbb{R}}^n$. Deduce that if the Möbius transformations $\phi_n$ converge uniformly to $I$ on some open subset of $\hat{\mathbb{R}}^n$, then they converge uniformly to $I$ on $\hat{\mathbb{R}}^n$.

# §3.7. The Topological Group Structure

There are several ways to give $GM(\hat{\mathbb{R}}^n)$ the structure of a topological group. The simplest construction is to observe that the elements of $GM(\hat{\mathbb{R}}^n)$ map the compact space $\hat{\mathbb{R}}^n$ onto itself so

$$D(\phi, \psi) = \sup\{d(\phi x, \psi x) : x \in \hat{\mathbb{R}}^n\},$$

(where $d$ is the chordal metric on $\hat{\mathbb{R}}^n$) is a metric on $GM(\hat{\mathbb{R}}^n)$. Clearly, $\phi_n \to \phi$ in this metric if and only if $\phi_n \to \phi$ uniformly on $\hat{\mathbb{R}}^n$.

**Theorem 3.7.1.** $GM(\hat{\mathbb{R}}^n)$ *is a topological group with respect to the topology induced by the metric* $D$.

PROOF. From Theorem 3.6.1, we see that for each $\phi$ in $GM(\hat{\mathbb{R}}^n)$ there is a positive constant $c(\phi)$ such that for all $x$ and $y$ we have

$$d(\phi x, \phi y) \le c(\phi)d(x, y).$$

Clearly, for any $\phi_1$, $\phi_2$ and $\psi$ we also have

$$D(\phi_1\psi, \phi_2\psi) = D(\phi_1, \phi_2),$$

so

$$\begin{aligned} D(\phi\psi, \phi_1\psi_1) &\le D(\phi\psi, \phi_1\psi) + D(\phi_1\psi, \phi_1\psi_1) \\ &\le D(\phi, \phi_1) + c(\phi_1)D(\psi, \psi_1). \end{aligned}$$

This shows that the composition map $(\phi, \psi) \mapsto \phi\psi$ is continuous at $(\phi_1, \psi_1)$. Similarly, the map $\psi \mapsto \psi^{-1}$ is continuous at $\phi$ as

$$\begin{aligned} D(\phi^{-1}, \psi^{-1}) &= D(\phi^{-1}\psi, I) \\ &\le c(\phi^{-1})D(\psi, \phi). \end{aligned} \qquad \square$$

For a different construction of the same topology we proceed as follows. The group $GM(\hat{\mathbb{R}}^n)$ is conjugate in $GM(\hat{\mathbb{R}}^{n+1})$ to the group $GM(B^{n+1})$ of all Möbius transformations preserving $B^{n+1}$. If $\phi$ in $GM(\hat{\mathbb{R}}^n)$ corresponds to $\phi_1$ in $GM(B^{n+1})$ then (by definition of the chordal metric)

$$D(\phi, \psi) = \sup\{|\phi_1 x - \psi_1 x| : x \in S^n\}.$$

Thus we may consider $GM(B^{n+1})$ instead of $GM(\hat{\mathbb{R}}^n)$ with the metric (which we continue to denote by $D$) of uniform convergence in Euclidean terms on $S^n$ and the conjugation is then an isometry between $GM(\hat{\mathbb{R}}^n)$ and $GM(B^{n+1})$.

For each non-zero $a$ in $B^{n+1}$ let $\sigma_a$ be the reflection in the sphere with centre $a^*$ that is orthogonal to $S^n$: thus $\sigma_a$ preserves $B^{n+1}$ and $\sigma_a(a) = 0$. Also, let $\tau_a$ denote the reflection in the plane $x \cdot a = 0$. Then, defining $T_a$ to be the composition $\tau_a\sigma_a$, we find that the isometry $T_a$ of $B^{n+1}$ leaves the Euclidean diameter through $a$ invariant and $T_a(a) = 0$. We call any isometry $T_a$ constructed in this way a *pure translation*: if $a = 0$ we define $T_a$ to be the identity.

**Lemma 3.7.2.** (i) *The map* $\phi \mapsto \phi(0)$ *of* $GM(B^{n+1})$ *onto* $B^{n+1}$ *is continuous*
(ii) *The map* $a \mapsto T_a$ *is a homeomorphism of* $B^{n+1}$ *onto the set of pure translations.*

PROOF. To prove (i) we suppose first that $D(\phi_n, I) < \varepsilon$. Each Euclidean diameter $L_j$ of $B^{n+1}$ is mapped by $\phi_n$ to a circular arc $\phi_n(L_j)$ (orthogonal to $S^n$) in $B^{n+1}$ whose end-points are at most a distance $\varepsilon$ from those of $L_j$. We deduce that the Euclidean cylinder $C_j$ with axis $L_j$ and radius of cross-section $\varepsilon$ contains $\phi_n(L_j)$. Thus

$$\phi_n(0) = \bigcap_j \phi_n(L_j)$$

$$\subset \bigcap_j C_j$$

$$= \{x \in B^{n+1} : |x| < \varepsilon\}.$$

This shows that if $\phi_n \to I$ uniformly on $S^n$, then $\phi_n(0) \to 0$: in fact,

$$|\phi_n(0)| < D(\phi_n, I).$$

Suppose now that $\phi_n \to \phi$ (as $\mathrm{GM}(B^{n+1})$ and $B^{n+1}$ are metric spaces, it is sufficient to consider sequential convergence). From Theorem 3.7.1 we have $\phi^{-1}\phi_n \to I$: thus (from above) $\phi^{-1}\phi_n(0) \to 0$ and hence $\phi_n(0) \to \phi(0)$. This proves (i).

To prove (ii) observe first that the map $T_a \mapsto T_a^{-1}$ is continuous (Theorem 3.7.1). By (i), the composite map

$$T_a \mapsto T_a^{-1} \mapsto T_a^{-1}(0),$$

namely, $T_a \mapsto a$, is continuous.

It remains to prove that the map $a \mapsto T_a$ is continuous: explicitly, as $b \to a$ so $T_b \to T_a$ uniformly on $S^n$. We have explicit formulae for $\phi_a$ and $\tau_a$ and the continuity follows from straightforward (if tedious) estimates: we omit the details. □

We know from Theorem 3.5.1 that every element $\phi$ of $\mathrm{GM}(B^{n+1})$ can be expressed uniquely as

$$\phi(x) = (\sigma_a x)A,$$

where $a = \phi^{-1}(0)$ and $A$ is an orthogonal matrix ($A$ acts after $\sigma_a$: it appears on the right because we are using row vectors). It follows that we can also write (uniquely)

$$\phi(x) = (T_a x)A_\phi,$$

where $A_\phi$ (namely, $\tau_a$ followed by $A$) is also an orthogonal matrix and this description establishes a natural bijection between $\mathrm{GM}(B^{n+1})$ and $O(n+1) \times B^{n+1}$ by the correspondence

$$\phi \mapsto (A_\phi, a), \qquad a = \phi^{-1}(0).$$

Now the group $O(n+1)$ of orthogonal matrices is itself a metric space. First, there is the natural metric

$$|(a_{ij}) - (b_{ij})| = \left[ \sum_{i,j} (a_{ij} - b_{ij})^2 \right]^{1/2},$$

and second, there is the metric $D$ induced by regarding $O(n + 1)$ as a subset of $GM(B^{n+1})$. In fact, these metrics yield the same topology because if $A = (a_{ij})$, $B = (b_{ij})$, $C = A - B$ and $x$ is on $S^n$, then

$$D(A, B)^2 = \sup_{|x|=1} |xA - xB|^2$$

$$= \sup_{|x|=1} \sum_{j=1}^{n} (x_1 c_{1j} + \cdots + x_n c_{nj})^2$$

$$\leq \sup_{|x|=1} \sum_{j=1}^{n} \left( \sum_{i=1}^{n} x_i^2 \right) \left( \sum_{i=1}^{n} c_{ij}^2 \right)$$

$$= |A - B|^2$$

$$= \sum_{i=1}^{n} \left( \sum_{j=1}^{n} c_{ij}^2 \right)$$

$$= \sum_{i=1}^{n} |e_i A - e_i B|^2$$

$$\leq nD(A, B)^2.$$

The space $O(n + 1) \times B^{n+1}$ now inherits a natural product topology and we have the following result.

**Theorem 3.7.3.** *The bijection* $\phi \mapsto (A_\phi, a)$ *is a homeomorphism of* $GM(B^{n+1})$ *onto* $O(n + 1) \times B^{n+1}$.

PROOF. The proof consists of repeated applications of Theorem 3.7.1 and Lemma 3.7.2. First, $a \mapsto T_a$ is continuous, hence so is the map $(A_\phi, a) \mapsto (A_\phi, T_a)$. Also the map of $(A_\phi, T_a)$ into their composition, namely $\phi$, is continuous thus so is the map $(A_\phi, a) \mapsto \phi$.

Next, $\phi \mapsto a\, (= \phi^{-1} 0)$ is continuous, as are the maps $a \mapsto T_a$ and $T_a \mapsto T_a^{-1}$: thus $\phi \mapsto T_a^{-1}$ is continuous. We deduce that the composition

$$\phi \mapsto (\phi, T_a^{-1}) \mapsto \phi T_a^{-1} = A_\phi$$

is continuous, hence so is $\phi \mapsto (A_\phi, a)$.                                           □

*Remark.* Theorem 3.7.3 simply means that the topology on $GM(B^{n+1})$ induced by the bijection from $O(n + 1) \times B^{n+1}$ coincides with the topology induced by the metric $D$. As $GM(\hat{\mathbb{R}}^n)$ has been identified isometrically with $GM(B^{n+1})$, this result provides a new construction for the topology induced on $GM(\hat{\mathbb{R}}^n)$ by the metric $D$.

For our third and final construction of the topology we need another model of hyperbolic space.

**Definition 3.7.4.** Let $Q$ be the *hyperboloid model* defined by

$$Q = \{(x_0, \ldots, x_n) \in \mathbb{R}^{n+1} : q(x, x) = 1, x_0 > 0\},$$

where

$$q(x, y) = x_0 y_0 - (x_1 y_1 + \cdots + x_n y_n).$$

Observe that $Q$ is one sheet of a hyperboloid of two sheets and that if $x \in Q$ then

$$x_0^2 = 1 + (x_1^2 + \cdots + x_n^2),$$

so, in fact, $x_0 \geq 1$.

Now let $\gamma = (\gamma_0, \ldots, \gamma_n)$ be any smooth curve on $Q$. Thus for all $t$,

$$\gamma_0(t)^2 = \gamma_1(t)^2 + \cdots + \gamma_n(t)^2 + 1,$$

so differentiating,

$$\gamma_0(t)\dot{\gamma}_0(t) = \gamma_1(t)\dot{\gamma}_1(t) + \cdots + \gamma_n(t)\dot{\gamma}_n(t),$$

(more briefly, $q(\gamma, \gamma) = 1$ so $q(\gamma, \dot{\gamma}) = 0$). We deduce that

$$
\begin{aligned}
q(\dot{\gamma}, \dot{\gamma}) &= \left( \frac{\gamma_1 \dot{\gamma}_1 + \cdots + \gamma_n \dot{\gamma}_n}{\gamma_0} \right)^2 - (\dot{\gamma}_1^2 + \cdots + \dot{\gamma}_n^2) \\
&\leq \left( \sum \gamma_j^2 \right) \left( \sum \dot{\gamma}_j^2 \right) / \gamma_0^2 - \left( \sum \dot{\gamma}_j^2 \right) \\
&= -\left( \sum \dot{\gamma}_j^2 \right) / \gamma_0^2 \\
&\leq 0,
\end{aligned}
$$

the summations being over $j = 1, \ldots, n$. Observe also that a strict inequality holds unless $\dot{\gamma}_1 = \cdots = \dot{\gamma}_n = 0$ in which case, $\dot{\gamma}_0 = 0$ also. It follows that we can construct a metric on $Q$ in the usual way by the line element

$$ds^2 = dx_1^2 + \cdots + dx_n^2 - dx_0^2, \tag{3.7.1}$$

the distance between two points on $Q$ being the infimum of

$$\int [-q(\dot{\gamma}, \dot{\gamma})]^{1/2} \, dt$$

over all curves joining the two points. The associated metric topology is the Euclidean topology on $Q$. We shall now compare $Q$ and this metric with the model $B^n$ and the metric

$$ds^2 = \frac{4 \, dx^2}{(1 - |x|^2)^2}. \tag{3.7.2}$$

**Theorem 3.7.5.** *The map*

$$F: (x_0, \ldots, x_n) \mapsto \left( \frac{x_1}{1 + x_0}, \ldots, \frac{x_n}{1 + x_0} \right)$$

*is an isometry of $Q$ with the metric (3.7.1) onto $B^n$ with the metric (3.7.2).*

PROOF. For brevity, we write

$$(y_1, \ldots, y_n) = \left( \frac{x_1}{1 + x_0}, \ldots, \frac{x_n}{1 + x_0} \right)$$

and denote the vectors by $x$ and $y$ in the obvious way. As $x \in Q$, a computation yields

$$|y|^2 = \frac{x_0 - 1}{x_0 + 1}, \tag{3.7.3}$$

so $0 \le |y| < 1$ and $F$ maps $Q$ into $B^n$.

By direct computation we find that the map

$$F^{-1}: (y_1, \ldots, y_n) \mapsto \left( \frac{1 + |y|^2}{1 - |y|^2}, \frac{2y_1}{1 - |y|^2}, \ldots, \frac{2y_n}{1 - |y|^2} \right) \tag{3.7.4}$$

is indeed the inverse of $F$ and so $F$ is a bijection of $Q$ onto $B^n$.

To verify that $F$ is an isometry, we observe that

$$dy_j = \frac{dx_j}{1 + x_0} - \frac{x_j \, dx_0}{(1 + x_0)^2}.$$

Thus, using this and (3.7.3) we have

$$\begin{aligned}
\frac{4(dy_1^2 + \cdots + dy_n^2)}{(1 - |y|^2)^2} &= (1 + x_0)^2 \sum_{j=1}^n \left( \frac{dx_j}{1 + x_0} - \frac{x_j \, dx_0}{(1 + x_0)^2} \right)^2 \\
&= \sum_{j=1}^n dx_j^2 + \frac{dx_0^2}{(1 + x_0)^2} \sum_{j=1}^n x_j^2 - \frac{2(\sum_{j=1}^n x_j \, dx_j) \, dx_0}{(1 + x_0)} \\
&= \sum_{j=1}^n dx_j^2 + \left( \frac{x_0 - 1}{x_0 + 1} \right) dx_0^2 - \frac{dx_0 \, d(x_0^2 - 1)}{1 + x_0} \\
&= \sum_{j=1}^n dx_j^2 - dx_0^2.
\end{aligned}$$ □

It is now clear that the group $G(Q)$ of isometries of $Q$ and the group $GM(B^n)$ of isometries of $B^n$ are isomorphic by virtue of the relation

$$GM(B^n) = F(G(Q))F^{-1}.$$

Our aim now is to prove an alternative characterization of $G(Q)$ and hence of $GM(B^n)$.

**Theorem 3.7.6.** *The isometries of $Q$ are precisely the $(n + 1) \times (n + 1)$ matrices which preserve both the quadratic form $q(x, x)$ and the half-space given by $x_0 > 0$.*

PROOF. First, let $A$ be any matrix with the prescribed properties. As $x_0 > 0$ is preserved and as

$$q(xA, xA) = q(x, x) = 1,$$

when $x \in Q$ we see that $A$ preserves $Q$. Moreover, for any curve $\gamma$ on $Q$, let $\Gamma = \gamma A$. Then $\dot{\Gamma} = \dot{\gamma} A$ so

$$q(\dot{\Gamma}, \dot{\Gamma}) = q(\dot{\gamma}, \dot{\gamma})$$

and this simply expresses the fact that $\gamma$ and $\gamma A$ have the same length. Thus each such $A$ is an isometry of $Q$ onto itself.

It remains to show that every $\phi$ in $\mathrm{GM}(B^n)$ is of the form $F(A)F^{-1}$ for some such matrix $A$ and to do this, we simply compute the action of $F(A)F^{-1}$ on $B^n$. Suppose then that $A = (a_{ij})$ where $i, j = 0, 1, \ldots, n$. With the obvious notation, we write

$$(y_1, \ldots, y_n) \overset{F^{-1}}{\longmapsto} (u_0, u_1, \ldots, u_n)$$
$$\overset{A}{\longmapsto} (v_0, v_1, \ldots, v_n)$$
$$\overset{F}{\longmapsto} (w_1, \ldots, w_n).$$

Now

$$(v_0, \ldots, v_n) = (u_0, \ldots, u_n)A,$$

so

$$v_j = u_0 a_{0j} + \cdots + u_n a_{nj}.$$

Using (3.7.4), this yields

$$(1 - |y|^2)v_j = (1 + |y|^2)a_{0j} + 2(y_1 a_{1j} + \cdots + y_n a_{nj}).$$

Thus

$$w_j = \frac{v_j}{1 + v_0}$$
$$= \frac{(1 - |y|^2)v_j}{(1 - |y|^2) + (1 - |y|^2)v_0}$$
$$= \frac{(1 + |y|^2)a_{0j} + 2(y_1 a_{1j} + \cdots + y_n a_{nj})}{|y|^2(a_{00} - 1) + 2(y_1 a_{10} + \cdots + y_n a_{n0}) + (a_{00} + 1)} \tag{3.7.5}$$

and this is the explicit expression for the map $F(A)F^{-1}$.

If $A_0$ is an orthogonal $n \times n$ matrix (viewed as an isometry of $B^n$), then

$$A = \begin{pmatrix} 1 & 0 & \cdots & 0 \\ 0 & & & \\ \vdots & & A_0 & \\ 0 & & & \end{pmatrix}$$

preserves $q$ and the condition $x_0 > 0$. In this case, (3.7.5) yields $w = yA_0$ and so every isometry of $B^n$ which fixes the origin does arise in the form $F(A)F^{-1}$.

It is only necessary to show now that the reflection in the sphere $S(\zeta, r)$ orthogonal to $S^{n-1}$ is of the form $F(A)F^{-1}$. Because orthogonal transformations are of this form, we need only consider the case when $\zeta$ is of the form $(s, 0, \ldots, 0)$. It is actually more convenient to introduce another positive parameter $t$ with

$$\zeta = (c(t), 0, \ldots, 0), \qquad c(t) = \frac{\cosh t}{\sinh t}$$

and

$$r = 1/\sinh t,$$

so the orthogonality requirement $|\zeta|^2 = 1 + r^2$ is satisfied.

Consider now the matrix

$$P = \begin{pmatrix} \cosh 2t & \sinh 2t & 0 & \cdots & 0 \\ -\sinh 2t & -\cosh 2t & 0 & \cdots & 0 \\ 0 & 0 & & & \\ \vdots & \vdots & & I_{n-1} & \\ 0 & 0 & & & \end{pmatrix} :$$

observe that $\det(P) = -1$ and that $P$ preserves both the quadratic form $q(x, x)$ and the half-space $x_0 > 0$. The effect $y \mapsto w$ of $F(A)F^{-1}$ on $B^n$ is given by (3.7.5) and the denominator of this expression can be simplified as follows:

$$\begin{aligned} |y|^2(a_{00} - 1) + 2(y_1 a_{10} + \cdots + y_n a_{n0}) + (a_{00} + 1) \\ = 2|y|^2 \sinh^2 t - 2y_1 \sinh(2t) + 2 \cosh^2 t \\ = 2|y - \zeta|^2 \sinh^2 t \\ = 2|y - \zeta|^2/r^2. \end{aligned}$$

Now for $j = 2, \ldots, n$ the formula (3.7.5) yields

$$w_j = \frac{r^2 y_j}{|y - \zeta|^2}.$$

Also,

$$\begin{aligned} w_1 &= \frac{(1 + |y|^2) \sinh(2t) - 2y_1 \cosh(2t)}{2|y - \zeta|^2 \sinh^2 t} \\ &= \frac{\sinh(2t)[|y - \zeta|^2 + 1 - |\zeta|^2 + 2(y \cdot \zeta)] - 2y_1[2 \cosh^2 t - 1]}{2|y - \zeta|^2 \sinh^2 t} \\ &= c(t) + \frac{r^2}{|y - \zeta|^2}(y_1 - c(t)). \end{aligned}$$

This proves that $F(P)F^{-1}$ is

$$y \mapsto \zeta + r^2(y - \zeta)^*,$$

that is, the reflection in $S(\zeta, r)$. □

In view of Theorem 3.7.6, we examine briefly the group $O(1, n)$ of matrices which preserve the quadratic form $q(x, x)$. If $A \in O(1, n)$, then

$$q(x, x) = q(xA, xA),$$

so

$$AJA^t = J, \qquad (3.7.6)$$

where

$$J = \begin{pmatrix} 1 & 0 & \cdots & 0 \\ 0 & & & \\ \vdots & & -I_n & \\ 0 & & & \end{pmatrix}.$$

We deduce that $\det(A)^2 = 1$: the subgroup of $O(1, n)$ with determinant 1 is $SO(1, n)$.

Next, we show that the set of matrices $A$ in $O(1, n)$ with $a_{00} > 0$ is also a subgroup. We denote this subgroup by $O^+(1, n)$ with

$$SO^+(1, n) = SO(1, n) \cap O^+(1, n).$$

Suppose that the matrices $A$, $B$ and $C$ satisfy $a_{00} > 0$, $b_{00} > 0$ and $C = AB$: then

$$\begin{aligned} c_{00} &= a_{00}b_{00} + \cdots + a_{0n}b_{n0} \\ &\geq a_{00}b_{00} - |a_{01}b_{10} + \cdots + a_{0n}b_{n0}| \\ &\geq a_{00}b_{00} - (a_{01}^2 + \cdots + a_{0n}^2)^{1/2}(b_{10}^2 + \cdots + b_{n0}^2)^{1/2}. \end{aligned}$$

Because of (3.7.6), we have

$$(a_{00}, -a_{01}, \ldots, -a_{0n}) \cdot (a_{00}, a_{01}, \ldots, a_{0n})^t = 1,$$

so

$$a_{00}^2 = a_{01}^2 + \cdots + a_{0n}^2 + 1.$$

Taking the transpose of both sides of (3.7.6) after replacing $A$ with $B$ yields

$$b_{00}^2 = b_{10}^2 + \cdots + b_{n0}^2 + 1,$$

so $c_{00} > 0$.

Finally, the inverse of $A\ (=(a_{ij}))$ is $(JAJ)^t$ because

$$\begin{aligned} A(JAJ)^t &= AJA^tJ \\ &= J^2 \\ &= I. \end{aligned}$$

Thus $A \mapsto A^{-1}$ preserves the condition $a_{00} > 0$ and so $O^+(1, n)$ is indeed a group. Observe that an element $A$ of $O(1, n)$ leaves the hyperboloid of two sheets $\{x : q(x, x) = 1\}$ invariant: the component $Q$ is $A$-invariant if and only if $a_{00} > 0$.

We have proved that the isometries of $Q$ are precisely the elements of $O^+(1, n)$ and that in the isomorphism $A \mapsto F(A)F^{-1}$ of $O^+(1, n)$ onto $GM(B^n)$ the subgroup $SO^+(1, n)$ corresponds exactly to the directly conformal elements of $GM(B^n)$ (in the proof of Theorem 3.7.6, each reflection corresponds to a matrix of determinant $-1$). We can now induce a topology on $GM(B^n)$ by transferring the natural topology from $O^+(1, n)$ to $GM(B^n)$ and it is not hard to see that convergence of matrices in $O^+(1, n)$ corresponds exactly to uniform convergence on $S^{n-1}$: thus this topology agrees with those previously constructed. Reverting back to $GM(\hat{\mathbb{R}}^n)$, we have proved the following result.

**Theorem 3.7.7.** $GM(\hat{\mathbb{R}}^n)$ *with the topology of uniform convergence in the chordal metric is isomorphic as a topological group to the group* $O^+(1, n+1)$ *of matrices.*

In particular, if we identify $\hat{\mathbb{R}}^2$ with the extended complex plane, then $M(\hat{\mathbb{R}}^2)$ is (as we shall see) the class of complex Möbius transformations

$$z \mapsto \frac{az + b}{cz + d}, \qquad ad - bc \neq 0,$$

and this is isomorphic to the Lorentz group of matrices preserving both the quadratic form $x_1^2 + x_2^2 + x_3^2 - t^2$ and the inequality $t > 0$.

EXERCISE 3.7

1. Show that if the Möbius transformations $\phi_m$ preserve $B^{n+1}$ and if $\phi_m \to I$ uniformly on some relatively open subset of $S^n$ then $\phi_m \to I$ uniformly on $B^{n+1}$ and on $S^n$. [Identify $S^n$ with $\hat{\mathbb{R}}^n$ and consider convergence on $\hat{\mathbb{R}}^n$ first.]

2. Suppose that $n = 2$ so that $Q$ in Definition 3.7.4 lies in $\mathbb{R}^3$. Show that the geodesics in $B^2$ through the origin correspond via $F$ and $F^{-1}$ to the intersections of $Q$ with certain planes through the origin in $\mathbb{R}^3$.

# §3.8. Notes

For recent treatments of Möbius transformations in $\hat{\mathbb{R}}^n$, see [5], [101] and [110]: for shorter works see (for example) [3], [33] and [108]. A more algebraic treatment based on quadratic forms is given in [19]. Theorem 3.1.5 is well documented: see, for instance, [36], p. 133.

The inversive product (Section 3.2) is discussed in [7], [21], [22], [110]: it can be derived from the metrical theory of the hyperboloid model (see [110]).

It is known that the only (smooth) conformal maps in $\hat{\mathbb{R}}^n$ (or in part of $\hat{\mathbb{R}}^n$) are the Möbius transformations: this is due originally to Liouville (1850) and has been considerably extended since then (by diminishing the degree of smoothness required). For further information see [105], pp. 15 and 43 and the references given there; also, see [88].

# CHAPTER 4

# Complex Möbius Transformations

## §4.1. Representation by Quaternions

In this chapter we shall examine the action of Möbius transformations in $\hat{\mathbb{R}}^2$ and their extensions to $\hat{\mathbb{R}}^3$. We identify $\mathbb{R}^2$ with the complex plane $\mathbb{C}$ and the algebraic structure of $\mathbb{C}$ then allows us to express the action of Möbius transformations algebraically. We shall also identify $(x, y, t)$ in $\mathbb{R}^3$ with the quaternion

$$x + yi + tj \tag{4.1.1}$$

(Section 2.4): this enables us to express the Poincaré extension of a Möbius transformation in terms of the algebra of quaternions. The extended complex plane $\hat{\mathbb{C}}$ is $\mathbb{C} \cup \{\infty\}$ and this is identified with $\hat{\mathbb{R}}^2$. In terms of quaternions,

$$H^3 = \{z + tj : z \in \mathbb{C}, t > 0\}$$

and the boundary of $H^3$ in $\hat{\mathbb{R}}^3$ is $\hat{\mathbb{C}}$.

Möbius transformations are usually encountered first as mappings of the form

$$g(z) = \frac{az + b}{cz + d}, \tag{4.1.2}$$

where $a, b, c$ and $d$ are given complex numbers with $ad - bc \neq 0$. This latter condition ensures that $g$ is not constant: it also ensures that $c$ and $d$ are not both zero and the algebra of $\mathbb{C}$ then guarantees that $g$ is defined on $\mathbb{C}$ if $c = 0$ or on $C - \{-d/c\}$ if $c \neq 0$. Now define $g(\infty) = \infty$ if $c = 0$ and

$$g(-d/c) = \infty, \qquad g(\infty) = a/c$$

if $c \neq 0$. With these definitions, $g$ is a 1–1 *map of $\hat{\mathbb{C}}$ onto itself.* In addition, $g^{-1}$ is of the same form.

Any finite composition $g_1 \cdots g_n$ of these maps can be computed alge-braically and the resulting map, say $g$, is again of the same form. Note, however, that the algebra is only valid on the complement of some finite set $E$ so $g = g_1 \cdots g_n$ on $\mathbb{C} - E$. Each map of the form (4.1.2) when extended as above, is a *continuous* map of $\hat{\mathbb{C}}$ onto itself (here, continuity is with respect to the chordal metric) and so by continuity, $g = g_1 \cdots g_n$ on $\hat{\mathbb{C}}$. These facts (which are left for the reader to check) show that the class $\mathcal{M}$ of maps of the form (4.1.2) is a group under the usual composition of functions. We must now show that $\mathcal{M} = M(\hat{\mathbb{R}}^2)$, the class of orientation preserving Möbius transformations of $\hat{\mathbb{C}}$ onto itself.

In the case of dimension two, the reflections (3.1.1) and (3.1.2) are both of the form

$$z \to \frac{a\bar{z} + b}{c\bar{z} + d}, \qquad ad - bc \neq 0.$$

The composition of two such maps is in $\mathcal{M}$ (again, we use algebra first and then appeal to continuity) and so $M(\hat{\mathbb{R}}^2) \subset \mathcal{M}$.

Now suppose that $g$ is in $\mathcal{M}$ and is given by (4.1.2). If $c = 0$ then $g$ is either a translation (if $a = d$) or a rotation and expansion, namely,

$$g(z) = \alpha + (a/d)(z - \alpha),$$

about some $\alpha$. In both cases, $g$ is a composition of an even number of reflections and so is in $M(\hat{\mathbb{R}}^2)$.

Now assume that $c \neq 0$. The *isometric circle* $Q_g$ of $g$ is (see Section 3.5)

$$Q_g = \{z \in \mathbb{C} : |cz + d| = |ad - bc|^{1/2}\}:$$

the significance of this lies in the fact that if $z$ and $w$ are on $Q_g$ then

$$|g(z) - g(w)| = \left| \frac{(ad - bc)(z - w)}{(cz + d)(cw + d)} \right| = |z - w|.$$

This property is also shared by the reflection $\sigma$ in $Q_g$ and so also by $\phi$ where $\phi = g\sigma$.

Now

$$\sigma(z) = \frac{-d}{c} + \frac{|ad - bc|}{|c|^2} \frac{(z + d/c)}{|z + d/c|^2}$$

and so

$$\phi(z) = g(\sigma(z))$$
$$= \frac{a\sigma(z) + b}{c\sigma(z) + d}$$
$$= \frac{a[c\sigma(z) + d] - (ad - bc)}{c[c\sigma(z) + d]}$$
$$= (a/c) - (u/c|u|)(\overline{cz + d}), \qquad (4.1.3)$$

where $u = ad - bc$. Any map

$$z \to \alpha \bar{z} + \beta, \qquad |\alpha| = 1,$$

is a composition of an odd number of reflections so again, $g \in M(\hat{\mathbb{R}}^2)$. This shows that $\mathcal{M} = M(\hat{\mathbb{R}}^2)$.

We shall use the notation $\mathcal{M}$ in preference to $M(\hat{\mathbb{R}}^2)$ for the remainder of the text. Also, there are many arguments which, strictly speaking, depend on an algebraic computation followed by an appeal to continuity: we shall not mention this again. The next result is well known.

**Theorem 4.1.1.** *Let* $z_1, z_2, z_3$ *be a triple of distinct points in* $\hat{\mathbb{C}}$ *and let* $w_1, w_2, w_3$ *be another such triple. Then there is a unique Möbius transformation which maps* $z_1, z_2, z_3$ *to* $w_1, w_2, w_3$ *respectively.*

We come now to the representation of $g$ in (4.1.2) in terms of quaternions. The quaternion (4.1.1) is $z + tj$ where $z = x + iy$ and *the Poincaré extension of* $g$ *is given by*

$$g(z + tj) = \frac{(az + b)\overline{(cz + d)} + a\bar{c}t^2 + |ad - bc|tj}{|cz + d|^2 + |c|^2 t^2}. \qquad (4.1.4)$$

Observe that this agrees with (4.1.2) if $t = 0$. We shall verify (4.1.4) when $c \neq 0$: the case $c = 0$ is easier and the proof is omitted.

The Poincaré extension of $\sigma$ is the reflection in the sphere in $\mathbb{R}^3$ with the same centre and radius as $Q_g$: thus the action of $\sigma$ in $\mathbb{R}^3$ is given by

$$\sigma(z + tj) = \frac{-d}{c} + \frac{|ad - bc|}{|c|^2} \frac{(z + (d/c) + tj)}{|z + (d/c) + tj|^2}$$

$$= \frac{-d}{c} + \frac{|u|}{cv}(cz + d + ctj),$$

where

$$u = ad - bc, \qquad v = |cz + d|^2 + |c|^2 t^2.$$

It is convenient to write

$$\sigma(z + tj) = z_1 + t_1 j,$$

so

$$cz_1 + d = \frac{|u|}{v}(cz + d), \qquad t_1 = \frac{|u|t}{v}. \qquad (4.1.5)$$

The Poincaré extension of $g$ is found by composing the extensions of $\phi$ and $\sigma$. The extension of $\sigma$ is given above and the extension of $\phi$ (and of any Euclidean isometry of $\mathbb{C}$) is given by

$$\phi(w + sj) = \phi(w) + sj.$$

Thus

$$g(z + tj) = \phi(\sigma(z + tj))$$
$$= \phi(z_1 + t_1 j)$$
$$= \phi(z_1) + t_1 j$$

and using (4.1.3) and (4.1.5) this simplifies to give (4.1.4).

If $ad - bc > 0$ we can describe the action of $g$ in $\mathbb{R}^3$ through the *algebra* of quaternions. Indeed,

$$[a(z + tj) + b] \cdot [c(z + tj) + d)]^{-1} = [(az + b) + atj] \cdot [(cz + d) + ctj]^{-1}$$
$$= \frac{[(az + b) + atj][\overline{cz + d} - tcj]}{|cz + d|^2 + |ct|^2}$$
$$= \frac{(az + b)(\overline{cz + d}) + a\bar{c}t^2 + (ad - bc)tj}{|cz + d|^2 + |c|^2 t^2}$$

and this is $g(z + tj)$ precisely when $ad - bc > 0$.

It is possible to write each transformation in $GM(\hat{\mathbb{R}}^3)$ in terms of quaternions. For example, the function

$$f(w) = (w - j)(w + j)^{-1}j, \qquad w = z + tj, \tag{4.1.6}$$

is the reflection in $x_3 = 0$ followed by reflection in $S(e_3, \sqrt{2})$ (note that $e_3 = j$). In fact, $f$ maps $H^3$ onto $B^3$ and the restriction of $f$ to $\mathbb{C}$ is simply the stereographic projection discussed in Section 3.1. In general,

$$f(z + tj) = (z + [t - 1]j)(z + [t + 1]j)^{-1}j$$
$$= \frac{(z + [t - 1]j)(\bar{z} - [t + 1]j)j}{|z|^2 + (t + 1)^2},$$

which simplifies to

$$f(z + tj) = \frac{2z + (|z|^2 + t^2 - 1)j}{|z|^2 + (t + 1)^2}. \tag{4.1.7}$$

For $t = 0$ this gives the formula for stereographic projection on $\mathbb{C}$: it also shows that $f(j) = 0$.

EXERCISE 4.1

1. Let $g$ be given by (4.1.2) with $c \neq 0$. Prove

   (i) $d_1(gz, a/c) \to 0$ as $d_1(z, \infty) \to 0$;
   (ii) $d_1(gz, \infty) \to 0$ as $d_1(z, -d/c) \to 0$

   where $d_1$ is the chordal metric on $\hat{\mathbb{C}}$.

2. Let $g$ be given by (4.1.2) and (4.1.4) with $ad - bc = 1$. Show that $g(j) = j$ if and only if

$$\begin{pmatrix} a & b \\ c & d \end{pmatrix} \in SU(2, \mathbb{C}).$$

3. Show that the Poincaré extension of any Euclidean isometry $g$ is given by
$$g(z + tj) = g(z) + tj.$$

Describe the action on $H^3$ of a Euclidean isometry $g$ of $\mathbb{C}$ which fixes $j$ in $H^3$.

4. Show that in terms of quaternions, the reflection in $S(\alpha, r)$, $\alpha \in \mathbb{R}^3$, is given by
$$w \mapsto (a\bar{w} + b)(c\bar{w} + d)^{-1}$$

for some suitable $a, b, c$ and $d$ where $\bar{w} = z - tj$ when $w = z + tj$.

5. Let $g$ be given by (4.1.2) with $c \neq 0$. Show that for quaternions $w$ and $w'$ of the form $x + iy + tj$,
$$g(w) - g(w') = (ad - bc)(wc + d)^{-1}(w - w')(cw' + d)^{-1}.$$

Deduce that if $ad - bc = 1$, then $g$ acts as a Euclidean isometry on the sphere $S(-d/c, 1/|c|)$ in $\mathbb{R}^3$.

## §4.2. Representation by Matrices

Any $2 \times 2$ matrix $A$ in $GL(2, \mathbb{C})$ induces a mapping $g$ in $\mathcal{M}$ by the formula $A \to g_A$ where
$$A = \begin{pmatrix} a & b \\ c & d \end{pmatrix}, \qquad g_A(z) = \frac{az + b}{cz + d}.$$

We denote the map $A \to g_A$ by $\Phi$ and this maps $GL(2, \mathbb{C})$ onto $\mathcal{M}$: we shall say that $A$ *projects to* or *represents* $g_A$.

An elementary computation shows that
$$g_A(g_B(z)) = g_{AB}(z), \qquad z \in \hat{\mathbb{C}},$$

where $AB$ is the matrix product and so $\Phi$ is a homomorphism. The kernel $K$ of $\Phi$ is easily found for $A \in K$ if and only if
$$\frac{az + b}{cz + d} = z$$

for all $z$ in $\hat{\mathbb{C}}$. If $A \in K$ we take $z = 0, \infty$ and 1 and find that
$$A = \begin{pmatrix} a & 0 \\ 0 & a \end{pmatrix}, \qquad a \neq 0.$$

Clearly any matrix of this form is in $K$ and so
$$K = \text{Ker } \Phi = \left\{ \begin{pmatrix} a & 0 \\ 0 & a \end{pmatrix} : a \neq 0 \right\}.$$

In particular, $\mathcal{M}$ is isomorphic to $GL(2, \mathbb{C})/K$: in less formal language, $g_A$ determines the matrix $A$ to within a non-zero multiple.

In general, we shall be more concerned with the restriction of $\Phi$ to $\mathrm{SL}(2, \mathbb{C})$. The kernel of this restriction is

$$K_0 = K \cap \mathrm{SL}(2, \mathbb{C}) = \{I, -I\}$$

and each $g$ in $\mathcal{M}$ is therefore the projection of exactly two matrices, say $A$ and $-A$, in $\mathrm{SL}(2, \mathbb{C})$. We deduce that $\mathcal{M}$ is isomorphic to $\mathrm{SL}(2, \mathbb{C})/\{I, -I\}$.

The two functions

$$\frac{\mathrm{tr}^2(A)}{\det(A)}, \qquad \frac{\|A\|^2}{|\det(A)|}, \qquad A \in \mathrm{GL}(2, \mathbb{C}),$$

are invariant under the transformation $A \mapsto \lambda A$, $\lambda \neq 0$, and so they induce corresponding functions on $\mathcal{M}$, namely

$$\mathrm{trace}^2(g) = \frac{\mathrm{tr}^2(A)}{\det(A)} \qquad (4.2.1)$$

and

$$\|g\| = \frac{\|A\|}{|\det(A)|^{1/2}},$$

where $A$ is any matrix which projects to $g$. We often abbreviate $\mathrm{trace}^2(g)$ to $\mathrm{tr}^2(g)$; also, we use $|\mathrm{trace}(g)|$ for the positive square root of $|\mathrm{tr}^2(g)|$. These functions are of great *geometric* significance: we shall consider $\|g\|$ now and discuss $\mathrm{tr}^2(g)$ in Section 4.3. Observe, however, that $\mathrm{trace}^2(\mathrm{g})$ is invariant under any conjugation $g \mapsto hgh^{-1}$.

**Theorem 4.2.1.** *For each $g$ in $\mathcal{M}$, we have*

$$\|g\|^2 = 2 \cosh \rho(j, gj).$$

PROOF. Write

$$g(z) = \frac{az + b}{cz + d}, \qquad ad - bc = 1;$$

then by (4.1.4) (with $z = 0$ and $t = 1$),

$$g(j) = \frac{(b\bar{d} + a\bar{c}) + j}{|c|^2 + |d|^2}.$$

According to (3.3.4), if $\zeta_1 = z_1 + t_1 j$ and $\zeta_2 = z_2 + t_2 j$, then

$$\frac{|z_1 - z_2|^2 + (t_1 - t_2)^2}{2t_1 t_2} + 1 = \cosh \rho(\zeta_1, \zeta_2).$$

The result now follows by substituting $z_1 = 0$, $t_1 = 1$ (so $\zeta_1 = j$), $\zeta_2 = g(j)$ and using the identity

$$|b\bar{d} + a\bar{c}|^2 + 1 = |b\bar{d} + a\bar{c}|^2 + |ad - bc|^2$$
$$= (|a|^2 + |b|^2)(|c|^2 + |d|^2). \qquad \square$$

We have already seen from (4.1.7) that

$$f(w) = (w - j)(w + j)^{-1}j, \qquad w = z + tj, \qquad (4.2.2)$$

is the reflection in $\hat{\mathbb{C}}$ followed by the reflection in $S(j, \sqrt{2})$ and that this transforms the hyperbolic metric in $H^3$ to the metric

$$ds = \frac{2|dx|}{1 - |x|^2}$$

in $B^3$. As another illustration of the use of quaternions let us consider an alternative proof of Theorem 4.2.1, this time the computations being carried out in $B^3$.

SECOND PROOF. Let $w = g(j)$ and $\zeta = f(w)$ so $\zeta \in B^3$. Now for any quaternions $\alpha$ and $\beta$,

$$|\alpha\beta| = |\alpha||\beta|, \qquad |\alpha^{-1}| = |\alpha|^{-1}$$

and so

$$\begin{aligned}
|\zeta| &= \frac{|(aj + b)(cj + d)^{-1} - j| \cdot |j|}{|(aj + b)(cj + d)^{-1} + j|} \\
&= \frac{|(aj + b) - j(cj + d)||(cj + d)^{-1}|}{|(aj + b) + j(cj + d)||(cj + d)^{-1}|} \\
&= \frac{|(b + \bar{c}) + (a - \bar{d})j|}{|(b - \bar{c}) + (a + \bar{d})j|}.
\end{aligned}$$

Thus

$$\begin{aligned}
|\zeta|^2 &= \frac{(b + \bar{c})(\bar{b} + c) + (a - \bar{d})(\bar{a} - d)}{(b - \bar{c})(\bar{b} - c) + (a + \bar{d})(\bar{a} + d)} \\
&= \frac{\|g\|^2 + (bc - ad) + (\overline{bc} - \overline{ad})}{\|g\|^2 + (ad - bc) + (\overline{ad} - \overline{bc})} \\
&= \frac{\|g\|^2 - 2}{\|g\|^2 + 2}. \qquad (4.2.3)
\end{aligned}$$

Using $\rho$ for both the metric in $H^3$ and the metric in $B^3$, we have

$$\begin{aligned}
\rho(j, g(j)) &= \rho(f(j), f(g(j))) \\
&= \rho(0, \zeta) \\
&= \log \frac{1 + |\zeta|}{1 - |\zeta|}.
\end{aligned}$$

Writing $\rho$ for $\rho(j, g(j))$ and using (4.2.3), this gives

$$\begin{aligned}
2 \cosh \rho &= e^\rho + e^{-\rho} \\
&= \frac{2(1 + |\zeta|^2)}{1 - |\zeta|^2} \\
&= \|g\|^2. \qquad \qquad \square
\end{aligned}$$

We can now review Theorems 2.5.1 and 2.5.2 in the light of the geometric action of Möbius transformations. Suppose that

$$A = \begin{pmatrix} a & b \\ c & d \end{pmatrix}, \qquad g(z) = \frac{az + b}{cz + d},$$

where $A$ is in $SL(2, \mathbb{C})$ and suppose also that $f$ is given by (4.2.2).

**Theorem 4.2.2.** *The following statements are equivalent.*

(i) $A \in SU(2, \mathbb{C})$;
(ii) $g(j) = j$;
(iii) $\|g\|^2 = 2$;
(iv) $fgf^{-1}$ *is a linear orthogonal transformation*;
(v) $g$ *is an isometry of the chordal metric space* $(\hat{\mathbb{C}}, d)$.

PROOF. The equivalence of (ii) and (iii) is a direct corollary of Theorem 4.2.1. As $A \in SL(2, \mathbb{C})$ we have $\|A\|^2 = \|g\|^2$ and the equivalence of (i) and (iii) is a direct consequence of Theorem 2.5.1.

Next, (ii) is equivalent to

$$fgf^{-1}(0) = 0$$

and by Theorem 3.4.1, this is equivalent to (iv).

Finally, the equivalence of (i) and (v) is established by observing that $g$ is an isometry if and only if for all $z$,

$$\frac{|g^{(1)}(z)|}{1 + |g(z)|^2} = \frac{1}{1 + |z|^2}.$$

Thus (v) holds if and only if for all $z$,

$$1 + |z|^2 = |az + b|^2 + |cz + d|^2,$$

or, equivalently,

$$1 + |z|^2 = (|a|^2 + |c|^2)|z|^2 + (|b|^2 + |d|^2) + 2 \operatorname{Re}(a\bar{b} + c\bar{d})z.$$

This is equivalent to

$$|a|^2 + |c|^2 = |b|^2 + |d|^2 = 1$$

and

$$a\bar{b} + c\bar{d} = 0,$$

which, in turn, is equivalent to $\bar{A}^t A = I$ and this is (i). $\qquad \square$

Of course, Theorem 4.2.2 shows that the classical symmetry groups of the regular solids (embedded in $B^3$) correspond to the finite subgroups of $SU(2, \mathbb{C})$: indeed, each rotation of $B^3$ is represented by a Möbius $g$ derived from a matrix in $SU(2, \mathbb{C})$ and the symmetry groups can be realized as finite Möbius groups.

EXERCISE 4.2

1. Show that if $g(j) = w + sj$ then

$$\|g\|^2 = (|w|^2 + s^2 + 1)/s.$$

2. Let a subgroup $\Gamma$ of SL(2, $\mathbb{C}$) project to a subgroup $G$ of $\mathcal{M}$. Show that if $\Gamma$ is discrete then for any compact subset $K$ of $H^3$, $g(j) \in K$ for only a finite number of $g$ in $G$.

3. Show that if a matrix $A$ in SL(2, $\mathbb{C}$) is of order two then $A = I$ or $-I$. Deduce that if $B$ is a matrix in SL(2, $\mathbb{C}$) representing a Möbius transformation of order two, then $B$ is of order four.

4. Show that $g: z \mapsto -z$ is not the projection of any matrix in SL(2, $\mathbb{R}$). Verify that the projection of SL(2, $\mathbb{R}$) consists of those Möbius transformations which preserve the extended real axis and the upper half-plane in $\mathbb{C}$.

5. Show that the transformations

$$z \mapsto z; \qquad z \mapsto \frac{3z - 1}{7z - 2}; \qquad z \mapsto \frac{2z - 1}{7z - 3};$$

$$z \mapsto \frac{2z - 1}{3z - 2}; \qquad z \mapsto \frac{z}{5z - 1}; \qquad z \mapsto \frac{3z - 1}{8z - 3}$$

form a group. Show that there is a unique point $w + tj$ in $H^3$ fixed by every element of this group and describe the corresponding group of rotations in $\mathbb{R}^3$.

# §4.3. Fixed Points and Conjugacy Classes

We begin with a brief discussion of the relationship between certain algebraic concepts and some geometric ideas concerning fixed points. Initially, the discussion will be quite general and there is no advantage to be gained by restricting ourselves to Möbius transformations (indeed, such a restriction may even deflect the reader from the central ideas).

Let $X$ be any non-empty set. A *permutation* of $X$ is a 1–1 mapping of $X$ onto itself: for example, a reflection in a sphere is a permutation of $\hat{\mathbb{R}}^n$. The *fixed points* of a permutation $g$ are those $x$ in $E$ which satisfy $g(x) = x$: if this is so we say that $g$ *fixes* $x$.

If $G$ is any group of permutations of $X$ then the *stabilizer* $G_x$ (in $G$) *of* $x$ is the subgroup of $G$ defined by

$$G_x = \{g \in G : g(x) = x\}.$$

Finally, the *orbit* (or $G$-orbit) $G(x)$ *of* $x$ is the subset of $X$ defined by

$$G(x) = \{g(x) \in X : g \in G\}.$$

Observe that *there is a natural one-to-one correspondence between the set* $G/G_x$ *of cosets and the orbit* $G(x)$. If $g$ and $h$ are in $G$, then $h(x) = g(x)$ if and only if $hG_x = gG_x$ and this shows that the map $hG \mapsto h(x)$ is both properly defined and one-to-one. It clearly maps $G/G_x$ onto $G(x)$ and this is the required correspondence. The same facts show that the coset $hG_x$ is the complete set of $g$ in $G$ which map $x$ to $h(x)$.

Two subgroups $G_0$ and $G_1$ of $G$ are *conjugate* if for some $h$ in $G$, $G_0 = hG_1h^{-1}$. As $g$ fixes $x$ if and only if $hgh^{-1}$ fixes $h(x)$, we see that

$$G_{h(x)} = hG_xh^{-1}:$$

thus *if $x$ and $y$ are in the same orbit then $G_x$ and $G_y$ are conjugate*.

Conjugate subgroups are, of course, isomorphic; however, they are also the same from a geometric point of view. This is not necessarily true of isomorphic subgroups, for example, the groups generated by $z \mapsto z + 1$ and $z \mapsto 3z$ are isomorphic but have quite different geometric actions. We are primarily interested in the geometric action of subgroups of $\mathcal{M}$ and we shall, in general, *state our results in a form which remains invariant under conjugation*.

Now let $F_g$ be the set of fixed points of $g$. If $gh = hg$ then

$$g(F_h) = F_h, \qquad h(F_g) = F_g, \qquad (4.3.1)$$

This is clear for if $x \in F_h$ then

$$h(g(x)) = g(h(x)) = g(x)$$

and so $g(x) \in F_h$: thus, $g(F_h) \subset F_h$. Replacing $g$ by $g^{-1}$ we obtain $g(F_h) = F_h$ and (similarly) $h(F_g) = F_g$. We shall see later (Theorem 4.3.6) that *the converse is also true when $G$ is a group of Möbius transformations*.

We return now to study the transformations in $\mathcal{M}$. In its action on $\hat{\mathbb{C}}$, a Möbius transformation $g$ has exactly one fixed point, exactly two fixed points or is the identity. This provides a rather primitive classification and we can obtain a finer classification based on the fixed points in $\hat{\mathbb{R}}^3$. This new classification is invariant under conjugation and so there is a still finer classification, namely the classification into conjugacy classes. One of our main results is that the function $\text{tr}^2$ defined by (4.2.1) actually parametrizes the conjugacy classes.

It is convenient to introduce certain normalized Möbius transformations. For each non-zero $k$ in $\mathbb{C}$ we define $m_k$ by

$$m_k(z) = kz \quad (\text{if } k \neq 1)$$

and

$$m_1(z) = z + 1:$$

we call these the *standard forms*. For future use, note that for all $k$ (including $k = 1$),

$$\text{tr}^2(m_k) = k + \frac{1}{k} + 2. \qquad (4.3.2)$$

If $g \ (\neq I)$ is any Möbius transformation then either $g$ has exactly two fixed points $\alpha$ and $\beta$ in $\hat{\mathbb{C}}$ or $g$ has a unique fixed point $\alpha$ in $\hat{\mathbb{C}}$ (in this case, we choose $\beta$ to be some point other than $\alpha$). Now let $h$ be any Möbius transformation with

$$h(\alpha) = \infty, \qquad h(\beta) = 0, \qquad h(g(\beta)) = 1 \quad \text{if } g(\beta) \neq \beta,$$

and observe that

$$hgh^{-1}(\infty) = \infty, \qquad hgh^{-1}(0) = \begin{cases} 0 & \text{if } g(\beta) = \beta, \\ 1 & \text{if } g(\beta) \neq \beta. \end{cases}$$

If $g$ fixes $\alpha$ and $\beta$, then $hgh^{-1}$ fixes $0$ and $\infty$ and so for some $k$ ($k \neq 1$), we have $hgh^{-1} = m_k$. If $g$ fixes $\alpha$ only then $hgh^{-1}$ fixes $\infty$ only and $hgh^{-1}(0) = 1$: thus $hgh^{-1} = m_1$. This shows that any *Möbius transformation* $g$ ($\neq I$) is *conjugate to one of the standard forms* $m_k$ and this leads to a simple proof of of the next result.

**Theorem 4.3.1.** *Let $f$ and $g$ be Möbius transformations, neither the identity. Then $f$ and $g$ are conjugate if and only if* $\text{tr}^2(f) = \text{tr}^2(g)$.

For brevity, we use $\sim$ to denote conjugacy in $\mathcal{M}$.

PROOF. We have already noted (following (4.2.1)) that if $f \sim g$ then $\text{tr}^2(f) = \text{tr}^2(g)$.

Now assume that $\text{tr}^2(f) = \text{tr}^2(g)$. We know that $f$ and $g$ are each conjugate to some standard form, say $f \sim m_p$ and $g \sim m_q$. Thus

$$\text{tr}^2(m_p) = \text{tr}^2(f) = \text{tr}^2(g) = \text{tr}^2(m_q)$$

and using (4.3.2), this shows that $p = q$ or $p = 1/q$. Now note that $m_p \sim m_{1/p}$: this is trivial if $p = 1$ while if $p \neq 1$, we have

$$hm_p h^{-1} = m_{1/p}, \qquad h(z) = -1/z.$$

We now have $f \sim m_p$, $g \sim m_q$ and (as $p = q$ or $p = 1/q$) $m_p \sim m_q$. As conjugacy is an equivalence relation, this shows that $f \sim g$ and the proof is complete. $\qquad\qquad\square$

We shall now classify Möbius transformations in terms of fixed points in $\hat{\mathbb{R}}^3$ and it is natural to begin by studying the fixed points of the standard forms. The action of $m_k$ in $\hat{\mathbb{R}}^3$ as given by (4.1.4) is

$$m_k(z + tj) = kz + |k|tj \qquad (k \neq 1);$$
$$m_1(z + tj) = z + 1 + tj,$$

and this enables one to find the fixed points of each $m_k$. Clearly:

(i) $m_1$ fixes $\infty$ but no other point in $\hat{\mathbb{R}}^3$;
(ii) if $|k| \neq 1$, then $m_k$ fixes $0$ and $\infty$ but no other points in $\hat{\mathbb{R}}^3$;
(iii) if $|k| = 1$, $k \neq 1$, then the set of fixed points of $m_k$ is

$$\{tj : t \in \mathbb{R}\} \cup \{\infty\}.$$

**Definition 4.3.2.** Let $g$ ($\neq I$) be any Möbius transformation. We say

(i) $g$ is *parabolic* if and only if $g$ has a unique fixed point in $\hat{\mathbb{C}}$ (equivalently, $g \sim m_1$);
(ii) $g$ is *loxodromic* if and only if $g$ has exactly two fixed points in $\hat{\mathbb{R}}^3$ (equivalently, $g \sim m_k$ for some $k$ satisfying $|k| \neq 1$);
(iii) $g$ is *elliptic* if and only if $g$ has infinitely many fixed points in $\hat{\mathbb{R}}^3$ (equivalently, $g \sim m_k$ for some $k$ satisfying $|k| = 1, k \neq 1$).

It is convenient to subdivide the loxodromic class by reference to invariant discs rather than invariant (fixed) points. Note, however, that the following usage is not universal: some authors use "loxodromic" for our "strictly loxodromic" and have no name for our loxodromic transformations.

**Definition 4.3.3.** Let $g$ be a loxodromic transformation. We say that $g$ is *hyperbolic* if $g(D) = D$ for some open disc (or half-plane) $D$ in $\hat{\mathbb{C}}$: otherwise $g$ is said to be *strictly loxodromic*.

The classification described in these definitions is invariant under conjugation and by virtue of Theorem 4.3.1, we must be able to classify $g$ according to the value of $\text{tr}^2(g)$. This is our next result.

**Theorem 4.3.4.** *Let $g$ ($\neq I$) be any Möbius transformation. Then*

(i) *$g$ is parabolic if and only if $\text{tr}^2(g) = 4$;*
(ii) *$g$ is elliptic if and only if $\text{tr}^2(g) \in [0, 4)$;*
(iii) *$g$ is hyperbolic if and only if $\text{tr}^2(g) \in (4, +\infty)$;*
(iv) *$g$ is strictly loxodromic if and only if $\text{tr}^2(g) \notin [0, +\infty)$.*

PROOF. We shall verify (i), (ii) and (iii): then (iv) will automatically be satisfied. Throughout the proof, we suppose that $g$ is conjugate to the standard form $m_p$ so by (4.3.2),

$$\text{tr}^2(g) = p + \frac{1}{p} + 2. \tag{4.3.3}$$

Recall that $g$ is conjugate to $m_p$ and to $m_{1/p}$ but to no other $m_q$.

If $g$ is parabolic, then $g$ is conjugate to $m_1$ only: so $p = 1$ and $\text{tr}^2(g) = 4$. Conversely, if $\text{tr}^2(g) = 4$, then $p = 1$ and $g$ is parabolic. This proves (i).

If $g$ is elliptic, then $p = e^{i\theta}$, say, with $\theta$ real and $\cos\theta \neq 1$. Then

$$\text{tr}^2(g) = 2 + 2\cos\theta \tag{4.3.4}$$

and so $\text{tr}^2(g) \in [0, 4)$. Conversely, suppose that $\text{tr}^2(g) \in [0, 4)$. Then we may write $\text{tr}^2(g)$ in the form (4.3.4) with $\cos\theta \neq 1$ and then (4.3.3) has solutions $p = e^{i\theta}, e^{-i\theta}$. Thus $|p| = 1, p \neq 1$ and we deduce that $g$ is elliptic. This proves (ii).

Finally, we prove (iii). Suppose first that $\text{tr}^2(g) \in (4, +\infty)$. Then (4.3.3) has solutions $p = k$, $1/k$ say, where $k > 0$. As both solutions are positive, $m_p$ necessarily preserves the upper half-plane and so is hyperbolic. This means that $g$ is hyperbolic. Now suppose that $g$, and hence $m_p$, is hyperbolic and let $D$ be a disc which is invariant under $m_p$. For any $z$ in $D$, the images of $z$ under the iterates of $m_p$ are in $D$ and so

$$\{p^n z : n \in \mathbb{Z}\} \subset D.$$

Because $|p| \neq 1$, this shows that $0$ and $\infty$ are in the closure of $D$. The same argument, but with $z$ chosen in the exterior of $D$, leads to the conclusion that $0$ and $\infty$ lie on the boundary of $D$. Thus $D$ is a half-plane and in order to preserve $D$, it is necessary that $m_p$ leaves invariant each of the half-lines from $0$ to $\infty$ on the boundary of $D$. Thus $p > 0$ and so $\text{tr}^2(g) > 4$. $\qquad\square$

We now prove three useful results concerning fixed points. Recall that in any group the commutator of $g$ and $h$ is

$$[g, h] = ghg^{-1}h^{-1} = g(hg^{-1}h^{-1}).$$

If $A$ and $B$ are matrices in $\text{SL}(2, \mathbb{C})$ representing Möbius transformations $g$ and $h$ then they are determined to within a factor of $-1$ and so

$$\text{tr}[g, h] = \text{tr}(ABA^{-1}B^{-1})$$

is uniquely determined, independently of the choice of $A$ and $B$.

**Theorem 4.3.5.**(i) *Two Möbius transformations $g$ and $h$ have a common fixed point in $\hat{\mathbb{C}}$ if and only if $\text{tr}[g, h] = 2$.*

(ii) *If $g$ and $h$ (neither the identity) have a common fixed point in $\hat{\mathbb{C}}$ then either:*

    (a) *$[g, h] = I$ (so $gh = hg$) and $F_g = F_h$; or*
    (b) *$[g, h]$ is parabolic (and $gh \neq hg$) and $F_g \neq F_h$.*

PROOF. The assertions in (i) remain invariant under conjugation so we may assume that in terms of matrices in $\text{SL}(2, \mathbb{C})$,

$$g = \begin{pmatrix} a & b \\ 0 & d \end{pmatrix}, \qquad h = \begin{pmatrix} \alpha & \beta \\ \gamma & \delta \end{pmatrix}.$$

A computation shows that

$$\text{tr}[g, h] = 2 + b^2\gamma^2 + b(a - d)\gamma(\alpha - \delta) - (a - d)^2\gamma\beta.$$

If $g$ and $h$ have a common fixed point, we may assume that it is $\infty$ so $\gamma = 0$ and $\text{tr}[g, h] = 2$.

Now suppose that $\text{tr}[g, h] = 2$. If $g$ is parabolic we can take $a = d = 1$ and $b \neq 0$: then $\gamma = 0$ so both $g$ and $h$ fix $\infty$. If $g$ is not parabolic we can take $b = 0$ so $ad = 1$ and $a \neq d$: then $\gamma\beta = 0$ so $h$ fixes one of $0$ and $\infty$. This proves (i).

To prove (ii) we may assume that $g$ and $h$ are as above with $\gamma = 0$. Then $[g, h] = I$ if and only if

$$\beta(a - d) = b(\alpha - \delta)$$

and this is equivalent to $F_g = F_h$ (consider the cases $a = d, a \neq d$).

For an alternative approach to (ii), suppose that the common fixed point is $\infty$ and so $g$ and also $h$ is of the form $z \mapsto az + b$. The map $g \mapsto a$ is a homomorphism of $\langle g, h \rangle$ to the group $\mathbb{C} - \{0\}$: as this group is abelian, every commutator is in the kernel of the homomorphism and so is a translation (or $I$). $\qquad\square$

A Euclidean similarity is a map $x \to r\phi(x) + x_0$ where $\phi$ is a Euclidean isometry and the above proof is concerned with such similarities. In fact, Theorem 4.3.5 is a theorem on Euclidean similarities but stated in a form that is invariant under conjugation.

**Theorem 4.3.6.** *Let $g$ and $h$ be Möbius transformations other than $I$. The following statements are equivalent:*

(i) $hg = gh$;
(ii) $h(F_g) = F_g, g(F_h) = F_h$;
(iii) *either $F_g = F_h$ or $g$ and $h$ have a common fixed point in $H^3$ with $g^2 = h^2 = (gh)^2 = I$ and $F_g \cap F_h = \varnothing$.*

PROOF. First, (4.3.1) shows that (i) implies (ii).

The proof that (iii) implies (i) is easy. If $F_g = F_h$ then $g$ and $h$ have a common fixed point and so by Theorem 4.3.5, $[g, h] = I$: thus in this case, $gh = hg$. The other alternative offered by (iii) also leads to $gh = hg$ as

$$hg = hg(ghgh) = gh$$

and so (iii) implies (i).

It remains to prove that (ii) implies (iii). We assume that (ii) holds and also that $F_g \neq F_h$ (else (iii) certainly holds). This means that there is some $w$ in exactly one of the sets $F_g, F_h$ and we may assume that $w \in F_g - F_h$: thus $g(w) = w$ and $h(w) \neq w$. By (ii), $F_g$ contains the points $w, h(w), h^2(w)$ and as these cannot be distinct (else $g = I$) we must have $h^2(w) = w$. This shows that $F_g$ has exactly two points and that these are interchanged by $h$. It also shows that $F_g \cap F_h = \varnothing$.

By conjugation, we may assume that $F_g = \{0, \infty\}$: thus for some $a$ and $b$,

$$g(z) = az, \qquad h(z) = b/z.$$

It is now clear that $h^2 = (gh)^2 = I$. Moreover, as $g(F_h) = F_h$, we must have $g(\sqrt{b}) = -\sqrt{b}$ so $a = -1$ and $g^2 = I$. Finally $g$ and $h$ have a common fixed point, namely $|b|^{1/2}j$, in $H^3$: this follows directly from (4.1.4). $\qquad\square$

Theorem 4.3.5 is concerned with two transformations with a common fixed point in $\hat{\mathbb{C}}$: the next result concerns a common fixed point in $H^3$.

**Theorem 4.3.7.** *A subgroup G of $\mathcal{M}$ contains only elliptic elements (and I) if and only if the elements of G have a common fixed point in $H^3$.*

It follows from Definition 4.3.2 that if $g(\neq I)$ is of finite order then $g$ is necessarily elliptic. As every element in a finite group has finite order we have the following corollary.

**Corollary.** *The elements in a finite subgroup of $\mathcal{M}$ have a common fixed point in $H^3$.*

To understand the geometric nature of the proof it is convenient to introduce the notion of the *axis* of an elliptic element $g$. If the fixed points of $g$ in $\hat{\mathbb{C}}$ are $\alpha$ and $\beta$, then (by considering a conjugation to one of the standard forms), the fixed points of $g$ in $\hat{\mathbb{R}}^3$ are precisely the points on the circle $\Gamma$ which is orthogonal to $\mathbb{C}$ and which passes through $\alpha$ and $\beta$. The *axis* $A_g$ of $g$ is the Euclidean semi-circle $\Gamma \cap H^3$ (in fact, this is a geodesic in the hyperbolic geometry of $H^3$). The condition that two elliptic elements $g$ and $h$ have a common fixed point in $H^3$ is simply that the two axes $A_g$ and $A_h$ are concurrent in $H^3$. Note that a *necessary and sufficient condition* for this is that the fixed points of $g$ and $h$ in $\hat{\mathbb{C}}$ lie on a circle $Q$ and separate each other on $Q$.

Parts of the proof of Theorem 4.3.7 are algebraic (the geometry is complicated) but even so, we shall stress the geometric interpretation. First, we prove a preliminary result.

**Lemma 4.3.8.** *Suppose that $g$, $h$ and $gh$ are elliptic. Then the fixed points of $g$ and $h$ in $\hat{\mathbb{C}}$ are concyclic. If, in addition, $[g, h]$ is elliptic or $I$, then the axes $A_g$ and $A_h$ are concurrent in $H^3$.*

PROOF. If $g$ and $h$ have a common fixed point in $\hat{\mathbb{C}}$, then $F_g \cup F_h$ has at most three points and so lies in some circle. If, in addition, $[g, h]$ is elliptic or $I$, then from Theorem 4.3.5, $F_g = F_h$ and so $A_g = A_h$: thus $g$ and $h$ have infinitely many common fixed points in $H^3$.

We may now assume that $g$ and $h$ have no common fixed points in $\hat{\mathbb{C}}$. By conjugation we may assume that

$$g(z) = \alpha^2 z, \qquad h(z) = \frac{az + b}{cz + d},$$

where $\alpha^2 \neq 1$, $|\alpha| = 1$ and $ad - bc = 1$. Now

$$\text{tr}^2(h) = (a + d)^2, \qquad \text{tr}^2(gh) = (\alpha a + \bar{\alpha} d)^2$$

and so by Theorem 4.3.4, the numbers

$$\lambda = a + d, \qquad \mu = \alpha a + \bar{\alpha} d$$

are in the interval $(-2, 2)$. Solving for $a$ and $d$ in terms of $\alpha, \lambda$ and $\mu$, we obtain

$$a = \bar{d} = u + iv,$$

say.

The fixed points of $h$ are (using $ad - bc = 1$)

$$\frac{a - d \pm i[4 - (a + d)^2]^{1/2}}{2c}$$

and these are the points

$$\xi, \zeta = (i/c)[v \pm (1 - u^2)^{1/2}].$$

As $|a + d| < 2$, we find that $u^2 < 1$ and so $\xi$ and $\zeta$ lie on a straight line $L$ through the origin: thus the fixed points of $g$ and $h$ are concyclic.

A computation (after writing $\alpha = e^{i\theta}$ and using $ad - bc = 1$) gives

$$\text{tr}^2([g, h]) = 4[1 + (|a|^2 - 1) \sin^2 \theta]^2$$

and so the additional hypothesis that $[g, h]$ is elliptic or $I$ implies that $|a| \leq 1$ because we must have

$$0 \leq \text{tr}^2([g, h]) \leq 4.$$

Now $|a| = 1$ implies that $u^2 + v^2 = 1$ and so one of the points $\xi, \zeta$ is zero. This is excluded as $g$ and $h$ are assumed to have no common fixed points: thus $|a| < 1$ and so (taking the positive root)

$$(1 - u^2)^{1/2} > v.$$

This means that

$$\xi = is/c, \qquad \zeta = it/c,$$

where $s$ and $t$ are real with $st < 0$. Thus the origin (fixed by $g$) lies between $\xi$ and $\zeta$ and so $A_g$ and $A_h$ are concurrent in $H^3$. $\qquad\square$

We now use Lemma 4.3.8 to obtain information about subgroups of $\mathcal{M}$ of the form $\langle g, h \rangle$ which contain only elliptic elements and $I$. First, by Lemma 4.3.8, $g$ and $h$ have a common fixed point $\zeta$, say, in $H^3$ and, of course, every element of $\langle g, h \rangle$ fixes $\zeta$. By considering a conjugate group, we may assume that $g$ and $h$ preserve $B^3$ and that $\zeta = 0$.

**Lemma 4.3.9.** *Let $g$ and $h$ be Möbius transformations $(\neq I)$ which preserve $B^3$ and fix the origin. Then*

(i) *the elements of $\langle g, h \rangle$ have the same axis and same fixed points or*
(ii) *there is some $f$ in $\langle g, h \rangle$ such that the three axes $A_g$, $A_h$, $A_f$ are not coplanar.*

Assuming the validity of Lemma 4.3.9 for the moment, we complete the proof of Theorem 4.3.7.

PROOF OF THEOREM 4.3.7. The conclusion is obviously true if all elliptic elements of $G$ have the same axis so we may assume that $G$ contains elements $g$ and $h$ with distinct axes. By Lemma 4.3.8, $g$ and $h$ have a common fixed point in $H^3$ and by considering a conjugate group we may assume that $G$ acts on $B^3$ and that Lemma 4.3.9 is applicable. By assumption, (i) fails so (ii) of Lemma 4.3.9 holds.

Every element of $\langle g, h \rangle$ fixes the origin so the axes $A_g, A_h, A_f$ are Euclidean diameters of $B^3$: moreover, by (ii), they are not coplanar. Now take any $q$ in $G$, $q \neq I$. We shall show that $q(0) = 0$ and this will complete the proof. By Lemma 4.3.8, the fixed points of $q$ and $g$ lie on some circle on $\partial B^3$ and so also lie on a Euclidean plane $\Pi_g$. As $\Pi_g$ contains the end-points of the diameter $A_g$, we see that $0 \in \Pi_g$: also $A_q \subset \Pi_g$. A similar definition and argument holds for $\Pi_h$ and $\Pi_f$: so

$$0 \in \Pi_g \cap \Pi_h \cap \Pi_f$$

and

$$A_q \subset \Pi_g \cap \Pi_h \cap \Pi_f. \tag{4.3.5}$$

The planes $\Pi_g, \Pi_h, \Pi_f$ cannot be the same plane (else $A_g, A_h, A_f$ would be coplanar) thus the intersection

$$\Pi_g \cap \Pi_h \cap \Pi_f$$

is either $\{0\}$ or is a diameter $D$ of $B^3$. Because this intersection contains the fixed points of $q$ on $\partial B^3$ it is a diameter $D$ and we conclude from (4.3.5) that $A_q = D$. In particular, $0 \in A_q$ and so $q(0) = 0$.                           □

PROOF OF LEMMA 4.3.9. Every element of $\langle g, h \rangle$ fixes the origin and so is elliptic or $I$. For each such elliptic $f$, let $A_f$ denote the axis (of fixed points) of $f$ in $B^3$. Note that by assumption, $A_g$ and $A_h$ are Euclidean diameters of $B^3$.

We shall assume that (i) fails so $A_g$ and $A_h$ are distinct diameters and hence determine a Euclidean plane $\Pi$. Let the normal to $\Pi$ through the origin be the diameter $D$ of $B^3$. If $h(A_g)$ does not lie in $\Pi$, then take $f = hgh^{-1}$ and this satisfies (ii) as then $A_f = h(A_g)$. A similar construction of $f$ is possible if $g(A_h)$ does not lie in $\Pi$. These attempts to construct $f$ can only fail if $g$ and $h$ preserve $\Pi$ in which case, they are both rotations of order two. Then both $g$ and $h$ interchange the end-points of $D$ and so (ii) is satisfied with $f = gh$.                           □

We end this section with a discussion of the iterates of a Möbius transformation.

If $g$ is *parabolic*, then for some $h$ we have

$$hgh^{-1}(z) = z + t \qquad (t \neq 0).$$

Thus

$$hg^n h^{-1}(z) = z + nt$$

and

$$g^n(z) = h^{-1}(hz + nt).$$

Observe that for each $z$ in $\hat{\mathbb{C}}$, $hg^n h^{-1}(z) \to \infty$ as $|n| \to \infty$: thus in general, if $g$ is parabolic then

$$g^n(z) \to \alpha,$$

where $\alpha$ is the fixed point of $g$.

If $g$ is not parabolic, then $g$ has two fixed points, say $\alpha$ and $\beta$, and for some $h$ we have

$$hgh^{-1}(z) = tz \qquad (t \neq 0, 1)$$

and hence

$$hg^n h^{-1}(z) = t^n z.$$

These facts show that if $g$ is *loxodromic* (equivalently, $|t| \neq 1$) and if $z$ is not $\alpha$ or $\beta$, then the images $g^n(z)$ are distinct and accumulate at $\alpha$ and $\beta$ only. If $g^n(z) \to \alpha$, say, as $n \to +\infty$, then $\alpha$ is called the *attractive fixed point* of $g$ while $\beta$ is called the *repulsive fixed point*. Then for all $z$ other than $\beta$, $g^n(z) \to \alpha$ as $n \to +\infty$.

If $g$ is *elliptic* (equivalently, $|t| = 1$), then $g$ has invariant circles: indeed each circle for which $\alpha$ and $\beta$ are inverse points is a $g$-invariant circle and so each orbit under iterates of $g$ is constrained to lie on such a circle. We collect these results together for future reference.

**Theorem 4.3.10.** (i) *Let $g$ be parabolic with fixed point $\alpha$. Then for all $z$ in $\hat{\mathbb{C}}$, $g^n(z) \to \alpha$ as $n \to +\infty$, the convergence being uniform on compact subsets of $C - \{\alpha\}$.*

(ii) *Let $g$ be loxodromic. Then the fixed points $\alpha$ and $\beta$ of $g$ can be labelled so that $g^n(z) \to \alpha$ as $n \to +\infty$ (if $z \neq \beta$); the convergence being uniform on compact subsets of $\hat{\mathbb{C}} - \{\beta\}$.*

(iii) *Let $ge$ be elliptic with fixed points $\alpha$ and $\beta$. Then $g$ leaves invariant each circle for which $\alpha$ and $\beta$ are inverse points.*

If a Möbius $g$ is of finite order $k$ (so $g^k$, but no smaller power, is $I$) then $g$ is necessarily elliptic. In this case we have

$$hgh^{-1}(z) = e^{i\theta} z,$$

say, and so

$$\theta = 2\pi m/k,$$

where $k$ and $m$ are coprime. We deduce that

$$\text{tr}^2(g) = 4\cos^2(\theta/2)$$
$$= 2[1 + \cos(2\pi m/k)].$$

Note that this can take different values depending on the prime factors of $k$. If $g$ is elliptic of order two, then $k = 2$ and necessarily, $\text{tr}^2(g) = 0$: the converse is also true. Observe that among all $g$ of order $k$, the largest value of $\text{tr}^2(g)$ occurs when $m = 1$ or $k - 1$,

$$\text{tr}^2(g) = 4\cos^2(\pi/k)$$

and $\theta = \pm 2\pi/k$. Again we record this for future reference.

**Theorem 4.3.11.** *Let $g$ be an elliptic transformation of order $k$. Then*

$$\text{tr}^2(g) \leq 4\cos^2(\pi/k),$$

*with equality if and only if $g$ is a rotation of angle $\pm 2\pi/k$.*

EXERCISE 4.3

1. Find Möbius transformations $g$ and $h$ such that

    (i) $\text{tr}[g, h] = -2$; and
    (ii) $g$ and $h$ have no common fixed point in $\hat{\mathbb{C}}$.

2. Let $g$ be any Möbius transformation which does not fix $\infty$. Show that $g = g_1 g_2 g_3$, where $g_1$ and $g_3$ are parabolic elements fixing $\infty$ and where $g_2$ is of order two.

3. An $n$th root of a Möbius transformation $g$ is any Möbius transformation $h$ satisfying $h^n = g$. Prove

    (i) if $g = I$ then $g$ has infinitely many $n$th roots;
    (ii) if $g$ is parabolic then $g$ has a unique $n$th root;
    (iii) in all other cases, $g$ has exactly $n$ $n$th roots.

4. Show that if $A$ and $B$ are in SL(2, $\mathbb{C}$) then

$$\det(A - I) = 2 - \text{tr}(A)$$

    and

$$\det(AB - BA) = 2 - \text{tr}[A, B]$$

    ($[A, B]$ is the commutator of $A$ and $B$). Deduce that if $A$ and $B$ viewed as Möbius transformations do not have a common fixed point in $\hat{\mathbb{C}}$, then $AB - BA$ is a non-singular matrix which represents a Möbius transformation or order two.

5. Let $g(z) = z/(cz + 1)$. Verify (i) by induction and (ii) by considering a suitable $hgh^{-1}$ that

$$g^n(z) = \frac{z}{ncz + 1}$$

    Find $f^n$ where $f(z) = 6z/(z + 3)$ and check your result by induction.

## §4.4. Cross Ratios

Given four distinct points $z_1, z_2, z_3, z_4$ of $\mathbb{C}$ we define the *cross-ratio* of these points as

$$[z_1, z_2, z_3, z_4] = \frac{(z_1 - z_3)(z_2 - z_4)}{(z_1 - z_2)(z_3 - z_4)}:$$

compare this with (3.2.5) where division is not permitted. The definition is extended by continuity to include the case when one of the $z_j$ is $\infty$ so, for example,

$$[z_1, z_2, z_3, \infty] = \frac{z_1 - z_3}{z_1 - z_2}.$$

Note that in particular,

$$[0, 1, z, \infty] = z. \tag{4.4.1}$$

If

$$g(z) = \frac{az + b}{cz + d} \quad (ad - bc \neq 0),$$

then

$$g(z) - g(w) = \frac{(z - w)(ad - bc)}{(cz + d)(cw + d)}$$

and it is immediate that *the cross-ratio is invariant under Möbius transformations*; that is,

$$[g(z_1), g(z_2), g(z_3), g(z_4)] = [z_1, z_2, z_3, z_4]. \tag{4.4.2}$$

This is a useful property which often leads to a considerable simplification. Moreover, the converse is also true: if

$$[w_1, w_2, w_3, w_4] = [z_1, z_2, z_3, z_4] \tag{4.4.3}$$

holds then there is a Möbius transformation $g$ with $g(z_j) = w_j$. To see this, let $f$ and $h$ be Möbius transformations which map $z_1, z_2, z_4$ to $0, 1, \infty$ and $w_1, w_2, w_4$ to $0, 1, \infty$ respectively: these exist by Theorem 4.1.1. Then by (4.4.1), (4.4.2) and (4.4.3),

$$\begin{aligned}
f(z_3) &= [0, 1, f(z_3), \infty] \\
&= [f(z_1), f(z_2), f(z_3), f(z_4)] \\
&= [z_1, z_2, z_3, z_4] \\
&= [w_1, w_2, w_3, w_4] \\
&= [h(w_1), h(w_2), h(w_3), h(w_4)] \\
&= [0, 1, h(w_3), \infty] \\
&= h(w_3).
\end{aligned}$$

It is now clear that $g(z_j) = w_j$ where $g = h^{-1} \circ f$. $\qquad\qquad\square$

We are now going to study how the cross ratio

$$\lambda = [z_1, z_2, z_3, z_4] \tag{4.4.4}$$

varies as we permute the $z_j$. With this in mind we let $S_n$ denote the permutation group of $\{1, \ldots, n\}$ and remark that (as with all functions) we regard permutations as acting on the *left*: for example, $(12)(13)$ maps 3 to 2.

Each $\sigma$ in $S_4$ induces a change in the value of the cross ratio by the formula

$$\lambda = [z_1, z_2, z_3, z_4] \mapsto [z_{\sigma 1}, z_{\sigma 2}, z_{\sigma 3}, z_{\sigma 4}]$$

and it is essential to realize that the resulting value depends on $\sigma$ and $\lambda$ *but not on the individual values* $z_j$. This is so because if

$$[z_1, z_2, z_3, z_4] = [w_1, w_2, w_3, w_4],$$

then there is some $g$ with $g(z_j) = w_j$ and so

$$[z_{\sigma 1}, z_{\sigma 2}, z_{\sigma 3}, z_{\sigma 4}] = [g(z_{\sigma 1}), g(z_{\sigma 2}), g(z_{\sigma 3}), g(z_{\sigma 4})]$$
$$= [w_{\sigma 1}, w_{\sigma 2}, w_{\sigma 3}, w_{\sigma 4}].$$

Because of this fact, we can introduce functions $f_\sigma$ ($\sigma \in S_4$) by the formula

$$f_\sigma(\lambda) = [z_{\sigma 1}, z_{\sigma 2}, z_{\sigma 3}, z_{\sigma 4}],$$

where $\lambda$ is given by (4.4.4). Because

$$f_\pi(f_\sigma(\lambda)) = [z_{\pi\sigma 1}, z_{\pi\sigma 2}, z_{\pi\sigma 3}, z_{\pi\sigma 4}]$$
$$= f_{\pi\sigma}(\lambda)$$

we have the important relation

$$f_\pi f_\sigma = f_{\pi\sigma}. \tag{4.4.5}$$

Now suppose that $\sigma$ is the transposition $(1, 2)$ and let $g$ be the Möbius transformation which maps $z_1, z_2, z_4$ to $0, 1, \infty$ respectively. Then

$$\lambda = [z_1, z_2, z_3, z_4]$$
$$= [0, 1, g(z_3), \infty]$$
$$= g(z_3)$$

and so

$$f_\sigma(\lambda) = [z_2, z_1, z_3, z_4]$$
$$= [1, 0, \lambda, \infty]$$
$$= 1 - \lambda.$$

A similar argument holds for all six transpositions in $S_4$ and we find

(i) *if* $\sigma = (1, 2)$ *or* $(3, 4)$ *then* $f_\sigma(\lambda) = 1 - \lambda$;
(ii) *if* $\sigma = (1, 3)$ *or* $(2, 4)$ *then* $f_\sigma(\lambda) = \lambda/(\lambda - 1)$;
(iii) *if* $\sigma = (1, 4)$ *or* $(2, 3)$ *then* $f_\sigma(\lambda) = 1/\lambda$.

This information leads to a determination of all $f_\sigma$. As $S_4$ is generated by transpositions, (i), (ii) and (iii) together with (4.4.5) suffice to give all $f_\sigma$. Note that for each transposition $\sigma$, the function $f_\sigma$ is actually a Möbius transformation which maps $\{0, 1, \infty\}$ onto itself. Thus if we denote by $\mathcal{M}_0$ the subgroup of Möbius transformations which map $\{0, 1, \infty\}$ onto itself we find from (4.4.5) that the map

$$\theta : \sigma \to f_\sigma$$

is actually a homomorphism of $S_4$ into $\mathcal{M}_0$ (which is isomorphic to $S_3$). In addition to this, it is clear from (i), (ii) and (iii) and (4.4.5) that the subgroup

$$K = \{I, (1, 2)(3, 4), (1, 3)(2, 4), (1, 4)(2, 3)\}$$

of $S_4$ is contained within the kernel of $\theta$. We can now describe the situation completely.

**Theorem 4.4.1.** *The map $\theta : S_4 \to \mathcal{M}_0$ is a homomorphism of $S_4$ onto $\mathcal{M}_0$ with kernel $K$.*

PROOF. Theorem 4.1.1. implies that $\mathcal{M}_0$ has exactly six elements: these are the functions

$$\lambda, \ 1 - \lambda, \ \lambda/(\lambda - 1), \ 1/\lambda, \ 1/(1 - \lambda), \ (\lambda - 1)/\lambda$$

of $\lambda$. There are six permutations $\sigma$ in $S_4$ with $\sigma(4) = 4$ and a straightforward computation shows that the corresponding $f_\sigma$ are precisely the six elements of $\mathcal{M}_0$. This shows that $\theta$ maps $S_4$ onto $\mathcal{M}_0$ and as this implies that the kernel of $\theta$ has exactly four elements, the kernel must be $K$. $\qquad\square$

Four distinct points $z_1, z_2, z_3, z_4$ in $\hat{\mathbb{C}}$ are concyclic if and only if they lie on some circle. Let $g$ be the Möbius transformation which maps $z_1, z_2, z_4$ to $0, 1, \infty$ respectively. Then the $z_j$ are concyclic if and only if the $g(z_j)$ are and this is so if and only if $g(z_3)$ is real. However,

$$g(z_3) = [0, 1, g(z_3), \infty]$$
$$= [z_1, z_2, z_3, z_4]:$$

thus $z_1, z_2, z_3, z_4$ *are concyclic if and only if* $[z_1, z_2, z_3, z_4]$ *is real.*

If $z_1, z_2, z_3, z_4$ lie on a circle $Q$ and are arranged in this order around $Q$, then $g(z_3) > 1$ and so

$$\lambda = [z_1, z_2, z_3, z_4] > 1.$$

EXERCISE 4.4

1. Show that the unique Möbius transformation which maps $z_1, z_2, z_4$ to $0, 1, \infty$ respectively is $g$ where

$$g(z) = [z_1, z_2, z, z_4].$$

2. Verify that $f_\sigma(\lambda) = \lambda/(\lambda - 1)$ when $\sigma = (2, 4)$.

3. Let $z_1, z_2, z_3, z_4$ be distinct points in $\hat{\mathbb{C}}$. Show that the circle through $z_1, z_2, z_4$ is orthogonal to the circle through $z_1, z_3, z_4$ if and only if

$$\text{Re}[z_1, z_2, z_3, z_4] = 0.$$

Generalize this to the case where the circles meet at an angle $\theta$ (note that the $z_j$ are concyclic if and only if $\theta = 0$).

4. Let $g$ be any Mobius transformation. Show that if $g$ does not fix $z$ then $[z, gz, g^2 z, g^3 z]$ is independent of $z$ and evaluate this in terms of $\text{tr}^2(g)$.

## §4.5. The Topology on $\mathcal{M}$

As described in Section 4.2, there is a homomorphism

$$\Phi: \text{SL}(2, \mathbb{C}) \to \mathcal{M},$$

which associates to each $g$ in $\mathcal{M}$ exactly two matrices $A$ and $-A$ in $\text{SL}(2, \mathbb{C})$. The group $\text{SL}(2, \mathbb{C})$ is a topological group with respect to the metric $\|A - B\|$ and the map $\Phi$ induces the quotient topology $\mathcal{T}$ on $\mathcal{M}$, namely the largest topology on $\mathcal{M}$ with respect to which, $\Phi$ is continuous. In addition, $\mathcal{M}$ has a topology $\mathcal{T}^*$, namely the topology of uniform convergence with respect to the chordal metric on $\hat{\mathbb{C}}$ (see Section 3.7) and it is essential to know that these topologies are the same. One method is to compare the action of $\text{SL}(2, \mathbb{C})$ through the action of $\mathcal{M}$ on $H^3$ (and then $B^3$) to the matrix group $O^+(1, 3)$. However, a more direct approach is not without interest.

**Theorem 4.5.1.** *The topology $\mathcal{T}$ induced on $\mathcal{M}$ by $\Phi$ coincides with the topology $\mathcal{T}^*$ of uniform convergence on $\hat{\mathbb{C}}$.*

PROOF. It is sufficient to show that the map

$$\Phi: \text{SL}(2, \mathbb{C}) \to (\mathcal{M}, \mathcal{T}^*) \tag{4.5.1}$$

is open and continuous: see Proposition 1.4.1.

Assuming that this has been established, observe that if $X$ is in $\text{SL}(2, \mathbb{C})$ then

$$\|X - (-X)\| = 2\|X\|$$
$$\geq 2\sqrt{2},$$

(see (x) of Section 2.2). This yields the next result.

**Corollary 4.5.2.** *The restriction of $\Phi$ to any open ball of radius $\sqrt{2}$ in $\text{SL}(2, \mathbb{C})$ is a homeomorphism: thus $\text{SL}(2, \mathbb{C})$ is a two-sheeted covering space of $\mathcal{M}$.*

It remains to prove that the map (4.5.1) is open and continuous. Define

$$\sigma(f, g) = \sup_{z \in \hat{\mathbb{C}}} d(fz, gz),$$

where $d$ is the chordal metric: thus $\mathcal{T}^*$ is the metric topology induced by the metric $\sigma$. We shall derive the continuity of $\Phi$ from the next result.

**Proposition 4.5.3.** *If $A$ in $\mathrm{SL}(2, \mathbb{C})$ represents $g$, then*

$$\sigma(g, I) \leq \sqrt{6}\, \|A - I\|.$$

Explicitly, if $B$ represents $f$, then

$$\begin{aligned}
\sigma(g, f) &= \sigma(gf^{-1}, I) \\
&\leq \sqrt{6}\, \|AB^{-1} - I\| \\
&\leq \sqrt{6}\, \|A - B\| \cdot \|B\|
\end{aligned}$$

and so $\Phi$ is continuous at the general element $B$ of $\mathrm{SL}(2, \mathbb{C})$.

PROOF OF PROPOSITION 4.5.3. There is a unitary matrix $B$ representing a Möbius map $h$ such that $hgh^{-1}$ fixes $\infty$ ($h$ corresponds to a rotation of the sphere moving a selected fixed point of $g$ to $\infty$). By Theorems 2.5.2 and 4.2.2 we have

$$\|A - I\| = \|BAB^{-1} - I\|$$

and

$$\begin{aligned}
\sigma(hgh^{-1}, I) &= \sigma(gh^{-1}, h^{-1}) \\
&= \sigma(g, I).
\end{aligned}$$

These remarks show that we may assume, without loss of generality, that $g$ fixes $\infty$. In addition, if $g$ is loxodromic we may assume that the repulsive fixed point of $g$ is $\infty$ (we simply choose $h$ appropriately).

Assume then that

$$A = \begin{pmatrix} \alpha & \beta \\ 0 & \delta \end{pmatrix}, \qquad \alpha\delta = 1:$$

the condition on the fixed point of $g$ in the loxodromic case means that in all cases,

$$|\alpha| \leq 1 \leq |\delta|.$$

Now

$$\begin{aligned}
d(z, gz) &\leq d\left(z, \frac{\alpha z}{\delta}\right) + d\left(\frac{\alpha z}{\delta}, \frac{\alpha z + \beta}{\delta}\right) \\
&\leq \frac{2|z| \cdot |1 - (\alpha/\delta)|}{(1 + |z|^2)^{1/2}(1 + |\alpha z/\delta|^2)^{1/2}} + 2|\beta/\delta| \\
&\leq \frac{2|z| \cdot |\alpha - \delta|}{|\delta| \cdot |2z|^{1/2}|2\alpha z/\delta|^{1/2}} + 2|\beta|,
\end{aligned}$$

the last line being an application of the Arithmetic–Geometric Mean inequality. This upper bound simplifies to a value independent of $z$ and using $\alpha\delta = 1$, we have

$$
\begin{aligned}
\sigma(g, I) &\leq |\alpha - \delta| + 2|\beta| \\
&\leq |\alpha - 1| + |1 - \delta| + 2|\beta| \\
&\leq (|a - 1|^2 + |1 - \delta|^2 + |\beta|^2)^{1/2}(1 + 1 + 4)^{1/2} \\
&\leq \sqrt{6}\,\|A - I\|.
\end{aligned}
$$
□

Finally, we must show that the map (4.5.1) is an open map and this will be derived from the next result.

**Proposition 4.5.4.** *Let* $g_1, g_2, \ldots$ *be Möbius transformations and suppose that* $g_n(w) \to w$ *for* $w = 0, 1, \infty$. *Then:*

(i) *there exist matrices* $A_n$ *representing* $g_n$ *which converge to* $I$; *and*
(ii) $g_n \to I$ *uniformly on* $\hat{\mathbb{C}}$.

PROOF. Choose matrices

$$
A_n = \varepsilon_n \begin{pmatrix} a_n & b_n \\ c_n & d_n \end{pmatrix}
$$

in $SL(2, \mathbb{C})$ representing $g_n$ where $\varepsilon_n$ is $1$ or $-1$ and is to be chosen later. In the following argument, trivial modifications are required if $g_n(\infty) = \infty$: we ignore these cases.

As

$$
\begin{aligned}
d_n^2 &= \frac{1}{g_n(1) - g_n(0)} - \frac{1}{g_n(\infty) - g_n(0)} \\
&\to 1,
\end{aligned}
$$

we can select $\varepsilon_n$ so that $\varepsilon_n d_n \to 1$. Next,

$$
\begin{aligned}
(\varepsilon_n a_n)(\varepsilon_n d_n) &= a_n d_n \\
&= \frac{g_n(\infty)}{g_n(\infty) - g_n(0)} \\
&\to 1,
\end{aligned}
$$

so $\varepsilon_n a_n \to 1$ also. As

$$
c_n = a_n/g_n(\infty), \qquad b_n = d_n g_n(0),
$$

we see that $c_n$ and $b_n$ tend to zero: thus $A_n \to I$. This proves (i). Observe that (ii) follows from (i) and Proposition 4.5.3.
□

Finally, we can complete the proof of Theorem 4.5.1. Let $\mathscr{B}$ be an open subset of $SL(2, \mathbb{C})$ and suppose that $\Phi(\mathscr{B})$ is not an open subset of $\mathscr{M}$ (with respect to the metric topology $\mathcal{T}^*$). Then there is some $g$ in $\Phi(\mathcal{B})$ and some $g_1, g_2, \ldots$ not in $\Phi(\mathcal{B})$ with

$$\sigma(g_n, g) \to 0.$$

As

$$\sigma(g_n, g) = \sigma(g_n g^{-1}, I),$$

we see from Proposition 4.5.4 that there are matrices $A_n$ representing $g_n g^{-1}$ with $A_n \to I$. If $B$ (in $\mathcal{B}$) represents $g$, then $A_n B \to B$ so $A_n B$ is in $\mathcal{B}$ for all large $n$. It follows that $g_n (= \Phi(A_n B))$ is in $\Phi(\mathcal{B})$ for these $n$ and this is a contradiction. $\qquad\square$

A subgroup $G$ of $\mathcal{M}$ is discrete if and only if the topology described by Theorem 4.5.1 induces the discrete topology on $G$. It is clear from Corollary 4.5.2 that if $G$ is discrete, then $\Phi^{-1}(G)$ is a discrete subgroup of $SL(2, \mathbb{C})$. Conversely, if $\Gamma$ is a discrete subgroup of $SL(2, \mathbb{C})$, then $\Phi(\Gamma)$ is a discrete subgroup of $\mathcal{M}$.

Of course, if $G$ is a discrete subgroup of $\mathcal{M}$, then $G$ is countable (see Section 2.3), say $G = \{g_1, g_2, \ldots\}$, and

$$\|g_n\| \to +\infty$$

as $n \to +\infty$. In view of this, the next result is of interest.

**Theorem 4.5.5.** *Suppose that $K$ is a compact subset of a domain $D$ in $\hat{\mathbb{C}}$ and that $g$ omits the values $0$ and $\infty$ in $D$. Then for some positive $m$ depending only on $D$ and $K$, we have*

$$d(gz, gw) \le \frac{md(z, w)}{\|g\|^2}$$

*for all $z$ and $w$ in $K$.*

PROOF. Define $m_1$ by

$$2m_1 = \inf\{d(z, w): z \in K, w \notin D\}$$

and suppose that

$$g(z) = \frac{az + b}{cz + d}, \qquad ad - bc = 1.$$

As $g^{-1}(\infty) \notin D$, we see that for $z$ in $K$,

$$2m_1 \le d(z, g^{-1}\infty)$$

$$\le \frac{2|cz + d|}{(1 + |z|^2)^{1/2}(|c|^2 + |d|^2)^{1/2}}.$$

A similar inequality holds for $g^{-1}0$ so

$$(1 + |z|^2)\|g\|^2 (m_1)^2 \le |az + b|^2 + |cz + d|^2.$$

As

$$\frac{d(gz, gw)}{d(z, w)} \le \left(\frac{1 + |z|^2}{|az + b|^2 + |cz + d|^2}\right)^{1/2} \left(\frac{1 + |w|^2}{|aw + b|^2 + |cw + d|^2}\right)^{1/2}$$

the result follows.                                                             $\square$

The implication of this is that if $G$ is discrete, then under the assumptions in Theorem 4.5.6, the chordal diameters of the sets $g_n(K)$ tend to zero.

EXERCISE 4.5

1. Prove that if $ad - bc = 1$ then for all $z$

$$(|a|^2 + |c|^2)(|az + b|^2 + |cz + d|^2) \ge 1$$

with equality if and only if $z = -(\bar{a}b + \bar{c}d)/(|a|^2 + |c|^2)$. Show that if $g(z) = (ab + b)(cz + d)^{-1}$ then for all $z$,

$$\frac{1}{\|g\|^2} \le \frac{|az + b|^2 + |cz + d|^2}{1 + |z|^2} \le \|g\|^2.$$

2. Let $G$ be a group of Möbius transformations preserving $H^2$. Show that each $g$ in $G$ can be written uniquely in the form $g = fh$ where $f(z) = az + b$ $(a > 0, b \in \mathbb{R})$ and $h(i) = i$. Deduce that $G$ is homeomorphic to $\mathbb{R}^2 \times S^1$.

3. Show that a sequence $g_n$ of loxodromic transformations can converge to an elliptic element but if this is so, then $g_n$ is strictly loxodromic for almost all $n$. Show that a sequence of elliptic elements cannot converge to a loxodromic element.

# §4.6. Notes

For a discussion of quaternions and Möbius transformations see [1], [5] and [26]. The problem of obtaining a subgroup of $SL(2, \mathbb{C})$ isomorphic to a given subgroup of $\mathcal{M}$ has been considered in [2] and [74]. For general information on Möbius transformations see [30] (especially for isometric circles), [51] and [52]. See [53] for Theorems 4.2.2 and 4.3.7.

# CHAPTER 5
# Discontinuous Groups

## §5.1. The Elementary Groups

In this section we shall define and describe a class of subgroups of $\mathscr{M}$ which have a particularly simple structure. This class contains all finite subgroups of $\mathscr{M}$, all abelian subgroups of $\mathscr{M}$ and the stabilizer of each point in $\mathbb{R}^3$.

**Definition 5.1.1.** A subgroup $G$ of $\mathscr{M}$ is said to be *elementary* if and only if there exists a finite $G$-orbit in $\mathbb{R}^3$.

Of course, the emphasis here is on the word *finite*. Also, note that this definition makes no reference to discreteness. The group $\mathscr{M}$ acts as the group of directly conformal isometries of $H^3$ and $G$ is elementary if there is a finite $G$-orbit in the closure of *hyperbolic* space.

Obviously, if a single point is $G$-invariant then $G$ is elementary. If $G$ is abelian, then either $G$ contains only elliptic elements and $I$ or $G$ contains some parabolic or loxodromic element $g$. In the first case (whether $G$ is abelian or not), $G$ is elementary by virtue of Theorem 4.3.7: in the second case, $G$ is elementary by Theorem 4.3.6(iii). Thus every abelian subgroup of $\mathscr{M}$ is elementary.

*Remark.* Elementary groups are sometimes defined by the condition that for every $g$ and $h$ in $G$ which are of infinite order, we have trace$[g, h] = 2$: equivalently, $g$ and $h$ have a common fixed point in $\hat{\mathbb{C}}$ (Theorem 4.3.5). However, with this definition, the stabilizer of a point in $H^3$ is not necessarily elementary.

Let us now assume that $G$ is an elementary group and examine the possibilities. Suppose that the finite orbit is $\{x_1, \ldots, x_n\}$. If $g$ is in $G$ then the points $g^m(x_j), m = 0, 1, 2, \ldots,$ cannot all be distinct so there is an integer $m_j$ with the property that $g^{m_j}$ fixes $x_j$. If $m$ is now the product of the $m_j$, then $g^m$ fixes each $x_j$. With this available we can now classify the elementary groups into three types.

*Type 1: suppose that $n \geq 3$ or that $\{x_1, \ldots, x_n\}$ is not in $\hat{\mathbb{C}}$.*

If the points $x_j$ are not in $\hat{\mathbb{C}}$ then each $g$ in $G$ has some power $g^m$ fixing $x_j$ and so $g^m$, and hence $g$ itself, is elliptic (or $I$). If $n \geq 3$ and the $x_j$ are in $\hat{\mathbb{C}}$, then $g^m$ has at least three fixed points and so is the identity: thus again, each non-trivial element of $G$ is elliptic. This shows that if $G$ is of Type 1, then $G$ contains only elliptic elements and $I$. By Theorem 4.3.7, there is some $x$ in $H^3$ which is fixed by every element of $G$ and by mapping $H^3$ onto $B^3$ and $x$ to $0$ we see that $G$ is conjugate in $GM(\hat{\mathbb{R}}^3)$ to a subgroup of the Special Orthogonal group $SO(3)$ (see Theorem 3.4.1).

*Type 2: suppose that $n = 1$ and $x_1$ is in $\hat{\mathbb{C}}$.*

In this case, $G$ is conjugate to a subgroup of $\mathcal{M}$, every element of which fixes $\infty$ and so is of the form $z \mapsto az + b$. Thus $G$ is conjugate to a group of Euclidean similarities of $\mathbb{C}$.

*Type 3: suppose that $n = 2$ and that $x_1, x_2$ are in $\hat{\mathbb{C}}$.*

In this case, $G$ is conjugate to a subgroup of $\mathcal{M}$, every element of which leaves $\{0, \infty\}$ invariant and is therefore of the form

$$z \mapsto az^s, \qquad a \neq 0, s^2 = 1.$$

Note that $G$ is then conjugate to a group of isometries of the space $\mathbb{C} - \{0\}$ with the metric derived from $|dz|/|z|$.

We shall now describe all *discrete elementary groups*. If $G$ is a discrete elementary group of Type 1 we may assume that every element of $G$ fixes the point $j$ in $H^3$. Thus by Theorem 4.2.1, $\|g\|^2 = 2$ for every $g$ in $G$ and (as $G$ is discrete) $G$ is necessarily finite. Thus $G$ is conjugate to a finite subgroup of $SO(3)$ and hence to one of the symmetry groups of the regular solids.

We can use the fact that $G$ is finite to obtain the possible structures of $G$ without reference to the regular solids. We say that $v$ in $\hat{\mathbb{C}}$ is a *vertex* if $v$ is fixed by some $g$ $(\neq I)$ in $G$ and we denote the set of vertices by $V$. Now consider the number $|E|$ of elements of the finite set

$$E = \{(g, v): g \in G, g \neq I, v \in V, g(v) = v\}.$$

As each $g$ in $G$ $(g \neq I)$ is elliptic it fixes exactly two vertices and we have

$$|E| = 2(|G| - 1).$$

The stabilizer of a vertex $v$ is $G_v$ so we also have

$$|E| = \sum_{v \in V} (|G_v| - 1).$$

The set $V$ is partitioned by $G$ into disjoint orbits $V_1, \ldots, V_s$ and as the stabilizers of each $v$ in $V_j$ have the same number, say $n_j$, of elements we have

$$|E| = \sum_{j=1}^{s} \sum_{v \in V_j} (|G_v| - 1)$$

$$= \sum_{j=1}^{s} |V_j|(n_j - 1).$$

Finally, each orbit $G(v)$ is in 1–1 correspondence with the class of cosets $G/G_v$ so for $v$ in $V_j$, we have $V_j = G(v)$ and

$$|V_j| = \frac{|G|}{|G_v|} = \frac{|G|}{n_j}.$$

Eliminating $|V_j|$ we obtain

$$2\left(1 - \frac{1}{|G|}\right) = \sum_{j=1}^{s} \left(1 - \frac{1}{n_j}\right). \tag{5.1.1}$$

We shall exclude the trivial group, so $|G| \geq 2$ and

$$1 \leq 2\left(1 - \frac{1}{|G|}\right) < 2.$$

By definition, $n_j \geq 2$ so

$$\tfrac{1}{2}s \leq \sum_{j=1}^{s} \left(1 - \frac{1}{n_j}\right) < s.$$

These inequalities together with (5.1.1) show that $s = 2$ or $s = 3$.

*Case 1: $s = 2$.*
In this case, (5.1.1) becomes

$$2 = \frac{|G|}{n_1} + \frac{|G|}{n_2}$$

and hence (as $|n_j| \leq G$),

$$|G| = n_1 = n_2, \qquad |V_1| = |V_2| = 1.$$

In this case there are only two vertices and each is fixed by every element of $G$. By conjugation, we may take the vertices to be $0$ and $\infty$ and $G$ is then a finite, cyclic group of rotations of $\mathbb{C}$.

*Case* 2: $s = 3$.

In this case, (5.1.1) becomes

$$\frac{1}{n_1} + \frac{1}{n_2} + \frac{1}{n_3} = 1 + \frac{2}{|G|}$$

and we may assume that $n_1 \le n_2 \le n_3$. Clearly $n_1 \ge 3$ leads to a contradiction: thus $n_1 = 2$ and

$$\frac{1}{n_2} + \frac{1}{n_3} = \frac{1}{2} + \frac{2}{|G|}.$$

If $n_3 \ge n_2 \ge 4$ we again obtain a contradiction, so $n_2 = 2$ or 3. The case $n_2 = 2$ leads to

$$(|G|, n_1, n_2, n_3) = (2n, 2, 2, n) \qquad (n \ge 2)$$

and this is isomorphic to the group of orientation preserving symmetries of a regular plane $n$-gon (the dihedral group $D_n$).

The remaining cases are those with $s = 3$, $n_1 = 2$, $n_2 = 3$ and

$$\frac{1}{n_3} = \frac{1}{6} + \frac{2}{|G|}, \qquad n_3 \ge 3,$$

and the (integer) solutions of this are

(i) $(|G|, n_1, n_2, n_3) = (12, 2, 3, 3)$;
(ii) $(|G|, n_1, n_2, n_3) = (24, 2, 3, 4)$;
(iii) $(|G|, n_1, n_2, n_3) = (60, 2, 3, 5)$.

These groups are isomorphic to $A_4$, $S_4$ and $A_5$ respectively and they correspond to the symmetry groups of the tetrahedron, the octahedron and the icosahedron respectively. For more details, see the references in Section 5.5.

We continue with our discussion of discrete, elementary groups. The next result essentially distinguishes between groups of Types 2 and 3.

**Theorem 5.1.2.** *Let $g$ be loxodromic and suppose that $f$ and $g$ have exactly one fixed point in common. Then $\langle f, g \rangle$ is not discrete.*

PROOF. As discreteness is preserved under conjugation we may assume that the common fixed point is $\infty$ and, say,

$$g(z) = \alpha z \quad (|\alpha| > 1), \qquad f(z) = az + b$$

(if necessary, we may replace $g$ by $g^{-1}$).

Then

$$g^{-n} f g^n(z) = az + \alpha^{-n} b.$$

As $f$ and $g$ have only one common fixed point, we see that $b \neq 0$. As $|\alpha| > 1$, we find that the sequence

$$\|g^{-n}fg^n\|, \qquad n = 1, 2, \ldots$$

is a convergent sequence of distinct terms: thus $\langle f, g \rangle$ is not discrete. For a much more illuminating proof, the reader need only draw a diagram and locate (for large $n$) the points $z$, $g^n z$, $fg^n z$ and $g^{-n}fg^n z$. $\qquad \square$

Suppose now that $G$ is elementary, discrete but not of Type 1. Then $G$ must contain parabolic or loxodromic elements. If $G$ contains a parabolic element $g$, fixing $\infty$ say, then every element of $G$ fixes $\infty$ (because all other orbits are infinite) and by Theorem 5.1.2, $G$ has no loxodromic elements. Such a group is of Type 2. If $G$ contains a loxodromic element $g$, fixing 0 and $\infty$ say, then every element of $G$ must leave the set $\{0, \infty\}$ invariant. This implies that $G$ cannot contain parabolic elements and such a group is of Type 2 or 3.

Let us now examine the structure of a discrete group of Type 2 with parabolic elements. Thus $G$ contains only $I$, parabolic elements and possibly some elliptic elements.

By conjugation, we may assume that every element of $G$ fixes $\infty$ and so is of the form $z \mapsto \alpha z + \beta$. As this is either elliptic or parabolic, we see that $|\alpha| = 1$: thus $G$ is conjugate to a group of Euclidean isometries of $\mathbb{C}$.

We call $\alpha$ the *multiplier* of the map $z \mapsto \alpha z + \beta$ and in general, we denote the multiplier of $g$ by $\alpha_g$. Note that $\alpha_g = 1$ if and only if $g$ is parabolic or $I$. It is a trivial matter to check that the set $S$ of multipliers of $g$ in $G$ is a (multiplicative) subgroup of $\{|z| = 1\}$ and that the map $\theta: G \to S$ defined by $\theta(g) = \alpha_g$ is a homomorphism of $G$ into $S$. The statement that $\alpha_g = 1$ if and only if $g$ is parabolic or $I$ is precisely the statement that the kernel, $T$, of $\theta$ is the subgroup of translations in $G$. As $G/T$ is isomorphic to $S \,(=\theta(G))$, we can describe $G$ by giving explicit descriptions of $S$ and $T$: this effectively separates the parabolic and elliptic elements.

First, we show that $S$ is a finite cyclic group. Now $G$ contains a translation, say $f(z) = z + \lambda$ and if $g(z) = \alpha z + \beta$ is in $G$, then so is

$$gfg^{-1}(z) = z + \alpha\lambda.$$

We deduce that $G$ contains $z \mapsto z + s\lambda$ for every $s$ in $S$ and as $G$ is discrete, $S$ cannot accumulate in $\mathbb{C}$. Thus $S$ is a finite subgroup of $\{|z| = 1\}$ and (as is easily seen) it is necessarily cyclic.

We can obtain even more information about $S$. With $f$ and $g$ as above,

$$f^{-1}(gfg^{-1})(z) = z + (\alpha - 1)\lambda$$

and so if $|\alpha - 1| < 1$, then there is a translation $z \mapsto z + \lambda_1$ in $G$ with

$$|\lambda_1| = |(\alpha - 1)\lambda| < |\lambda|.$$

The same argument yields translations $z \mapsto z + \lambda_n$ in $G$ with

$$|\lambda_n| = |\alpha - 1|^n |\lambda| \to 0$$

as $n \to +\infty$ and this violates the discreteness of $G$. It follows that for every $\alpha$ in $S$, $|\alpha - 1| \geq 1$. As $S$ is a cyclic group, say,

$$S = \{1, \omega, \omega^2, \ldots, \omega^{q-1}\},$$

where

$$\omega = \exp(2\pi i / q),$$

we see that $q \leq 6$. In fact, $q \neq 5$. Indeed

$$fgfg^{-1}(z) = z + (\alpha + 1)\lambda$$

and for exactly the same reason as above, we must have $|\alpha + 1| \geq 1$. This implies that $q \neq 5$ for if $q = 5$, then $|\omega^2 + 1| < 1$. The remaining possibilities, namely $q = 1, 2, 3, 4$ and $6$ can all occur.

We must now describe $T$. Let $\Lambda$ be the set of $\lambda_1$ for which $z \mapsto z + \lambda_1$ is in $G$ and let $\Lambda^* = \Lambda - \{0\}$. As $G$ is discrete, $\Lambda$ cannot accumulate in $\mathbb{C}$ and so $\Lambda^*$ contains an element $\lambda$ of smallest (positive) modulus. If $\Lambda = \{n\lambda : n \in \mathbb{Z}\}$, then

$$T = \{z \mapsto z + n\lambda : n \in \mathbb{Z}\}. \tag{5.1.2}$$

If this is not so, there is an element $\mu$ of smallest (positive) modulus in $\Lambda^* - \{n\lambda : n \in \mathbb{Z}\}$: note that $|\mu| \geq |\lambda|$. The translations

$$z \mapsto z + n\lambda + m\mu; \qquad n, m \in \mathbb{Z}, \tag{5.1.3}$$

are in $G$ and we shall show that $T$ consists precisely of these translations. It is clear that $\mu$ is not a real multiple of $\lambda$ (else we write $\mu = (k + \delta)\lambda$ where $k \in \mathbb{Z}, 0 \leq \delta < 1$, and consider $\delta\lambda$). Thus $\lambda$ and $\mu$ span the vector space $\mathbb{C}$ (over $\mathbb{R}^1$) and if $z \mapsto z + \gamma$ is in $G$ we may write

$$\gamma = (n_1 + x)\lambda + (m_1 + y)\mu,$$

where $n_1, m_1 \in \mathbb{Z}$ and $x, y \in [-\frac{1}{2}, \frac{1}{2}]$. Now $\gamma - n_1\lambda - m_1\mu$ is in $\Lambda$ and

$$|\gamma - n_1\lambda - m_1\mu| = |x\lambda + y\mu| < |\mu|,$$

a strict inequality holding because $\lambda$ and $\mu$ are linearly independent. We deduce that

$$\gamma - n_1\lambda - m_1\mu \in \{n\lambda : n \in \mathbb{Z}\}$$

and so $T$ is precisely the set of translations (5.1.3).

We can now describe $G$. We select $g$ in $G$ with multiplier $\omega$ which generates $S$. Then $g, g^2, \ldots, g^{q-1}$ have multipliers $\omega, \omega^2, \ldots, \omega^{q-1}$ ($\omega^q = 1, q \leq 6$) and so $G$ has the coset decomposition

$$G = T \cup Tg \cup \cdots \cup Tg^{q-1}.$$

This shows that every element of $G$ is of the form

$$z \mapsto \omega^k z + n\lambda + m\mu,$$

where $k, m, n$ are integers, $0 \le k \le q$ and $q \le 6$, $q \ne 5$.

Next, we suppose that $G$ is discrete, elementary with loxodromic elements. First we suppose that every element of $G$ fixes both $0$ and $\infty$ and so is of the form

$$g(z) = \alpha z, \qquad \alpha \ne 0.$$

The map $\theta: G \to \{x \in R^1 : x > 0\}$ defined by $\theta(g) = |\alpha_g|$ is a homomorphism of $G$ into the multiplicative group of positive numbers and this time the kernel $E$ of $\theta$ consists of $I$ and all elliptic elements of $G$. Because $G$ and hence $E$, is discrete we see that $E$ is a finite cyclic group generated by, say, $z \to \omega z$ where $\omega^q = 1$.

The image $\theta(G)$ is the set $\{|\alpha_g| : g \in G\}$ and this set cannot accumulate at 1 else there are distinct elements $g_n$ in $G$ with

$$\|g_n\|^2 = |\alpha_n|^2 + \frac{1}{|\alpha_n|^2} \to 2 \qquad (g_n(z) = \alpha_n^2 z)$$

and this violates discreteness. It is now very easy to see that the multiplicative group $\theta(G)$ is of the form

$$\theta(G) = \{\lambda^n : n \in \mathbb{Z}\}$$

for some positive $\lambda$. We may assume that $g(z) = \alpha z$ where $|\alpha| = \lambda$: then $G$ has the coset decomposition

$$G = \bigcup_{n \in \mathbb{Z}} Eg^n$$

and each element of $G$ is of the form

$$z \mapsto \omega^k \alpha^n z, \qquad\qquad (5.1.4)$$

where $n \in \mathbb{Z}$, $k \in \mathbb{Z}$ and $0 \le k < q$. If $|\alpha| = 1$, then $\theta(G)$ is the trivial group and $G$ is a finite cyclic group of Type 1. Otherwise, $G$ is infinite and contains loxodromic elements but in any event, $G$ has no parabolic elements.

Finally, we consider the general discrete, elementary group of this type. We may assume that $\{0, \infty\}$ is the $G$-invariant and we denote by $G_0$ the elements in $G$ which fix both $0$ and $\infty$ so $G_0$ is of the form given by (5.1.4). If $G_0$ is a proper subgroup of $G$, then $G$ necessarily contains some element $h$ with

$$h(0) = \infty, \qquad h(\infty) = 0.$$

By a further conjugation (leaving $0$ and $\infty$ fixed) we may assume that $h(1) = 1$: thus $h(z) = 1/z$. If $f$ in $G$ interchanges $0$ and $\infty$, then $fh \in G_0$ and so $G_0$ is of index two in $G$: this shows that all elements of $G$ are of the form (5.1.4) or of the form

$$z \mapsto \omega^k \alpha^n / z.$$

This completes our discussion of all elementary discrete groups. In general, we shall be more interested in the *non-elementary* subgroups of $\mathcal{M}$. We end with two results which give necessary conditions for a group to be non-elementary: these results make no reference to discreteness. The first of these results gives some insight into the complexity of such groups.

**Theorem 5.1.3.** *Every non-elementary subgroup $G$ of $\mathcal{M}$ contains infinitely many loxodromic elements, no two of which have a common fixed point.*

PROOF. We begin by showing that $G$ has some loxodromic elements. Suppose, then, that $G$ has no loxodromic elements. If $G$ contains only $I$ and elliptic elements then $G$ is elementary. It follows that $G$ contains a parabolic element which we may take to be

$$f(z) = z + 1.$$

For any $g$ in $G$, say

$$g(z) = \frac{az + b}{cz + d}, \qquad ad - bc = 1,$$

we find that

$$f^n g(z) = \frac{(a + nc)z + (b + nd)}{cz + d}$$

and

$$\operatorname{tr}^2(f^n g) = (a + d + nc)^2.$$

As $f^n g$ is not loxodromic, we see that for all integers $n$,

$$0 \le (a + d + nc)^2 \le 4$$

and so $c = 0$. This implies that every element in $G$ fixes $\infty$ and so $G$ is elementary, a contradiction. Thus every non-elementary group contains loxodromic elements.

Now consider any non-elementary group $G$ and let $g$ be a loxodromic element of $G$ fixing, say, $\alpha$ and $\beta$. As $G$ is non-elementary, there is some $f$ in $G$ which does not leave $\{\alpha, \beta\}$ invariant and two cases arise:

(i) $\{\alpha, \beta\}$, $\{f\alpha, f\beta\}$ are disjoint;
(ii) $\{\alpha, \beta\}$, $\{f\alpha, f\beta\}$ have exactly one element in common.

In case (i), $g$ and $g_1 = fgf^{-1}$ are loxodromic with no common fixed points. It is now easy to see that the elements $g^n g_1 g^{-n}$ $(n \in \mathbb{Z})$ contain the desired loxodromic elements because the fixed points of $g^n g_1 g^{-n}$ are $g^n f\alpha$, $g^n f\beta$ and these are distinct from but converge towards $\alpha$ or $\beta$ (see Theorem 4.3.10).

In case (ii), $g$ and $g_1$ have exactly one common fixed point, say $\alpha$, so by Theorem 4.3.5, $p = [g, g_1]$ is parabolic and also fixes $\alpha$. As $\{\alpha\}$ cannot be

$G$-invariant, there is some $h$ in $G$ not fixing $\alpha$ so $q = hph^{-1}$ is parabolic and does not fix $\alpha$. Thus $q$ and $g$ (or $q$ and $g_1$) have no common fixed points. Then for suitably large $n$, the elements $g$ and $q^n g q^{-n}$ are loxodromic with no common fixed points and case (i) is applicable. $\qquad\square$

**Theorem 5.1.4.** *Let $f(\neq I)$ be a Möbius transformation not of order two and define the map $\theta: \mathcal{M} \to \mathcal{M}$ by $\theta(g) = gfg^{-1}$. If for some $n$, we have $\theta^n(g) = f$, then $\langle f, g \rangle$ is elementary and $\theta^2(g) = f$.*

PROOF. Define $g_0 = g$ and $g_n = \theta^n(g)$ so for $m \geq 0$,

$$g_{m+1} = g_m f(g_m)^{-1}.$$

Suppose first that $f$ is parabolic; then without loss of generality, $f(z) = z + 1$. As $g_1, \ldots, g_n$ are conjugate to $f$, they are each parabolic and so have a unique fixed point. Now for $r \geq 0$, $g_{r+1}$ fixes $g_r(\infty)$. Thus if $g_{r+1}$ fixes $\infty$, then so does $g_r$. As $g_n(=f)$ fixes $\infty$, we deduce that each $g_j$ (including $g_0$) fixes $\infty$. This shows that $\langle f, g \rangle$ is elementary as both elements fix $\infty$. Also, $g_1$ is parabolic and fixes $\infty$ and so commutes with $f$: thus $g_2 = f$.

Suppose now that $f$ has exactly two fixed points: then we may assume that $f(z) = kz$. Clearly $g_1, \ldots, g_n$ each have exactly two fixed points. Now suppose that $g_{r+1}$ fixes 0 and $\infty$ (as does $g_n$): then

$$\{0, \infty\} = \{g_r(0), g_r(\infty)\}.$$

Now $g_r$ cannot interchange 0 and $\infty$ $(r \geq 1)$ else $(g_r)^2$ fixes 0, $\infty$ and other points too and so $g_r$, and hence $f$ (which is conjugate to $g_r$), is of order two. We deduce that if $g_{r+1}$ fixes both points 0 and $\infty$, then so does $g_r$ for $r \geq 1$. It follows that $g_1, \ldots, g_n$ each fix 0 and $\infty$. This shows that $f$ and $g$ leave the set $\{0, \infty\}$ invariant and so $\langle f, g \rangle$ is elementary. Again, $g_1$ and $f$ commute so $g_2 = f$. $\qquad\square$

The reader may wish to relate this result to the discussion in Section 1.5.

EXERCISE 5.1

1. Let $G$ be an elementary group containing a parabolic element which fixes $\infty$. Show that if the group of all such parabolic elements is cyclic then any elliptic element in $G$ is of order two.

2. Show that a group $G$ is elementary if and only if for all $f$ and $g$ in $G$, $\langle f, g \rangle$ is elementary.

3. Show that if $g$ and $h$ are of order two, then $\langle g, h \rangle$ is elementary. Is $\langle g, h \rangle$ necessarily discrete?

4. Show that the map

$$z \mapsto (z/|z|, \log|z|)$$

is an isometry of $\mathbb{C} - \{0\}$ with the metric $|dz|/|z|$ onto the cylinder $S^1 \times \mathbb{R}^1$ with the Euclidean metric. Deduce that an elementary group leaving $\{0, \infty\}$ invariant is isomorphic to a group of isometries of the cylinder. Find the Euclidean isometry corresponding to the group element $z \mapsto az^p$ where $p = 1$ or $-1$.

5. Let

$$f(z) = -z, \qquad g(z) = \frac{(1 + t)z - (1 + t)}{(1 - t)z + (1 - t)}$$

where $t = 1/\sqrt{2}$. Show that $g$ is parabolic with fixed point $w$, say, where $w \neq 0$. Deduce that $fgf^{-1}$ is parabolic with fixed point $-w (\neq w)$ so $\langle f, g \rangle$ is non-elementary Show however that in the notation of Theorem 5.1.4, $\theta^2(g) = f$. (The assumption that $f$ is not of order two in Theorem 5.1.4 is necessary.)

## §5.2. Groups with an Invariant Disc

Later, we shall be interested in those subgroups of $\mathcal{M}$ which have an invariant disc: here, we characterize such groups.

**Theorem 5.2.1.** *Let $G$ be a non-elementary subgroup of $\mathcal{M}$. Then there exists a G-invariant disc if and only if $G$ has no strictly loxodromic elements. If $D$ is a G-invariant open disc, then $D$ and its exterior are the only G-invariant discs.*

Note that we do not require $G$ to be discrete. The restriction to non-elementary groups is necessary: for example, if

$$p(z) = z + 1, \qquad q(z) = z + i,$$

then $\langle p, q \rangle$ has no loxodromic elements and no invariant disc and $\langle p \rangle$ has infinitely many invariant discs.

PROOF. Directly from Definition 4.3.3, if a $G$-invariant disc exists then $G$ has no strictly loxodromic elements.

To prove the converse, suppose that $G$ is non-elementary and has no strictly loxodromic elements. By Theorem 5.1.3, we can find loxodromic, and therefore hyperbolic, elements $g$ and $h$ in $G$ with no common fixed points. By conjugation, we may assume that $g$ fixes 0 and $\infty$.

Now select any $f$ in $G$. In terms of matrices we can write

$$g = \begin{pmatrix} u & 0 \\ 0 & 1/u \end{pmatrix}, \qquad f = \begin{pmatrix} \alpha & \beta \\ \gamma & \delta \end{pmatrix},$$

where each matrix is in $SL(2, \mathbb{C})$. As $g$ is hyperbolic, we find that $u$ is real. Next, write

$$t_1 = \text{trace}(f) = \alpha + \delta$$

and

$$t_2 = \text{trace}(gf) = u\alpha + \delta/u.$$

Because $f$ and $gf$ are not strictly loxodromic, $t_1$ and $t_2$ are real. Solving for $\alpha$ and $\delta$, we find that $\alpha$ and $\delta$ are real. This shows that *every element of $G$ has real diagonal elements*.

Now let

$$h = \begin{pmatrix} a & b \\ c & d \end{pmatrix}, \qquad ad - bc = 1,$$

so $a$ and $d$ are real. Also $(a + d)^2 > 4$ because $h$ is hyperbolic. The fixed points of $h$ are the points

$$w_1, w_2 = \frac{(a - d) \pm [(a + d)^2 - 4]^{1/2}}{2c}$$

and as $c \neq 0$, the ratio $w_1/w_2$ is real. This implies that the fixed points of $g$ and $h$ are collinear. In an invariant formulation, the absence of strictly loxodromic elements implies that the fixed points of every pair $g$ and $h$ of hyperbolic elements are concyclic. One can proceed by geometry but the algebraic proof seems simpler.

We may assume that the fixed points of $g$ and $h$ lie on the real axis. Then $g$ and $h$ leave $H^2$ invariant and all entries of $h$ are real. Now

$$fh = \begin{pmatrix} \alpha a + \beta c & * \\ * & \gamma b + \delta d \end{pmatrix}$$

and these diagonal elements are real. As $a, b, c, d, \alpha$ and $\delta$ are real and $bc \neq 0$, we find that $\beta$ and $\gamma$ are real so $f$ is in $SL(2, \mathbb{R})$. This shows that every element of $G$ preserves $H^2$.

Finally, let $D$ be an invariant disc. For any hyperbolic $h$ in $G$, the points $h^n(z)$ accumulate at the fixed points of $h$ (Theorem 4.3.10). By taking $z$ in $D$ and then in the exterior of $D$ we see that all hyperbolic fixed points must lie in the boundary of $D$: thus there are precisely two $G$-invariant discs, the common boundary containing all hyperbolic fixed points (see Theorem 5.1.3). $\qquad \square$

The argument given in the last part of this proof shows that *if $g$ is parabolic or hyperbolic with an invariant disc $D$, then the fixed points of $g$ lie on $\partial D$*. If $g$ is elliptic with an invariant disc $D$, then the fixed points of $g$ cannot lie on $\partial D$ (consider $g(z) = e^{i\theta}z$). If $w$ is a fixed point of $g$, then so is the inverse point of $w$ with respect to $\partial D$ because inverse points and $\partial D$ are preserved by $g$. Thus *if $g$ is elliptic with invariant disc $D$ then the fixed points of $g$ are inverse points with respect to $\partial D$ and are not on $\partial D$.*

EXERCISE 5.2

1. Verify the statements regarding the location of the fixed points of $g$ with invariant disc $D$ by taking $D$ to be $H^2$ and regarding $g$ as a matrix in SL(2, $\mathbb{R}$).

## §5.3. Discontinuous Groups

We begin with a general definition.

**Definition 5.3.1.** Let $X$ be any topological space and $G$ a group of homeomorphisms of $X$ onto itself. We say that $G$ *acts discontinuously on $X$* if and only if for every compact subset $K$ of $X$,

$$g(K) \cap K = \emptyset,$$

except for a finite number of $g$ in $G$.

In our applications, $X$ will always be a subset of $\hat{\mathbb{R}}^3$ with the usual topology. There are, however, several useful results which, even in the general situation, follow easily from this definition. Suppose now that $G$ acts discontinuously on $X$: then the following statements are true.

<blockquote>

*Every subgroup of $G$ acts discontinuously on $X$.*                (5.3.1)

*If $\phi$ is a homeomorphism of $X$ onto $Y$, then $\phi G \phi^{-1}$ acts discontinuously on $Y$.*                (5.3.2)

*If $Y$ is a $G$-invariant subset of $X$, then $G$ acts discontinuously on $Y$.*                (5.3.3)

*If $x \in X$ and if $g_1, g_2, \ldots$ are distinct elements of $G$, then the sequence $g_1(x), g_2(x), \ldots$ cannot converge to any $y$ in $X$.*                (5.3.4)

*If $x \in X$, then the stabilizer $G_x$ is finite.*                (5.3.5)

*If (for example) $X \subset \hat{\mathbb{R}}^3$, then $G$ is countable.*                (5.3.6)

</blockquote>

PROOFS. Clearly (5.3.1) and (5.3.2) are true. If $Y \subset X$, then any compact subset of $Y$ is also a compact subset of $X$ and (5.3.3) follows. To prove (5.3.4), observe that if the given sequence converges to $y$, then

$$K = \{y, x, g_1(x), g_2(x), \ldots\}$$

is a compact set. As $g_n(K) \cap K \neq \emptyset$ $(n = 1, 2, \ldots)$ and as the $g_n$ are distinct, $G$ cannot act discontinuously on $X$: thus (5.3.4) follows. For each $x$ in $X$, $\{x\}$ is compact; thus (5.3.5) is a direct consequence of Definition 5.3.1.

Finally, we have seen (in Section 4.3) that there is a 1–1 correspondence between $G/G_x$ and the orbit $G(x)$ and so by (5.3.5), $G$ is countable if and only if $G(x)$ is countable. Now any uncountable set in $\hat{\mathbb{R}}^3$ contains a limit point of itself and so by (5.3.4), $G(x)$ must be countable. This proves (5.3.6). □

Our aim is to study the relationship between discreteness and discontinuity as applied to subgroups of $\mathcal{M}$. First, we consider the action of $G$ in $H^3$.

**Theorem 5.3.2.** *A subgroup $G$ of $\mathcal{M}$ is discrete if and only if it acts discontinuously in $H^3$.*

PROOF. Suppose first that $G$ is discrete. As $G$ is the homomorphic image of a discrete (and therefore countable) subgroup of $\text{SL}(2, \mathbb{C})$, we see that $G$ is countable, say

$$G = \{g_1, g_2, \ldots\}.$$

As $G$ is discrete, $\|g_n\| \to +\infty$ and so using Theorem 4.2.1, we see that as $n \to +\infty$, so

$$\rho(j, g_n(j)) \to +\infty. \tag{5.3.7}$$

It is clear from (3.3.5) that a compact subset $K$ of $H^3$ lies in some hyperbolic ball

$$B = \{x \in H^3 : \rho(x, j) < k\}.$$

If $g(K) \cap K \neq \varnothing$, then $g(B) \cap B \neq \varnothing$ and so

$$\rho(j, g(j)) < 2k$$

By (5.3.7) this can only happen for a finite number of $g$ in $G$ and so $G$ acts discontinuously in $H^3$.

Now suppose that $G$ acts discontinuously in $H^3$ (or in any subdomain of $\hat{\mathbb{C}}$). If $G$ is not discrete, we can find distinct matrices $A_1, A_2, \ldots$ in $\text{SL}(2, \mathbb{C})$ projecting to $g_1, g_2, \ldots$ in $G$ with $A_n \to I$ as $n \to \infty$. Using (4.1.4), we see that $g_n(x) \to x$ as $n \to \infty$ for every $x$ in $\hat{\mathbb{R}}^3$. Clearly this violates (5.3.4) and so we deduce that $G$ is necessarily discrete. □

We now turn our attention to the extended complex plane and we seek to understand the relationship between discreteness and discontinuity in open subsets of $\hat{\mathbb{C}}$. Of course, the proof of Theorem 5.3.2 shows that *if $G$ acts discontinuously in some non-empty open subset of $\hat{\mathbb{C}}$, then $G$ is discrete.* The converse is false: it is possible for $G$ to be discrete yet not act discontinuously in any open subset of $\hat{\mathbb{C}}$. In order to give a simple example of this, we establish a criterion which excludes the possibility of a discontinuous action.

**Lemma 5.3.3.** *Let G be any subgroup of $\mathcal{M}$ and let D be an open subset of $\hat{\mathbb{C}}$ which contains a fixed point $v$ of some parabolic or loxodromic element $g$ of G. Then G does not act discontinuously in D.*

PROOF. This is trivial as the stabilizer $G_v$ contains the distinct iterates of $g$. If $g$ is parabolic or loxodromic, then $G_v$ is infinite and this violates (5.3.5). □

**Example 5.3.4.** Let $G$ be Picard's group, namely the group of transformations of the form

$$g(z) = \frac{az + b}{cz + d}, \tag{5.3.8}$$

where $a, b, c$ and $d$ are Gaussian integers (of the form $m + in$ where $m$, $n \in \mathbb{Z}$) and $ad - bc = 1$. Obviously $G$ is discrete.

By Lemma 5.3.3 it is sufficient to show that the parabolic fixed points of $G$ are dense in $\hat{\mathbb{C}}$. Let $w = (p + iq)/r$ where $p, q$ and $r$ are integers: obviously, the set of such $w$ is dense in $\hat{\mathbb{C}}$. Now simply observe that

$$h(z) = \frac{(1 - wr^2)z + r^2w^2}{-r^2z + (1 + wr^2)}$$

is a parabolic element of $G$ that fixes $w$. □

Our aim now is to understand the situation in which a discrete group does act discontinuously on some open subset of $\hat{\mathbb{C}}$. The exposition will be clearer if *we restrict our attention to the non-elementary groups*: the case of the elementary groups is rather easy and are left to the reader. Note, however, that once again we do not begin with the assumption of discreteness.

The discussion will be based on the fixed points of loxodromic elements of $G$ and we begin with a preliminary result which enables us to locate these fixed points.

**Lemma 5.3.5.** *Let $\Sigma$ be an open disc and suppose that $g \in \mathcal{M}$ and $g(\bar{\Sigma}) \subset \Sigma$. Then $g$ is loxodromic and has a fixed point in $g(\bar{\Sigma})$.*

PROOF. We may assume that $g(\infty) = \infty$. With this assumption, $\partial\Sigma$ is a Euclidean circle (and not a straight line) as clearly, no fixed point of $g$ is on the boundary of $\Sigma$. If $g$ is elliptic or parabolic then (as $g$ fixes $\infty$) $g$ is a Euclidean isometry and this is not compatible with $g(\bar{\Sigma}) \subset \Sigma$. Thus $g$ is loxodromic. For any $w$ not fixed by $g$, the images $g^n(w), n = 1, 2, \ldots$, accumulate at a point $v$ fixed by $g$. If $w \in \Sigma$, these images are in $g(\bar{\Sigma})$ and so $v \in g(\bar{\Sigma})$. □

We now begin our study of discontinuity in subsets of $\hat{\mathbb{C}}$.

**Definition 5.3.6.** Let $G$ be a non-elementary subgroup of $\mathcal{M}$ ($G$ need not be discrete) and let $\Lambda_0$ denote the set of points fixed by some loxodromic element in $G$. The *limit set* $\Lambda(G)$ of $G$ is the closure of $\Lambda_0$ in $\hat{\mathbb{C}}$: the *ordinary set* $\Omega(G)$ of $G$ is the complement of $\Lambda$ in $\hat{\mathbb{C}}$.

In general, we shall write $\Lambda$ and $\Omega$ without explicit mention of $G$. Note that if $G \subset G_1$ then

$$\Lambda(G) \subset \Lambda(G_1), \qquad \Omega(G) \supset \Omega(G_1).$$

We shall study $\Lambda$ first and then $\Omega$.

**Theorem 5.3.7.** *For any non-elementary group $G$, the limit set $\Lambda$ is the smallest non-empty $G$-invariant closed subset of $\hat{\mathbb{C}}$. In addition, $\Lambda$ is a perfect set and is therefore uncountable.*

PROOF. As $\Lambda_0$ is $G$-invariant, so is $\Lambda$. By definition, $\Lambda$ is closed and by Theorem 5.1.3, $\Lambda \neq \varnothing$. Now let $E$ be any non-empty, closed $G$-invariant subset of $\hat{\mathbb{C}}$. As $G$ is non-elementary, every orbit is infinite, thus $E$ is infinite. Now take any point $v$ fixed by a loxodromic element $g$ in $G$. There is some $w$ in $E$ not fixed by $g$ and the set $\{g^n(w): n \in \mathbb{Z}\}$ accumulates at $v$ (and at the other fixed point of $g$). As $E$ is closed, $v \in E$. This shows that $\Lambda_0 \subset E$; hence $\Lambda \subset E$.

This argument also shows that $\Lambda_0$ has no isolated points (we simply choose $w$ in $\Lambda_0$ but not fixed by $g$): hence $\Lambda$ has no isolated points. A set is *perfect* if it is closed and without isolated points and as is well known any non-empty perfect set is uncountable. As $\Lambda$ is perfect, the proof is complete.

$\square$

Theorem 5.3.7 shows that the countable set $\Lambda_0$ is dense in the uncountable set $\Lambda$ but we can say even more than this.

**Theorem 5.3.8.** *Let $G$ be a non-elementary subgroup of $\mathcal{M}$ and let $O_1$ and $O_2$ be disjoint open sets both meeting $\Lambda$. Then there is a loxodromic $g$ in $G$ with a fixed point in $O_1$ and a fixed point in $O_2$.*

PROOF. Recall that if $f$ is loxodromic with an attractive fixed point $\alpha$ and a repulsive fixed point $\beta$, then as $n \to +\infty$, $f^n \to \alpha$ uniformly on each compact subset of $\hat{\mathbb{C}} - \{\beta\}$ and $f^{-n} \to \beta$ uniformly on each compact subset of $\hat{\mathbb{C}} - \{\alpha\}$ (Theorem 4.3.10). The repulsive fixed point of $f$ is the attractive fixed point of $f^{-1}$.

Now consider $G, O_1$ and $O_2$ as in the theorem. It follows (Definition 5.3.6) that there is a loxodromic $p$ with attractive fixed point in $O_1$ and a loxodromic $q$ with attractive fixed point in $O_2$. By Theorem 5.1.3, there is a loxodromic $f$ with attractive fixed point $\alpha$ and repulsive fixed point $\beta$, neither fixed by $p$. Now choose (and then fix) some sufficiently large value of $m$ so that

$$g = p^m f p^{-m}$$

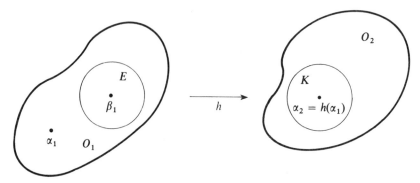

Figure 5.3.1

has its attractive fixed point $\alpha_1$ $(=p^m\alpha)$ and repulsive fixed point $\beta_1$ $(=p^m\beta)$ in $O_1$. Then choose (and fix) some sufficiently large value of $r$ so that

$$h = q^r$$

maps $\alpha_1$ into $O_2$: put $\alpha_2 = h(\alpha_1)$. See Figure 5.3.1

Next, construct open discs $E$ and $K$ with the properties

$$\beta_1 \in E \subset \bar{E} \subset O_1,$$
$$\alpha_2 \in K \subset \bar{K} \subset O_2.$$

As $\beta_1 \notin \bar{K}$ we see that $g^n \to \alpha_1$ uniformly on $\bar{K}$ as $n \to +\infty$. As $h^{-1}(K)$ is an open neighbourhood of $\alpha_1$ we see that for all sufficiently large $n$,

$$g^n(\bar{K}) \subset h^{-1}(K)$$

and so

$$hg^n(\bar{K}) \subset K. \tag{5.3.9}$$

As $h(\alpha_1) \notin \bar{E}$ so $\alpha_1$ is not in $h^{-1}(\bar{E})$ and so $g^{-n} \to \beta_1$ uniformly on $h^{-1}(\bar{E})$ as $n \to +\infty$. Thus for all sufficiently large $n$,

$$g^{-n}h^{-1}(\bar{E}) \subset E. \tag{5.3.10}$$

Choose a value of $n$ for which (5.3.9) and (5.3.10) hold. By Lemma 5.3.5, $hg^n$ is loxodromic with a fixed point in $K$: also, $g^{-n}h^{-1}$, which is $(hg^n)^{-1}$, has a fixed point in $E$, hence so does $hg^n$. $\qquad\square$

Theorems 5.3.7 and 5.3.8 do not require $G$ to be discrete. If we add the extra condition that $G$ is discrete, we can describe $\Lambda$ in terms of any one orbit. For any $z$ in $\hat{\mathbb{C}}$, let $\Lambda(z)$ be the set of $w$ with the property that there are distinct $g_n$ in $G$ with $g_n(z) \to w$ (the points $g_n(z)$ need not be distinct).

**Theorem 5.3.9.** *Let $G$ be a non-elementary discrete subgroup of $\mathscr{M}$. Then for all $z$ in $\hat{\mathbb{C}}$, we have $\Lambda = \Lambda(z)$.*

*Remark.* The group generated by $z \mapsto 2z$ shows that the conclusion may fail if $G$ is only discrete. The group of Möbius transformations preserving the unit disc shows that the conclusion may fail if $G$ is only non-elementary.

PROOF OF THEOREM 5.3.9. Each $\Lambda(z)$ is closed, non-empty and $G$-invariant so by Theorem 5.3.7, we have

$$\Lambda \subset \Lambda(z).$$

If $z \in \Lambda$, then $G(z) \subset \Lambda$ and so

$$\Lambda(z) \subset \overline{G(z)} \subset \Lambda:$$

in this case, then we have $\Lambda = \Lambda(z)$.

Now suppose that $z$ is in $\Omega$ and select any $w$ in $\Lambda(z)$: we must show that $w \in \Lambda$. Suppose not, then $w \in \Omega$ and there is a disc $Q$ with centre $w$ whose closure $\overline{Q}$ lies in $\Omega$. We may suppose that $0$ and $\infty$ are in $\Lambda$ so taking $K = \overline{Q} \cup \{z\}$ we deduce from Theorem 4.5.6 that for all $g$ in $G$ and all $z'$ in $Q$,

$$d(gz, gz') \le m/\|g\|^2.$$

As $w \in \Lambda(z)$, there are distinct $g_n$ with $g_n(z) \to w$: as $\|g_n\|^2 \to +\infty$, we deduce that $g_n \to w$ uniformly on $\overline{Q}$. This implies that for large $n$,

$$g_n(\overline{Q}) \subset Q:$$

hence for Lemma 5.3.5 we have $Q \cap \Lambda \ne \emptyset$ and this contradicts $Q \subset \Omega$. $\square$

We now turn our attention to the open set $\Omega$.

**Theorem 5.3.10.** *Suppose that $G$ is a discrete non-elementary subgroup of $\mathcal{M}$. Then $\Omega$ is the maximal domain of discontinuity in $\hat{\mathbb{C}}$ of $G$: precisely,*

(i) *$G$ acts discontinuously in $\Omega$; and*
(ii) *if $G$ acts discontinuously in an open subset $D$ of $\hat{\mathbb{C}}$, then $D \subset \Omega$.*

*Remark.* Traditionally, a discrete group $G$ was called *Kleinian* if $\Omega \ne \emptyset$. More recently, *Kleinian* is used synonomously with *discrete*.

PROOF OF THEOREM 5.3.10. If $G$ does not act discontinuously in $\Omega$, then there is a compact subset $K$ of $\Omega$ and distinct $g_1, g_2, \ldots$ in $G$ such that $g_n(K) \cap K \ne \emptyset$. Thus there are points $z_1, z_2, \ldots$ in $K$ with $g_n(z_n) \in K$. By taking a subsequence, we may assume that $g_n(z_n) \to w$ in $K$ and so $w \in \Omega$. However, exactly as in the proof of Theorem 5.3.9, we now see that $g_n \to w$ uniformly on $K$ and so $w \in \Lambda$, a contradiction. This proves (i).

It is easy to prove (ii). By Lemma 5.3.3, $D \cap \Lambda_0 = \emptyset$. As $D$ is open, this implies that $D \cap \Lambda = \emptyset$ so $D \subset \Omega$. $\square$

Theorem 5.3.10 has an interesting corollary.

**Corollary.** *Let G be discrete and non-elementary. Then $\Omega \neq \varnothing$ if and only if for some z, G(z) is not dense in $\hat{\mathbb{C}}$.*

PROOF. By Theorem 5.3.9, $\Omega \neq \varnothing$ if and only if $\Lambda(z)(=\Lambda)$ is not $\hat{\mathbb{C}}$ and this is the assertion in the corollary.                                                    $\square$

Lemma 5.3.3 shows that the fixed points of parabolic and loxodromic elements of $G$ lie in $\Lambda$ and hence not in $\Omega$. It is not hard to see that there can be fixed points of elliptic elements of $G$ both in $\Lambda$ and in $\Omega$. However, if an elliptic fixed point lies in $\Omega$, the stabilizer of that point must be cyclic.

**Theorem 5.3.11.** *Suppose that G is non-elementary and that $\Omega \neq \varnothing$. If $z \in \Omega$ then the stabilizer $G_z$ is cyclic and finite.*

PROOF. By virtue of Lemma 5.3.3, if $z \in \Omega$ then every element of the stabilizer $G_z$ is either elliptic or $I$. Thus by Theorem 4.3.7, there is some $\zeta$ in $H^3$ which is fixed by every $g$ in $G_z$. Now let $A$ be the unique semi-circle in $H^3$ which has end-point $z$, which passes through $\zeta$ and which is orthogonal to $\hat{\mathbb{C}}$. Every elliptic element of $G_z$ fixes $z$ and $\zeta$ and so has the axis $A$. This means that every element of $G_z$ fixes *both* end-points of $A$ and an examination of the discrete elementary groups listed in Section 5.1 shows that $G_z$ is necessarily a finite cyclic group.

For an alternative proof, suppose that $g$ and $h$ fix $z$ in $\Omega$. As both $g$ and $h$ are elliptic they each have another fixed point. If these other fixed points are distinct, then by Theorem 4.3.5, $[g, h]$ is parabolic and also fixes $z$ and this violates Lemma 5.3.3.                                                    $\square$

We can use Theorem 5.3.11 to obtain a result concerning the local behaviour of a discrete group $G$ near a point in $\Omega$ or $H^3$.

**Theorem 5.3.12.** *Let G be a discrete non-elementary subgroup of $\mathcal{M}$. Then (considering only g in G):*

(i) *each x in $H^3$ is the centre of an open hyperbolic ball N such that $g(N) = N$ if $g(x) = x$ and $g(N) \cap N = \varnothing$ otherwise;*

(ii) *if $\Omega \neq \varnothing$, each z in $\Omega$ has an open neighbourhood N in $\Omega$ such that $g(N) = N$ if $g(x) = x$ and $g(N) \cap N = \varnothing$ otherwise.*

PROOF. First, (i) is a direct consequence of the fact that $G$ is a group of isometries acting discontinuously in $H^3$.

To prove (ii), we may assume that $z = 0$ and that every $g$ in $G_z$ also fixes $\infty$ (use Theorem 5.3.11). Now select a disc

$$N = \{z : |z| < r\}$$

whose closure is contained in $\Omega$. As $G$ acts discontinuously in $\Omega$, $g(N) \cap N \neq \varnothing$ for only a finite set of $g$ in $G$. By continuity, for a sufficiently small $r$ (depending in this *finite* set) $g(N) \cap N = \emptyset$ unless $g(0) = 0$ in which case, $g(N) = N$.  $\square$

If $G$ is a discrete group, then $G = \{g_1, g_2, \ldots\}$ say, and

$$\|g_n\| \to +\infty \quad \text{as } n \to +\infty.$$

We now show this convergence cannot be too slow.

**Theorem 5.3.13.** *Let $G$ be a discrete subgroup of $\mathcal{M}$. Then:*

(i) *the number $n(t)$ of elements $g$ in $G$ with $\|g\| \leq t$ is $O(t^4)$;*
(ii) *for any $s > 4$, the series $\sum \|g\|^{-s}$ converges;*
(iii) *if $\Omega \neq \emptyset$, then the series $\sum \|g\|^{-4}$ converges.*

PROOF. The stabilizer $G_j$ of $j$ in $H^3$ is finite with, say, $k$ elements. Let $N$ be a hyperbolic ball in $H^3$ with centre $j$ and radius $r$, say, such that $g(N) \cap N = \emptyset$ when $g \in G - G_j$. Let $V(R)$ be the hyperbolic volume of a hyperbolic ball of radius $R$.

Now $\|g\| \leq t$ is equivalent to

$$2 \cosh \rho(j, gj) \leq t^2,$$

(Theorem 4.2.1) and so if $\|g\| \leq t$, then

$$g(N) \subset \{x \in H^3 : \rho(x, j) \leq r + \cosh^{-1}(\tfrac{1}{2}t^2)\}.$$

By adding the volumes of the disjoint images $g(N)$ of $N$ with $\|g\| \leq t$ and by taking into account the order of the stabilizer of $j$, we obtain

$$n(t)/k \leq V(r + \cosh^{-1}(\tfrac{1}{2}t^2)). \qquad (5.3.11)$$

Now (see [5], p.61)

$$V(R) = \pi[\sinh(2R) - 2R]$$
$$< \pi e^{2R}/2$$

and

$$\cosh^{-1}(y) = \log(y + [y^2 - 1]^{1/2})$$
$$< \log(2y).$$

Thus

$$n(t) \leq k(\pi/2) \exp[2r + 2\log(t^2)]$$
$$= (k\pi e^{2r}/2)t^4.$$

To prove (ii) simply observe that $n(1) = 0$ so

$$\sum_{g \in G, \|g\| \leq t} \|g\|^{-s} = \int_1^t \frac{dn(x)}{x^s}$$
$$= \frac{n(t)}{t^s} + s \int_1^t \frac{n(x)\, dx}{x^{s+1}} \qquad (5.3.12)$$

and so (i) implies (ii). Note that in general, this yields

$$\sum_{\|g\| \le t} \|g\|^{-4} = O(\log t)$$

and indeed, an estimate of the partial sums (5.3.12) for any positive $s$.

To prove (iii) we can use a similar argument but in $\Omega$ and with the chordal metric. We can find an open disc $N$ in $\Omega$ such that for all $g$ in $G$, $g \ne I$, we have $g(N) \cap N = \emptyset$. Then the sum of the areas of the $g(N)$ measured in the chordal metric converges to at most $4\pi$ (the chordal area of $\hat{\mathbb{C}}$) and it is only necessary to estimate this area of $g(N)$. Let

$$g(z) = \frac{az + b}{cz + d}, \qquad ad - bc = 1.$$

Then the chordal area of $g(N)$ is

$$\iint\limits_{g(N)} \frac{4dx\, dy}{(1 + |z|^2)^2} = 4 \iint\limits_{N} \frac{|g^{(1)}(z)|^2\, dx\, dy}{(1 + |g(z)|^2)^2}$$

$$= \iint\limits_{N} \frac{4dx\, dy}{(|az + b|^2 + |cz + d|^2)^2}$$

$$\ge \|g\|^{-4} \text{ (chordal area of } N),$$

the last line being an application of the Cauchy–Schwarz inequality, namely

$$|az + b|^2 + |cz + d|^2 \le (|a|^2 + |b|^2)(1 + |z|^2) + (|c|^2 + |d|^2)(1 + |z|^2).$$

$\square$

We end with two result which imply that $\Omega \ne \emptyset$.

**Theorem 5.3.14.** *Let $G$ be a discrete non-elementary subgroup of $\mathcal{M}$.*

(i) *If $D$ is a non-empty open $G$-invariant set which is not $\hat{\mathbb{C}}$, then $G$ acts discontinuously in $D$;*

(ii) *if $D$ is a non-empty open set such that $g(D) \cap D = \emptyset$ for all $g$ in $G$ except $I$, then $G$ acts discontinuously in $\bigcup_{g \in G} g(D)$.*

PROOF. The set $E = \hat{\mathbb{C}} - D$ is non-empty, closed and $G$-invariant and so by Theorem 5.3.7, $\Lambda \subset E$. Thus $G$ acts discontinuously in $D$ (Theorem 5.3.10).

By definition, $\bigcup g(D)$ is disconnected and so is not $\hat{\mathbb{C}}$: now apply (i) to $\bigcup g(D)$. $\square$

Referring to (ii) in the previous theorem, we say that a subdomain $D$ of $\hat{\mathbb{C}}$ is a $G$-packing if $g(D) \cap D = \emptyset$ whenever $g \in G$ and $g \ne I$. This terminology enables us to state our next result easily.

**Theorem 5.3.15.** *Let $G_1, G_2, \ldots$ be subgroups of $\mathcal{M}$ whose union generates the group $G$. Let $D_j$ be a $G_j$-packing and suppose that $D_i \cup D_j = \hat{\mathbb{C}}$ when $i \neq j$. Suppose also that $D^*(= \bigcap D_j)$ is nonempty. Then $G$ is the free product of the $G_j$, $D^*$ is a $G$-packing and $G$ acts discontinuously on $\bigcup_g g(D^*)$.*

PROOF. Consider any element $g_n \cdots g_1$ of $G$ where $g_k \in G_{i_k}$, $g_k \neq I$ and $i_k \neq i_{k+1}$ for any $k$. First, because $D_{i_1}$ is a $G_{i_1}$-packing, we have

$$g_1(D^*) \subset g_1(D_{i_1}) \subset \hat{\mathbb{C}} - D_{i_1}.$$

In fact, it follows (by induction) that

$$g_m \cdots g_1(D^*) \subset \hat{\mathbb{C}} - D_{i_m}$$

for if this is so, then

$$\begin{aligned} g_{m+1}(g_m \cdots g_1)(D^*) &\subset g_{m+1}(\hat{\mathbb{C}} - D_{i_m}) \\ &\subset g_{m+1}(D_{i_{m+1}}) \\ &\subset \hat{\mathbb{C}} - D_{i_{m+1}}. \end{aligned}$$

We deduce that

$$g_m \cdots g_1(D^*) \subset \hat{\mathbb{C}} - D_{i_m} \subset \hat{\mathbb{C}} - D^*$$

so $D^*$ is a $G$-packing. Because $D^* \neq \varnothing$ we must have $g_m \cdots g_1 \neq I$ so $G$ is the free product of the $G_j$. The last assertion follows from Theorem 5.3.14(ii). □

As an application of Theorem 5.3.15, consider $G_1 = \langle g \rangle$ and $G_2 = \langle h \rangle$ where

$$g(z) = z + 6, \qquad h(z) = z/(z + 1).$$

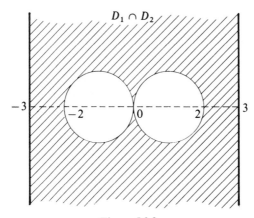

Figure 5.3.2

Let
$$D_1 = \{x + iy : |x| < 3\}$$
and
$$D_2 = z : |z + 1| > 1\} \cap \{z : |z - 1| > 1\}:$$

see Figure 5.3.2.

Clearly, $D_1$ is a $G$-packing: as $h$ maps the domain $|z + 1| > 1$ onto the disc $|z - 1| < 1$ we see that $D_2$ is a $G_2$-packing. Obviously $D^* \neq \varnothing$ and $D_1 \cup D_2 = \hat{\mathbb{C}}$: thus Theorem 5.3.15 is applicable.

EXERCISE 5.3

1. Verify the details in the Remark following Theorem 5.3.9.

2. Let $g$ and $\Sigma$ be as in Lemma 5.3.5. Show that for some $w$,

$$\bigcap_{n=1}^{\infty} g^n(\bar{\Sigma}) = \{w\}$$

   and that $w$ is the unique fixed point of $g$ in $\Sigma$.

3. Suppose that $G$ is discrete and non-elementary. Show that $\Omega$ is the largest domain in $\hat{\mathbb{C}}$ in which $G$ is a normal family.

4. Suppose that $G$ is non-elementary and contains parabolic elements. Show that $\Lambda$ is also the closure of the set of parabolic fixed points of $G$.

5. Let $G_j$, $D_j$ and $D^*$ be as in the application of Theorem 5.3.15 and let $G = \langle g, h \rangle$. Prove that $\Lambda \subset \mathbb{R}^1 \cup \{\infty\}$ so $G$ acts discontinuously in the upper and lower half-planes. Deduce that $\Omega$ is connected.

   Let $D$ be the set obtained by removing the origin from the closure of $D^*$. Prove that $D \subset \Omega$ and deduce that

$$\bigcup_{f \in G} f(D) = \Omega.$$

6. Let $Q_1, Q_{-1}, Q_2, Q_{-2}$ be four mutually exterior circles in $\mathbb{C}$. For $j = 1, 2$, let $g_j$ map the exterior of $Q_{-j}$ onto the interior of $Q_j$. Deduce that $G = \langle g_1, g_2 \rangle$ acts discontinuously on

$$\bigcup_{g \in G} g(\bar{D})$$

   where $D$ is the domain lying exterior to all four circles. This is called a *Schottky group* on two generators.

## §5.4. Jorgensen's Inequality

We end our general discussion of discreteness and discontinuity with an account of Jørgensen's inequality. Later, we shall examine the geometric interpretation in greater detail in the special case of isometries of the hyperbolic plane.

Let $A$ and $B$ be matrices in $SL(2, \mathbb{C})$ representing the Möbius transformations $f$ and $g$ respectively. As $A$ and $B$ are determined by $f$ and $g$ to within a factor of $-1$, we see that the commutator $ABA^{-1}B^{-1}$ is *uniquely* determined by $f$ and $g$. Thus we may (unambiguously) write

$$\operatorname{tr}(fgf^{-1}g^{-1}) = \operatorname{tr}(ABA^{-1}B^{-1}).$$

**Theorem 5.4.1.** (*Jørgensen's Inequality*). *Suppose that the Möbius transformations $f$ and $g$ generate a discrete non-elementary group. Then*

$$|\operatorname{tr}^2(f) - 4| + |\operatorname{tr}(fgf^{-1}g^{-1}) - 2| \geq 1. \tag{5.4.1}$$

*The lower bound is best possible.*

The inequality (5.4.1) can be interpreted in terms of the metric on $SL(2, \mathbb{C})$ for if $\langle f, g \rangle$ is non-elementary and discrete, then

$$|\operatorname{tr}^2(A) - 4| + |\operatorname{tr}(ABA^{-1}B^{-1}) - 2| \geq 1 \tag{5.4.2}$$

and so $A$ and $B$ cannot both be close to $I$. Thus (5.4.1) represents a quantitative statement about the isolated nature of $I$ within a discrete group.

It is easy to obtain an explicit numerical bound by writing

$$A = I + X, \qquad A^{-1} = I + X^*$$

and noting that

$$\|X\| = \|X^*\|, \qquad X + X^* + XX^* = 0:$$

similar expressions hold for $B = I + Y$, say. The Cauchy–Schwarz inequality yields

$$|\operatorname{tr}(X)| \leq \sqrt{2}\|X\|$$

and a computation shows that $[A, B] - I$ reduces to a sum of six terms, each being a product of at least two of the matrices $X$, $X^*$, $Y$ and $Y^*$. If $\|X\| < \varepsilon$ and $\|Y\| < \varepsilon$ then (5.4.2) yields

$$1 \leq \sqrt{2}\varepsilon(4 + \sqrt{2}\varepsilon) + 6\varepsilon^2$$
$$= 4\sqrt{2}\varepsilon + 8\varepsilon^2$$

so $\varepsilon > 0.146$. Thus we have the following (presumably) crude but explicit estimate.

**Corollary.** *If $A$ and $B$ generate a non-elementary discrete group then*

$$\max\{\|A - I\|, \|B - I\|\} > 0.146.$$

To show that the lower bound in (5.4.1) is best possible, consider the group generated by

$$f(z) = z + 1, \qquad g(z) = -1/z.$$

In this case, $G$ is the Modular group arising from $SL(2, \mathbb{Z})$: it is obviously non-elementary and equality holds in (5.4.1).

PROOF OF THEOREM 5.4.1. The idea of the proof is contained in Section 1.5 and Theorem 5.1.4. We know that $\langle f, g \rangle$ is discrete and non-elementary. Now (5.4.1) holds if $f$ is of order two (because then, $\text{tr}^2(f) = 0$) so we may assume that $f$ is not of order two. Select matrices $A$ and $B$ representing $f$ and $g$ respectively in $SL(2, \mathbb{C})$ and define

$$B_0 = B, \qquad B_{n+1} = B_n A B_n^{-1}. \tag{5.4.3}$$

It follows that $B_n$ represents $g_n$ as defined in the proof of Theorem 5.1.4, hence (by that Theorem) $B_n \neq A$ for any $n$. It remains only to show that if (5.4.2) fails, then for some $n$ we have

$$B_n = A \tag{5.4.4}$$

and we consider two cases.

  *Case 1: $f$ is parabolic.*
As the trace is invariant under conjugation we may assume that

$$A = \begin{pmatrix} 1 & 1 \\ 0 & 1 \end{pmatrix}, \qquad B = \begin{pmatrix} a & b \\ c & d \end{pmatrix},$$

where $c \neq 0$ (else $\langle A, B \rangle$ is elementary). We are assuming that (5.4.2) fails and this is the assumption that

$$|c| < 1.$$

The relation (5.4.3) yields

$$\begin{pmatrix} a_{n+1} & b_{n+1} \\ c_{n+1} & d_{n+1} \end{pmatrix} = \begin{pmatrix} 1 - a_n c_n & a_n^2 \\ -c_n^2 & 1 + a_n c_n \end{pmatrix}.$$

From this we deduce (by induction) that

$$c_n = -(-c)^{2^n}$$

(which is $-c^{2^n}$ except when $n = 0$) and as $|c| < 1$ we see that

$$c_n \to 0.$$

As $|c_n| < 1$, we have (by induction)

$$|a_n| \leq n + |a_0|$$

so $a_n c_n \to 0$ and

$$a_{n+1} \to 1.$$

This proves that

$$B_{n+1} \to A,$$

which, by discreteness, yields (5.4.4) for all large $n$.

*Case* 2: *f is loxodromic or elliptic.*
Without loss of generality,

$$A = \begin{pmatrix} u & 0 \\ 0 & 1/u \end{pmatrix},$$

where $B$ is as in Case 1 and $bc \neq 0$ (else $\langle A, B \rangle$ is elementary). The assumption that (5.4.2) fails is

$$\mu = |\text{tr}^2(A) - 4| + |\text{tr}(ABA^{-1}B^{-1}) - 2|$$
$$= (1 + |bc|)|u - 1/u|^2$$
$$< 1.$$

The relation (5.4.3) yields

$$\begin{pmatrix} a_{n+1} & b_{n+1} \\ c_{n+1} & d_{n+1} \end{pmatrix} = \begin{pmatrix} a_n d_n u - b_n c_n/u & a_n b_n(1/u - u) \\ c_n d_n(u - 1/u) & a_n d_n/u - b_n c_n u \end{pmatrix}$$

so

$$b_{n+1} c_{n+1} = -b_n c_n(1 + b_n c_n)(u - 1/u)^2.$$

We now obtain (by induction)

$$|b_n c_n| \leq \mu^n |bc| \leq |bc|$$

so

$$b_n c_n \to 0$$

and

$$a_n d_n = 1 + b_n c_n \to 1.$$

Also, we obtain

$$a_{n+1} \to u, \qquad d_{n+1} \to 1/u.$$

Now

$$|b_{n+1}/b_n| = |a_n(1/u - u)|$$
$$\to |u(1/u - u)|$$
$$\leq \mu^{1/2}|u|$$

so

$$\left| \frac{b_{n+1}}{u^{n+1}} \right| < \left( \frac{1 + \mu^{1/2}}{2} \right) \left| \frac{b_n}{u^n} \right|$$

for all sufficiently large $n$. Thus

$$b_n/u^n \to 0$$

and similarly, $c_n u^n \to 0$. It follows that

$$A^{-n} B_{2n} A^n = \begin{pmatrix} a_{2n} & b_{2n}/u^{2n} \\ u^{2n} c_{2n} & d_{2n} \end{pmatrix}$$

$$\to A.$$

As $\langle A, B \rangle$ is discrete, we must have

$$A^{-n} B_{2n} A^n = A$$

for all sufficiently large $n$ so for these $n$, $B_{2n} = A$ which is (5.4.4). $\qquad\square$

We end this chapter with several applications of Jørgensen's inequality.

**Theorem 5.4.2.** *A non-elementary group $G$ of Möbius transformations is discrete if and only if for each $f$ and $g$ in $G$, the group $\langle f, g \rangle$ is discrete.*

PROOF. If $G$ is discrete, then so is every subgroup of $G$. Now suppose that every subgroup $\langle f, g \rangle$ is discrete: we suppose that $G$ is not discrete and our aim is to reach a contradiction.

As $G$ is not discrete we can find distinct $f_1, f_2, \ldots (\neq I)$ in $G$ represented by matrices $A_1, A_2, \ldots$ in SL$(2, \mathbb{C})$ which converge to $I$. By considering traces, we may assume that no $f_n$ is of order two.

For any $g$ in $G$ with matrix $B$, say, we have

$$|\mathrm{tr}^2(A_n) - 4| + |\mathrm{tr}[A_n, B] - 2| \to 0$$

and so by Theorem 5.4.1, for $n \geq n(g)$ say, the group $\langle f_n, g \rangle$ is elementary.

Now $G$ contains two loxodromic elements $g$ and $h$ with no common fixed points (Theorem 5.1.3). For $n$ greater than $n(g)$ and $n(h)$, both groups

$$\langle g, f_n \rangle, \qquad \langle h, f_n \rangle$$

are elementary and discrete and, according to the discussion of such groups in Section 5.1, we deduce that $f_n$ must leave the fixed point pair of $g$ and of $h$ invariant. As $f_n$ is not elliptic of order two, it cannot interchange a pair of points so $f_n$ must fix each individual fixed point of $g$ and of $h$. We deduce that $g$ and $h$ have a common fixed point and this is the required contradiction. $\qquad\square$

Next, we give alternative formulations of (5.4.1) in the particular case when $f$ is parabolic ($\rho$ is the hyperbolic metric in $H^3$).

**Theorem 5.4.3.** *Let $f$ be parabolic and suppose that $\langle f, g \rangle$ is discrete and non-elementary. Then*

(i)                         $\|f - I\| \cdot \|g - I\| \geq 1$
   *and this is best possible;*

(ii) *if g is also parabolic, then for all x in $H^3$ we have*

$$\sinh \tfrac{1}{2}\rho(x, fx)\sinh \tfrac{1}{2}\rho(x, gx) \geq \tfrac{1}{4}$$

*and this is best possible.*

*Remark.* In (i), $\|f - I\|$ is to be interpreted as $\|A - I\|$ for either choice of the matrix $A$ representing $f$ and similarly for $g$.

PROOF. There is a Möbius $h$ corresponding to a unitary matrix $U$ such that $hfh^{-1}$ fixes $\infty$. If $A$ corresponds to $f$, then

$$\|UAU^{-1} - I\| = \|A - I\|$$

and similarly for $g$: thus we may assume that $f$ fixes $\infty$. Then

$$A = \begin{pmatrix} \varepsilon & \lambda \\ 0 & \varepsilon \end{pmatrix}, \qquad B = \begin{pmatrix} a & b \\ c & d \end{pmatrix} \qquad (ad - bc = 1),$$

where $\varepsilon^2 = 1$ and where $B$ represents $g$. Jørgensen's inequality (5.4.2) yields

$$|c\lambda| \geq 1$$

and (i) follows as

$$\|A - I\| \geq |\lambda|, \qquad \|B - I\| \geq |c|.$$

To prove (ii), select matrices $A$ and $B$ for $f$ and $g$ respectively with

$$\mathrm{tr}(A) = \mathrm{tr}(B) = 2.$$

Then using Theorem 4.2.1, we have

$$\begin{aligned} \|A - I\|^2 &= \|A\|^2 + 2 - 2\,\mathrm{Re}[\mathrm{tr}(A)] \\ &= \|A\|^2 - 2 \\ &= 4 \sinh^2 \tfrac{1}{2}\rho(j, gj), \end{aligned}$$

where $j = (0, 0, 1)$ in $H^3$. This verifies (ii) when $x = j$.

The general case of (ii) follows easily. If $x \in H^3$, choose a Möbius $h$ mapping $x$ to $j$. Now apply (ii) with $f, g$ and $x$ replaced by $hfh^{-1}, hgh^{-1}$ and $j$. The maps $f: z \mapsto z + 1, g(z): z \mapsto z/(z + 1)$ show that both bounds are best possible.  $\square$

Theorem 5.4.3 has an interesting geometric interpretation. A *horoball* $\Sigma$ in $H^3$ is an open Euclidean ball in $H^3$ which is tangent to $\hat{\mathbb{C}}$. If the point of tangency is $w$, we say that $\Sigma$ is *based* at $w$: the boundary $\partial\Sigma$ of $\Sigma$ (in $\mathbb{R}^3$) is a *horosphere*. A horoball based at $\infty$ is a set of the form

$$\{(x_1, x_2, x_3) \in H^3 : x_3 > k\},$$

where $k > 0$. Thus in this case, and hence in general, a horosphere is a surface in $H^3$ which is orthogonal to all hyperbolic planes containing the point $w$ on the sphere at $\infty$, namely $\hat{\mathbb{C}}$. This characterizes horoballs and horospheres in terms of the geometry of $H^3$ alone.

If $g$ is a parabolic element of $\mathcal{M}$ fixing $w$, then for all positive $k$,

$$\Sigma[g, k] = \{x \in H^3 : \sinh \tfrac{1}{2}\rho(x, gx) < k\}$$

is a horoball based at $\infty$. Indeed, if $g(z) = z + 1$, then using (3.3.4) we obtain

$$\sinh \tfrac{1}{2}\rho(x, gx) = 1/2x_3$$

and hence

$$\Sigma[g, k] = \{x \in H^3 : x_3 > 1/2k\}:$$

the general case follows because for all Möbius $h$,

$$h(\Sigma[g, k]) = \Sigma[hgh^{-1}, k].$$

Now define, for each parabolic $g$, the horoball

$$\Sigma_g = \{x \in H^3 : \sinh \tfrac{1}{2}\rho(x, gx) < \tfrac{1}{2}\}. \tag{5.4.5}$$

Obviously, for any Möbius $h$ we have

$$h(\Sigma_g) = \Sigma_{hgh^{-1}}. \tag{5.4.6}$$

It is clear from Theorem 5.4.3(ii) that if $\Sigma_g$ meets $\Sigma_h$, then $\langle g, h \rangle$ cannot be both discrete and non-elementary. In particular, if $g$ and $h$ are known to be in a discrete group, then $g$ and $h$ must have a common fixed point. This proves the next result.

**Theorem 5.4.4.** *Let $G$ be a discrete non-elementary subgroup of $\mathcal{M}$ with parabolic elements. For each parabolic $g$ in $G$, let $\Sigma_g$ be the horoball defined by (5.4.5). Then the family*

$$\{\Sigma_g : g \text{ parabolic in } G\}$$

*is permuted by $G$ according to (5.4.6) and $\Sigma_g \cap \Sigma_h = \varnothing$ unless $g$ and $h$ have a common fixed point.*

Our last application of Jørgensen's inequality relates Theorem 5.4.3(ii) to non-parabolic elements: for completeness, we include this in the statement of the next result.

**Theorem 5.4.5.** *Suppose that $\langle g, h \rangle$ is discrete and non-elementary.*

(i) *if $g$ is parabolic, then for all $x$ in $H^3$,*

$$\sinh \tfrac{1}{2}\rho(x, gx) \sinh \tfrac{1}{2}\rho(x, hgh^{-1}x) \geq \tfrac{1}{4};$$

(ii) *if $g$ is hyperbolic, then for all $x$ in $H^3$,*

$$\sinh \tfrac{1}{2}\rho(x, gx) \sinh \tfrac{1}{2}\rho(x, hgh^{-1}x) \geq \tfrac{1}{8};$$

(iii) *if g is elliptic or strictly loxodromic and if* $|\mathrm{tr}^2(g) - 4| < \frac{1}{4}$ *(which defines an open neighbourhood of I) then for all x in* $H^3$

$$\max\{\sinh \tfrac{1}{2}\rho(x, gx), \sinh \tfrac{1}{2}\rho(x, hgh^{-1}x)\} \geq \tfrac{1}{4}.$$

If

$$\rho(x, gx) < \varepsilon, \qquad \rho(x, hx) < \varepsilon,$$

then

$$
\begin{aligned}
\rho(x, hgh^{-1}x) &= \rho(h^{-1}x, gh^{-1}x) \\
&\leq \rho(h^{-1}x, x) + \rho(x, gx) + \rho(gx, gh^{-1}x) \\
&< 3\varepsilon:
\end{aligned}
$$

thus we obtain the following corollary of Theorem 5.4.5.

**Corollary 5.4.6.** *Let N be the open neighbourhood of I in $\mathcal{M}$ defined by* $\{f : |\mathrm{tr}^2(f) - 4| < \frac{1}{4}\}$. *If g is in N and if $\langle g, h \rangle$ is discrete and non-elementary, then for all x in* $H^3$,

$$\max\{\rho(x, gx), \rho(x, hx)\} \geq 0\cdot38 \ldots .$$

The proof of Theorem 5.4.5 requires details of the geometry of the action of loxodromic and elliptic elements. Suppose first that

$$g = \begin{pmatrix} u & 0 \\ 0 & 1/u \end{pmatrix}, \qquad u = |u|e^{i\theta}, \tag{5.4.7}$$

is loxodromic (this includes hyperbolic) or elliptic. Observe that

$$
\begin{aligned}
|u - 1/u|^2 &= (u - 1/u)(\bar{u} - 1/\bar{u}) \\
&= (|u| - 1/|u|)^2 + 4\sin^2\theta. \tag{5.4.8}
\end{aligned}
$$

Next, for all $x$ and $y$ in $H^3$, (3.3.4) yields

$$4\sinh^2 \tfrac{1}{2}\rho(x, y) = \frac{|x - y|^2}{x_3 y_3}.$$

The transformation $g$ acts on $\mathbb{R}^3$ (viewed as $\mathbb{C} \times \mathbb{R}^1$) by the formula

$$g : (z, t) \mapsto (u^2 z, |u|^2 t)$$

and so with $x = (z, t)$ we have

$$
\begin{aligned}
4\sinh^2 \tfrac{1}{2}\rho(x, gx) &= \frac{|z - u^2 z|^2 + (t - |u|^2 t)^2}{|u|^2 t^2} \\
&= \left(|u| - \frac{1}{|u|}\right)^2 + \left(\frac{|z|}{t}\right)^2 \left(u - \frac{1}{u}\right)^2. \tag{5.4.9}
\end{aligned}
$$

The *axis* $A$ of $g$ is, by definition, the geodesic joining the fixed points of $g$. In the particular case (5.4.7), the axis is given by $z = 0$ and it is clear from (5.4.9) that the displacement

$$T_g = \rho(x, gx)$$

is independent of $x$ on $A$: we call $T_g$ the *translation length* of $g$. The identity (5.4.9) shows that

$$4 \sinh^2(\tfrac{1}{2}T_g) = (|u| - 1/|u|)^2:  \qquad (5.4.10)$$

in particular, $T_g = 0$ if $g$ is elliptic. Note that the two terms involving $u$ in (5.4.8) are invariant under conjugation (they can be expressed in terms of trace($g$) and $T_g$), hence so is $\sin^2 \theta$. In particular, $\sin \theta = 0$ if $g$ is hyperbolic.

The next task is to express $|z|/t$ geometrically. The reader is referred forward to Section 7.9 where it is shown that

$$|z|/t = \sinh \rho(x, A): \qquad x = (z, t).$$

With this available, (5.4.9), (5.4.8) and (5.4.10) yield

$$\sinh^2 \tfrac{1}{2}\rho(x, gx) = \sinh^2(\tfrac{1}{2}T_g) \cosh^2 \rho(x, A) + \sinh^2 \rho(x, A) \sin^2 \theta. \quad (5.4.11)$$

Thus the displacement by $g$ arises out of a contribution corresponding to the shift $T_g$ along the axis and a contribution arising out of the rotational effect of $\theta$ and each contribution is adjusted according to the distance of $x$ from the axis.

PROOF OF THEOREM 5.4.5. We need only prove (ii) and (iii) and by considering conjugate elements we may suppose that $g$ is given by (5.4.7). As Jørgensen's inequality is applicable, we write

$$h = \begin{pmatrix} a & b \\ c & d \end{pmatrix}, \qquad ad - bc = 1,$$

and so

$$(1 + |bc|)|u - 1/u|^2 \geq 1: \qquad (5.4.12)$$

see the proof of Theorem 5.4.1, Case 2.

In order to interpret the term $|bc|$, we seek a Möbius transformation $f$ taking $0, \infty, h0, h\infty$ to $1, -1, w, -w$ respectively. Such a transformation exists if and only if we have equality of cross-ratios, namely

$$[1, -1, w, -w] = [0, \infty, b/d, a/c],$$

or, equivalently,

$$bc = (1 - w)^2/4w. \qquad (5.4.13)$$

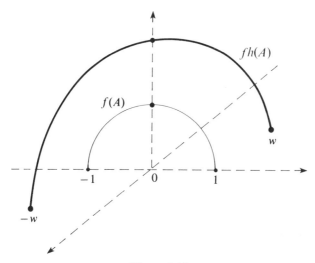

Figure 5.4.1

Now $A$ is the geodesic with end-points $0$ and $\infty$; $hA$ is the geodesic with end-points $h0$ and $h\infty$. As $\langle g, h \rangle$ is non-elementary, the geodesics $A$ and $hA$ cannot have a common end-point: thus $bc \neq 0$. It follows that there are two solutions in $w$ of (5.4.13), each solution being the reciprocal of the other. Let $w$ be such a solution and we may suppose that $|w| \geq 1$: the location of $f(A)$ and $f(hA)$ is illustrated in Figure 5.4.1.

It is an easy deduction from (3.3.4) that

$$
\begin{aligned}
\rho(A, hA) &= (\rho fA, fhA) \\
&= \inf\{\rho(x, y): x \in fA, \, y \in fhA\} \\
&= \rho(e_3, |w|e_3) \\
&= \log|w|,
\end{aligned}
$$

because if $(x, y, t) \in f(A)$ and $(u, v, s) \in fh(A)$ then

$$
\frac{(x - u)^2 + (y - v)^2 + (t - s)^2}{ts} = \frac{1 + |w|^2 - 2(xu + yv + st)}{ts}
$$

and the Cauchy–Schwarz inequality is applicable. We now write

$$
w = \exp 2(\alpha + i\beta)
$$

so

$$
\rho(A, hA) = 2\alpha.
$$

Also,

$$
bc = \sinh^2(\alpha + i\beta),
$$

hence

$$4|bc|^2 = |\cosh 2(\alpha + i\beta) - 1|^2$$
$$= (\cosh 2\alpha \cos 2\beta - 1)^2 + (\sinh 2\alpha \sin 2\beta)^2$$
$$= (\cosh 2\alpha - \cos 2\beta)^2$$
$$\leq (1 + \cosh 2\alpha)^2$$
$$= (2 \cosh^2 \alpha)^2.$$

Thus for all $x$ in $H^3$,

$$|bc| \leq \cosh^2 \alpha$$
$$= \cosh^2 \tfrac{1}{2}\rho(A, hA)$$
$$\leq \cosh^2 \tfrac{1}{2}[\rho(A, x) + \rho(x, hA)].$$

Now (by elementary means or because log cosh is a convex function) we have

$$\cosh^2\left(\frac{p + q}{2}\right) \leq \cosh p \cosh q, \qquad (p, q \text{ real})$$

thus

$$|bc| \leq \cosh \rho(x, A) \cosh \rho(x, hA). \qquad (5.4.14)$$

Finally, observe that the conjugate elements $g$ and $hgh^{-1}$ have the same trace[2], the same translation length and hence the same value of $\sin^2 \theta$. With this in mind, we combine (5.4.12), (5.4.14), (5.4.8), (5.4.10) and (5.4.11) to obtain

$$(\sinh^2 \tfrac{1}{2}\rho(x, gx) + \sin^2 \theta)(\sinh^2 \tfrac{1}{2}\rho(x, hgh^{-1}x) + \sin^2 \theta)$$
$$\geq [\cosh \rho(x, A) \cosh \rho(x, hA)|u - 1/u|^2/4]^2.$$

Because of (5.4.12) and (5.4.14) we have

$$2 \cosh \rho(x, A) \cosh \rho(x, hA)|u - 1/u|^2 \geq 1$$

so in all cases

$$(\sinh^2 \tfrac{1}{2}\rho(x, gx) + \sin^2 \theta)(\sinh^2 \tfrac{1}{2}\rho(x, hgh^{-1}x) + \sin^2 \theta) \geq \tfrac{1}{64}$$

If $g$ is hyperbolic, then $\sin \theta = 0$ and we obtain (ii). In all other cases write

$$m = \max\{\sinh \tfrac{1}{2}\rho(x, gx), \sinh \tfrac{1}{2}\rho(x, hgh^{-1}x)\}.$$

Then

$$m^2 + \sin^2 \theta \geq \tfrac{1}{8}.$$

The hypotheses of (iii) together with (5.4.8) yields

$$\sin^2 \theta \leq \tfrac{1}{16}$$

and so $m \geq \tfrac{1}{4}$.                                                                          $\square$

## §5.5. Notes

For a discussion of the elementary groups given in Section 5.1, see, for example, [30], [51] and [107]. Discrete Euclidean groups in $R^n$ are discussed in [91] and [111].

For a selection of papers concerned with the geometric action of discontinuous groups acting in plane domains or in $H^3$, see [8], [9], [13], [14], [15], [65], [108] and [109]. Theorem 5.3.15 and extensions of it can be found in [30], [51], [54], [60] and [61]. As more comprehensive accounts, we cite [5], [25], [30], [35], [50], [51], [52], [57] and [114].

Jørgensen's inequality (Theorem 5.4.1) appears in [41]: for related material, see [14], [40], [44], [45] and [89].

# CHAPTER 6
# Riemann Surfaces

## §6.1. Riemann Surfaces

Briefly, a Riemann surface is a topological space which, when viewed locally, is essentially the same as the complex plane. The formal definition is constructed so that the concept of an analytic function and complex analytic function theory extend without difficulty to a Riemann surface. The function theory will not concern us here and we shall confine our discussion to the relationship between Riemann surfaces and the quotient by a discontinuous group action. We shall develop these ideas only as far as is necessary to interpret results on discontinuous groups in terms of Riemann surfaces.

A Hausdorff connected topological space $X$ is a Riemann surface if there exists a family

$$\{(\phi_j, U_j) : j \in J\},$$

called an atlas (each $(\phi_j, U_j)$ is called a chart) such that

(i) $\{U_j : j \in J\}$ is an open cover of $X$;
(ii) each $\phi_j$ is a homeomorphism of $U_j$ onto an open subset of the complex plane; and
(iii) if $U = U_i \cap U_j \neq \emptyset$, then

$$\phi_i(\phi_j)^{-1} : \phi_j(U) \to \phi_i(U)$$

is an analytic map between the plane sets $\phi_j(U)$ and and $\phi_i(U)$,

Clearly, (i) is saying that $X$ is covered by a collection of "distinguished" open sets, each of which (by (ii)) is homeomorphic to an open subset of $\mathbb{C}$. Two distinguished sets may overlap but then by (iii), the corresponding homeomorphisms are related by an analytic homeomorphism.

It is now possible to define analytic functions between Riemann surfaces. If $X$ and $Y$ are Riemann surfaces with atlases $\{(\phi_j, U_j): j \in J\}$ and $\{(\psi_k, V_k): k \in K\}$ respectively, then a continuous map $f: X \to Y$ is analytic if each map

$$\psi_k f(\phi_j)^{-1}: \phi_j(U_j \cap f^{-1}(V_k)) \to \mathbb{C} \qquad (6.1.1)$$

is analytic. The domain of this map is a subset of $\mathbb{C}$ and the assumed continuity of $f$ guarantees that this set is open. Of course by (iii), it is only necessary to check that the maps (6.1.1) are analytic for subatlases which still provide an open cover of $X$ and $Y$ respectively.

We can also talk of the angle between (smooth) curves $\gamma$ and $\sigma$ on $X$ which cross at some point $x$. If $x \in U_j$, we can measure the angle $\theta$ between the curves $\phi_j(\gamma)$, $\phi_j(\sigma)$ which cross at $\phi_j(x)$ in the complex plane. If $x \in U_i$, also, then $\phi_i(\gamma)$ and $\phi_i(\sigma)$ will cross at the same angle $\theta$ because, being an analytic homeomorphism, the map $\phi_j(\phi_i)^{-1}$ is conformal. It follows that $\theta$ is defined independently of the choice of $j$ and this is then taken to be the angle between $\gamma$ and $\sigma$ at $x$.

The simplest non-planar example of a Riemann surface is $X = \mathbb{C} \cup \{\infty\}$ with the atlas given by $J = \{1, 2\}$ and

$$\phi_1(z) = z, \qquad U_1 = \mathbb{C};$$
$$\phi_2(z) = 1/z, \qquad U_2 = \{\infty\} \cup \{z \in \mathbb{C}: z \neq 0\}:$$

obviously, $\phi_2(\phi_1)^{-1}$ is analytic on $\phi_1(U_1 \cap U_2)$.

We say that two Riemann surfaces $R_1$ and $R_2$ are conformally equivalent if there is an analytic bijection $f$ of $R_1$ onto $R_2$ (then $f^{-1}$ is also analytic). This is an equivalence relation on the class of all Riemann surfaces and in general, we do not distinguish between conformally equivalent surfaces.

EXERCISE 6.1

1. Prove that a Riemann surface is arcwise connected.

2. Show that if $R$ is a Riemann surface containing points $w_j$, then $R - \{w_1, \ldots, w_n\}$ is also a Riemann surface.

3. Let $f: R \to S$ be a non-constant analytic map between the Riemann surfaces $R$ and $S$. Prove that $f$ maps open subsets of $R$ onto open subsets of $S$. Deduce that if $R$ is compact, then $f$ is surjective and so $S$ is compact.

# §6.2. Quotient Spaces

One method of constructing Riemann surfaces is by forming the quotient space with respect to a discontinuous group action. In fact, it is known that every Riemann surface arises in this way.

**Theorem 6.2.1.** *Let $D$ be a subdomain of $\hat{\mathbb{C}}$ and let $G$ be a group of Möbius transformations which leaves $D$ invariant and which acts discontinuously in $D$. Then $D/G$ is a Riemann surface.*

PROOF. We know that $D/G$ is a topological space with the quotient topology and that the quotient map $\pi: D \to D/G$ is continuous. As $D$ is connected and $\pi$ is continuous, it follows that $D/G$ is connected (in fact, arcwise connected). It is also clear that $\pi$ is an open map for if $A \subset D$, then

$$\pi^{-1}(\pi A) = \bigcup_{g \in G} g(A):$$

thus if $A$ (and therefore $g(A)$) is open, then so is $\pi(A)$.

We now show that $D/G$ is Hausdorff. First, choose distinct $z_1$ and $z_2$ in $D$ and choose a positive $r$ so that the discs

$$K_1 = \{z: |z - z_1| \leq r\}, \qquad K_2 = \{z: |z - z_2| \leq r\}$$

lie in $D$. For $n \geq 1$, define

$$A_n = \{z: |z - z_1| < r/n\},$$
$$B_n = \{z: |z - z_2| < r/n\}.$$

If for every $n$,

$$\pi(A_n) \cap \pi(B_n) \neq \varnothing,$$

then there is some $w_n$ in $A_n$ and some $g_n$ in $G$ with $g_n(w_n) \in B_n$. This implies that

$$g_n(K) \cap K \neq \varnothing,$$

where $K = K_1 \cup K_2$ (which is compact) and it follows (from discontinuity) that the set $\{g_1, g_2, \ldots\}$ is finite. On a suitable subsequence, $g_n = g$, say, and

$$g(z_1) = \lim_n g_n(w_n)$$

$$= z_2.$$

To prove that $D/G$ is Hausdorff, consider two distinct points, say $\pi(z_1)$ and $\pi(z_2)$ in $D/G$. Thus $z_1$ and $z_2$ are in $D$ but not equivalent under $G$.

It follows that for some $n$, the disjoint sets $\pi(A_n)$ and $\pi(B_n)$ separate $\pi(z_1)$ and $\pi(z_2)$ and these sets are open as $\pi$ is an open map.

Our last task is to construct an atlas for $D/G$. For each $z$ in $D$, we select an open disc $N_z$ (whose closure lies in $D$) with the properties

$$g(N_z) \quad\;\; = N_z \quad \text{if } g(z) = z;$$
$$g(N_z) \cap N_z = \varnothing \quad \text{if } g(z) \neq z:$$

see Theorems 5.3.11 and 5.3.12.

Observe that $N_z - \{z\}$ contains no fixed points of $G$. Indeed, if $h\,(\neq I)$ fixes a point in $N_z$, then (because of the definition of $N_z$) $h$ fixes $z$. The inverse point of $z$ with respect to $N_z$ is also fixed by $h$ so there are no fixed points of $h$ in $N_z - \{z\}$. Recall that if $h$ fixes $z$, then $h$ is elliptic.

For each $w$ in $D$, let $\sigma$ be a Möbius transformation which maps $w$ to zero and $N_w$ to the unit disc $\Delta$. The stabilizer of $w$ in $G$ is of order $n$, say, and is generated by some elliptic $g$ where

$$\sigma g \sigma^{-1}(z) = z \exp(2\pi i/n), \qquad z \in \Delta.$$

Now let $q(z) = z^n$: this maps $\Delta$ onto itself and has the property that for all $k$ and for all $z$ in $N_w$, we have

$$\begin{aligned} q\sigma g^k(z) &= [\sigma g^k \sigma^{-1}(\sigma z)]^n \\ &= [\sigma(z) \exp(2\pi i k/n)]^n \\ &= \sigma(z)^n. \end{aligned} \qquad (6.2.1)$$

Observe that this is independent of the integer $k$.

We shall take as charts for $D/G$ the pairs

$$(q\sigma(\pi_w)^{-1}, \pi_w(N_w)),$$

where $\pi_w$ is the restriction of $\pi$ to $N_w$: see Figure 6.2.1.

Each point in $\pi_w(N_w)$ is mapped by $(\pi_w)^{-1}$ into $n$ points $g^k(z)$, say, where $k = 0, 1, \ldots, n - 1$ in $N_w$. According to (6.2.1), these map under $q\sigma$ to the same point in $\Delta$, thus

$$\phi_w = q\sigma(\pi_w)^{-1}$$

is a bijection of $\pi_w(N_w)$ onto $\Delta$. As the maps $q$, $\sigma$ and $\pi_w$ are both open and continuous, we see that each $\phi_w$ is a homeomorphism.

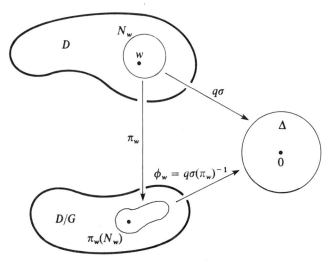

Figure 6.2.1

In order to check that the transition maps are analytic we must first study the maps

$$(\pi_v)^{-1}\pi_u, \qquad u \neq v. \tag{6.2.2}$$

Suppose that $\zeta_u \in N_u$, $\zeta_v \in N_v$ and

$$\pi(\zeta_u) = \pi(\zeta_v) = \zeta,$$

say: then for some $g$ in $G$, we have

$$\zeta_v = g(\zeta_u).$$

Suppose now that $\zeta_u$, and hence $\zeta_v$, are not elliptic fixed points. Then $\pi_v$ is 1–1 in some neighbourhood of $\zeta_v$ and therefore there is a local inverse $(\pi_v)^{-1}$ mapping $\zeta$ to $\zeta_v$. The two maps

$$\pi_v g, \quad \pi_u$$

agree with $\pi$, and hence with each other, on some neighbourhood of $\zeta_u$ and take values in $\pi_v(N_v)$. Applying $(\pi_v)^{-1}$ we see that

$$g = (\pi_v)^{-1}\pi_u$$

near $\zeta_u$. We deduce that the maps (6.2.2) are analytic near points which are not elliptic fixed points of $G$.

We now show that the transition maps

$$\phi_v(\phi_u)^{-1} \qquad (u \neq v)$$

are analytic (where defined): writing

$$\phi_v = q_v \sigma_v(\pi_v)^{-1}$$

and similarly for $u$, the situation is illustrated in Figure 6.2.2. At points corresponding to the non-fixed points of $G$, we can compute $\phi_v(\phi_u)^{-1}$ by choosing a single valued branch of $(q_u)^{-1}$ and the map $\phi_v(\phi_u)^{-1}$ is a composition of analytic maps. At points corresponding to elliptic fixed points the homeomorphism $\phi_v(\phi_u)^{-1}$ is analytic in a deleted neighbourhood of the point in question (by the previous remark) and hence has a removable singularity at this point. □

There is a converse to Theorem 6.2.1 (which we shall not prove here). Given any Riemann surface $R$ one can construct a simply connected Riemann surface $\hat{R}$ and a mapping $\pi: \hat{R} \to R$ with the properties

(i) each $\hat{z}$ in $\hat{R}$ has a neighborhood $\hat{N}$ such that $\pi$ restricted to $\hat{N}$ is a homeomorphism onto an open subset of $R$;
(ii) Given any curve $\gamma: [0, 1] \to R$ and any $\hat{z}$ on $\hat{R}$ with $\pi(\hat{z}) = \gamma(0)$, then there is a unique curve $\hat{\gamma}: [0, 1] \to \hat{R}$ such that $\pi\hat{\gamma} = \gamma$ and $\hat{\gamma}(0) = \hat{z}$ (we say that $\hat{\gamma}$ projects to $\gamma$ or that $\gamma$ lifts to $\hat{\gamma}$ from $\hat{z}$).

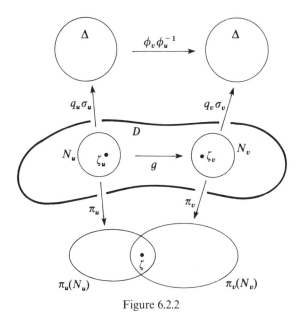

Figure 6.2.2

These properties are expressed by saying that $(\hat{R}, \pi)$ is a smooth unlimited covering surface of $R$. By the Riemann Mapping Theorem (for Riemann surfaces) $\hat{R}$ is conformally equivalent to one of the standard Riemann surfaces

$$\Delta = \{|z| < 1\}, \quad \mathbb{C}, \quad \mathbb{C} \cup \{\infty\},$$

(with the trivial atlases) so without loss of generality, we may assume that $\hat{R}$ is one of these.

It can now be shown that there is a group $G$ of Möbius transformations preserving $\hat{R}$ such that the given surface $R$ is conformally equivalent to $\hat{R}/G$. Writing the quotient map as $\pi$, this means that $\pi g = \pi$ for all $g$ in $G$. Further, one can show that $G$ acts discontinuously in $\hat{R}$ and has no elliptic elements.

If $\hat{R} = \mathbb{C} \cup \{\infty\}$, these restrictions imply that $G = \{I\}$ (the trivial group) so essentially, $R = \mathbb{C} \cup \{\infty\}$. If $\hat{R} = \mathbb{C}$, the only possibilities for $G$ are: (i) the trivial group; (ii) a cyclic group generated by some $z \mapsto z + \lambda$; (iii) a group generated by two translations $z \mapsto z + \lambda$, $z \mapsto z + \mu$ where $\lambda$, $\mu$ are linearly independent over the real numbers. These cases show that $R$ is either $\mathbb{C}$, $\mathbb{C}^* = \{z \in \mathbb{C} : z \neq 0\}$ or a torus. In all other cases, $R$ is of the form $\Delta/G$ where $G$ acts discontinuously in $\Delta$ and has no elliptic elements. If $R$ is compact, say with genus $g$, then $\hat{R} = \hat{\mathbb{C}}$ when $g = 0$, $\hat{R} = \mathbb{C}$ when $g = 1$ and $\hat{R} = \Delta$ when $g \geq 2$.

In view of these remarks, we can see the importance of groups acting discontinuously in $\Delta$ (or in some conformal image of $\Delta$).

**Definition 6.2.2.** A group $G$ of Möbius transformations is a *Fuchsian group* if and only if there is some $G$-invariant disc in which $G$ acts discontinuously.

A Riemann surface $R$ is said to be of hyperbolic type if it is of the form $\Delta/G$ where $G$ acts in $\Delta$. In this case, we can view the differential

$$ds = \frac{2|dz|}{1 - |z|^2}$$

as acting on $R$ and each curve on $R$ can be partitioned into small segments and the length of these segments then computed (in an invariant manner) in $\Delta$. In this way we can talk of the hyperbolic metric on $R$ and so compute lengths and areas on $R$.

If we join $z$ to $g(z)$ in $\Delta$ ($g \in G$) and project this to $R$ ($= \Delta/G$) we obtain a closed curve on $R$ for $\pi g(z) = \pi(z)$. Conversely, if we select a closed curve $\gamma: [0, 1] \to R$ and $z$ in $\Delta$ with $\pi(z) = \gamma(0)$, then there is a unique curve $\hat{\gamma}: [0, 1] \to \Delta$ with $\pi\hat{\gamma} = \gamma$ and $\hat{\gamma}(0) = z$. Note that

$$\pi\hat{\gamma}(1) = \gamma(1) = \gamma(0) = \pi(z),$$

so for some $h$ in $G$, $\hat{\gamma}(1) = h(z)$: thus $\hat{\gamma}$ is a curve from $z$ to $hz$. If $\gamma$ is homotopic to the point $z$ on $R$ then, by the Monodromy Theorem, $\hat{\gamma}$ is a closed curve on $\Delta$ and $h = I$ (because $h$ is not elliptic).

More generally, one can consider $n$-dimensional manifolds: in the definition of a Riemann surface, we replace $\mathbb{C}$ in (ii) by $\mathbb{R}^n$ and we delete (iii) (or replace "analytic" by some other smoothness condition). If $G$ is any discrete Möbius group, then $G$ acts discontinuously in $H^3$ and one can study $H^3/G$: this topic has attracted much attention in recent years.

EXERCISE 6.2

1. Let $G$ be generated by $g: z \mapsto z + 1$. Prove that $H^2/G$ is (conformally equivalent to) $\Delta^* = \{z: 0 < |z| < 1\}$. [Consider the map $z \mapsto \exp(2\pi i z)$.]

   Show how to project the metric $|dz|/\text{Im}[z]$ from $H^2$ to a metric $\mu(w)|dw|$ in $\Delta^*$. Find $\mu$ and show that in this metric, the area of $\{z: 0 < |z| < \frac{1}{2}\}$ is finite.

## §6.3. Stable Sets

Suppose that a domain $D$ (a subset of $\hat{\mathbb{C}}$) is $G$-invariant and that $G$ acts discontinuously in $D$. We need to consider the following type of invariance.

**Definition 6.3.1.** A subset $D_0$ of $D$ is said to be *stable* (or *precisely invariant*) with respect to $G$ if and only if for all $g$ in $G$, either

$$g(D_0) = D_0 \quad \text{or} \quad g(D_0) \cap D_0 = \varnothing.$$

The set of $g$ with $g(D_0) = D_0$ is the *stabilizer* of $D_0$.

For examples of stable sets, see Theorem 5.3.12.

Let $D_0$ be stable with stabilizer $G_0$: it is natural to form the quotient space $D_0/G_0$ and in general (for example, if $G_0$ is cyclic) this is easier to discuss than the projection $\pi(D_0)$ of $D_0$ into $D/G$. Unfortunately, the two spaces

$$D_0/G_0, \quad \pi(D_0)$$

need not be homeomorphic as the next example shows.

**Example 6.3.2.** Take $D = \mathbb{C}$, let $G$ be generated by $g(z) = z + 1$ and let $D_0 = \{x + iy : 0 \leq x < 1\}$. Clearly $D_0/G_0$ $(=D_0)$ is simply connected whereas $\pi(D_0)$ $(=\pi(\mathbb{C}) = \{z : z \neq 0\})$ is not.

There are important cases when $D_0/G_0$ and $\pi(D_0)$ are homeomorphic and we need explicit conditions which guarantee that this is so.

**Theorem 6.3.3.** *Suppose that $G$ acts discontinuously in $D$ and that $D_0$ is stable with stabilizer $G_0$. If either*

(i) *$D_0$ is open in $D$; or*
(ii) *$D_0/G_0$ is compact;*

*then $D_0/G_0$ (with the quotient topology) and $\pi(D_0)$ (with the subspace topology from $D/G$) are homeomorphic.*

PROOF. Both quotient maps

$$\pi : D \to D/G, \qquad \phi : D_0 \to D_0/G_0$$

are continuous and open as the respective groups are groups of homeomorphisms of the corresponding spaces. The restriction $\pi_0$ of $\pi$ to $D_0$ is continuous so the natural bijection

$$\theta = \pi_0 \phi^{-1} : D_0/G_0 \to \pi(D_0)$$

given by

$$G_0(x) \to G(x),$$

(where, for example, $G(x)$ is the $G$-orbit of $x$) is continuous: see Figure 6.3.1.

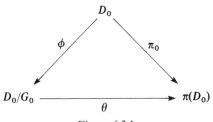

Figure 6.3.1

If (i) holds, then $\pi_0$ is an open map (because $\pi$ is an open map) and $\theta^{-1}$ is continuous. If (ii) holds, then $\theta$ is a continuous bijection from a compact space to a Hausdorff space and so is a homeomorphism (see Section 1.4). □

*Remark.* If $D$ is a subdomain of $\hat{\mathbb{C}}$, then $\phi$ and $\pi$ are analytic and $D_0/G_0$ and $\pi(D_0)$ are then conformally equivalent.

We end this chapter with some examples illustrating the hypotheses (i) and (ii) in Theorem 6.3.2.

**Example 6.3.4.** Suppose that $G$ preserves and acts discontinuously in the upper half-plane $H^2$ of $\mathbb{C}$ and let $g$ be a hyperbolic element of $G$. We may assume that $g$ fixes $0$ and $\infty$ so the positive imaginary axis, say $L$, is invariant under $g$.

Suppose now that for all $h$ in $G$, either $h(L) = L$ or $h(L) \cap L = \varnothing$ and suppose also that $G$ has no elliptic elements of order two (which might leave $L$ invariant and interchange the end-points of $L$). This situation will be discussed in detail later in the book. Then $h(L) = L$ only if $h$ lies in a cyclic subgroup of $G$ generated by a hyperbolic element (which we may assume is $g$) fixing $0$ and $\infty$. Now $g(z) = kz$, say, where $k > 1$, and $L/\langle g \rangle$ is compact and, in fact, is a simple closed curve. According to Theorem 6.3.3, the projection of $L$ into $H^2/G$ is also a simple closed curve.

**Example 6.3.5.** Suppose that a group $G$ acts discontinuously in a subdomain $D$ of $\hat{\mathbb{R}}$ and that there is an open disc $Q$ which is stable with stabilizer $\langle g \rangle$ where $g$ is parabolic. As $Q$ is open, Theorem 6.3.3 implies that the projection of $Q$ in $D/G$ is conformally equivalent to $Q/\langle g \rangle$.

By conjugation, we may assume that $g(z) = z + 1$ so that for some $y_0$,

$$Q = \{x + iy: y > y_0\}.$$

It is clear that the quotient space $Q/\langle g \rangle$ is conformally equivalent to the image of $Q$ under the map $z \mapsto \exp(2\pi i z)$: thus the projection of $Q$ in $D/G$ is conformally equivalent to a punctured disc and hence to

$$\{z \in \mathbb{C}: 0 < |z| < 1\}.$$

Now adjoin $\infty$ (the fixed point of $g$) and all of its $G$-images to $D$ to form the larger space $D^*$. We generate a topology on $D^*$ from the open subsets of $D$ together with sets of the form $\{\infty\} \cup \{x + iy: y > t\}$ and their $G$-images and the quotient space $D^*/G$ is also a Riemann surface: the adjoining of $\infty$ to $D$ corresponds to the addition of the origin to the punctured disc. Note, however, that the sequence $n + iy, n \geq 1$, does not converge in the topology of $D^*$ so $\infty$ does not have a compact neighbourhood in $D^*$. Of course, we may adjoin different orbits of parabolic fixed points to $D$ provided that in each case, a corresponding disc $Q$ exists. For more details and a converse result, see [50], Chapter 2.

EXERCISE 6.3

1. Let $G$ be generated by $g: z \mapsto z + 1$ and $h: z \mapsto z + i$ and let

$$D = \{x + iy: 0 \le y < 1\}.$$

Show that $D$ is stable under $\langle g \rangle$. Let $\pi$ be the natural projection of $\mathbb{C}$ onto $\mathbb{C}/G$. Show that $\pi(D)$ is compact whereas $D/\langle g \rangle$ is not compact.

# CHAPTER 7
# Hyperbolic Geometry

## §7.1. The Hyperbolic Plane

From the outset we have assumed both an acceptance and understanding of Euclidean geometry: we have not entered into a discussion of the axiomatic foundations of the geometry and we shall not do so. The question now arises as to how we should treat hyperbolic geometry. We must not assume that the reader is as familiar with this as with Euclidean geometry yet it is necessary to have available some of the more basic and elementary results in hyperbolic geometry for we shall be using this (rather than Euclidean geometry) for the remainder of the text. Indeed, we have already seen the importance of hyperbolic geometry in the earlier chapters.

We shall describe hyperbolic geometry in terms of Euclidean geometry, thus it can be thought of here as being subordinate to Euclidean geometry. The points, lines and other configurations will be defined as subsets of the Euclidean plane and in this way we avoid the need to discuss the axioms for hyperbolic geometry. Of course, appropriate sets of axioms do exist and once we have verified that these axioms hold in our model we are entitled to use those theorems which are derivable from these axioms: we shall not, however, follow this path. Within the limitations of Euclidean geometry we shall be as rigorous and complete as possible.

We have seen in Section 3.3 that we may use the *upper half-plane*

$$H^2 = \{x + iy: y > 0\}$$

as a model for the hyperbolic plane and that this supports a metric $\rho$ derived from the differential

$$ds = \frac{|dz|}{\text{Im}[z]}. \tag{7.1.1}$$

We have also seen that reflections in circles of the form $|z - x_0| = r$ ($x_0$ real, $r > 0$) and reflections in "vertical" lines of the form $x = x_1$ ($x_1$ real) are isometries of $(H^2, \rho)$. We shall return to these facts in the next few sections.

There is a parallel development in terms of the *unit disc*

$$\Delta = \{z \in \mathbb{C}: |z| < 1\}.$$

The results in Section 3.4 are applicable and the metric $\rho$ in $H^2$ transfers to a metric in $\Delta$ which is derived from the differential

$$ds = \frac{2|dz|}{1 - |z|^2}. \tag{7.1.2}$$

*Throughout the remainder of the book we shall use $\rho$ for both the metric in $H^2$ and the metric in $\Delta$*: no confusion should arise, indeed the reader must become adept at frequently changing from one model to the other as each has its own particular advantage.

One of the principal benefits of discussing hyperbolic geometry in Euclidean terms is that we can easily introduce *the circle of points at infinity*: by this we mean $\mathbb{R}^1 \cup \{\infty\}$ for $H^2$ and $\{z: |z| = 1\}$ for $\Delta$. These are not points in the hyperbolic plane, nevertheless they play a vital part in any discussion of hyperbolic geometry and Fuchsian groups. The union of the hyperbolic plane and the circle at infinity is called the *closed hyperbolic plane*.

We shall refer to the two models of hyperbolic geometry described above as the Poincaré models. There are other models available (see Section 3.7) and we shall discuss (briefly) one alternative, namely the Klein model. The reader should note, however, that apart from one result (in Section 7.5) and occasional remarks and exercises, we shall not use the Klein model.

We have seen in Section 3.4 that the reflection in the plane $x_3 = 0$ followed by stereographic projection maps $H^3$ isometrically onto $B^3$, the metrics being those analogous to (7.1.1) and (7.1.2). Let this composite map be denoted by $s$. It follows that the upper hemisphere

$$Q = \{(x_1, x_2, x_3): x_1^2 + x_2^2 + x_3^2 = 1, x_3 > 0\},$$

(which is a model of the hyperbolic plane embedded in hyperbolic space $H^3$) is mapped by $s$ isometrically onto $\Delta$ ($= B^2$) embedded in $B^3$. Observe that as $s$ is conformal, arcs of circles in $Q$ orthogonal to $\partial H^3$ map to arcs of circles in $\Delta$ orthogonal to $\partial B^3$.

We can also map $Q$ onto $\Delta$ by vertical projection, namely

$$v: (x_1, x_2, x_3) \mapsto x_1 + ix_2.$$

Thus under the map $F$ ($= vs^{-1}$) of $\Delta$ onto $\Delta$, arcs of circles in $\Delta$ orthogonal to $\partial \Delta$ (the geodesics in $\Delta$) map to Euclidean segments with the same end-points on $\partial \Delta$. The significance of this is that $F$ is a homeomorphism of the closed

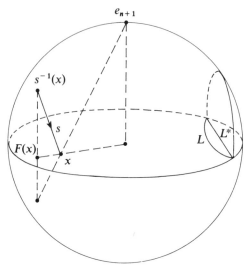

Figure 7.1.1

unit disc $\bar{\Delta}$ onto itself which maps each geodesic $L$ in the Poincaré model onto the Euclidean straight line segment $L^*$ of $\Delta$ with the same end-points as $L$: see Figure 7.1.1.

The effect of $F$ can easily be verified analytically and the preceding discussion is equally valid in $n$ dimensions. If $x \in B^n$, then

$$\begin{aligned} F(x) &= vs^{-1}(x), \\ &= v\pi^{-1}(x) \\ &= v\pi(x), \end{aligned}$$

where $\pi$ is stereographic projection (or, more properly, reflection in the sphere $S(e_{n+1}, \sqrt{2})$). The formula for $\pi$ given in Section 3.1 now yields the explicit formula for $F$, namely

$$F(x) = \frac{2x}{1 + |x|^2}.$$

Given that the sphere $S(a, r)$ is orthogonal to $\partial B^n$, the orthogonality implies that $|a|^2 = 1 + r^2$ and so $S$ has equation

$$|x|^2 + 1 = 2(x \cdot a).$$

Thus $F$ maps $S(a, r)$ onto the Euclidean hyperplane

$$S^* = \{y : y \cdot a = 1\},$$

which meets $\partial B^n$ at the same set of points as does $S(a, r)$.

The Klein model of hyperbolic geometry (that is, the model $\Delta$ with geodesics represented by the Euclidean segments $L^*$) is a useful model for establishing properties of incidence and convexity in that it transfers problems in hyperbolic geometry to corresponding problems in Euclidean geometry.

## §7.2. The Hyperbolic Metric

Our first task is to give a careful description of the construction of the metric $\rho$ from the differential (7.1.1). To each piecewise continuously differentiable curve in $H^2$, say $\gamma: [a, b] \to H^2$, we assign a "length" $\|\gamma\|$ by the formulae

$$\|\gamma\| = \int_a^b \frac{|\gamma^{(1)}(t)|}{\text{Im}[\gamma(t)]} \, dt.$$

The function $\rho$ is now defined by

$$\rho(z, w) = \inf \|\gamma\| \qquad (z, w \in H^2),$$

where the infimum is taken over all $\gamma$ which join $z$ to $w$ in $H^2$. It is clear that $\rho$ is non-negative, symmetric and satisfies the Triangle Inequality

$$\rho(z_1, z_3) \leq \rho(z_1, z_2) + \rho(z_2, z_3):$$

indeed, $\rho$ is a metric on $H^2$ (see Section 1.6).
    Now let

$$g(z) = \frac{az + b}{cz + d}, \tag{7.2.1}$$

where $a$, $b$, $c$ and $d$ are real and $ad - bc > 0$: thus $g$ maps $H^2$ onto itself. An elementary computation yields

$$\frac{|g^{(1)}(z)|}{\text{Im}[g(z)]} = \frac{1}{\text{Im}[z]}$$

and so

$$\|g\gamma\| = \int_a^b \frac{|g^{(1)}(\gamma(t))| \cdot |\gamma^{(1)}(t)|}{\text{Im}[g(\gamma(t))]} \, dt = \|\gamma\|.$$

Because of this invariance we immediately obtain the invariance of $\rho$, namely

$$\rho(gz, gw) = \rho(z, w) \tag{7.2.2}$$

and this proves that each such $g$ is an isometry of $(H^2, \rho)$. This will now be used to obtain an explicit expression for $\rho(z, w)$.

**Theorem 7.2.1.** *With $\rho$ as above, and with $z$, $w$ in $H^2$,*

(i)
$$\rho(z, w) = \log \frac{|z - \bar{w}| + |z - w|}{|z - \bar{w}| - |z - w|};$$

(ii)
$$\cosh \rho(z, w) = 1 + \frac{|z - w|^2}{2 \operatorname{Im}[z] \operatorname{Im}[w]};$$

(iii)
$$\sinh[\tfrac{1}{2}\rho(z, w)] = \frac{|z - w|}{2(\operatorname{Im}[z] \operatorname{Im}[w])^{1/2}};$$

(iv)
$$\cosh[\tfrac{1}{2}\rho(z, w)] = \frac{|z - \bar{w}|}{2(\operatorname{Im}[z] \operatorname{Im}[w])^{1/2}};$$

(v)
$$\tanh[\tfrac{1}{2}\rho(z, w)] = \left| \frac{z - w}{z - \bar{w}} \right|.$$

PROOF OF THEOREM 7.2.1. It is easy to see that the five equations are equivalent to each other: we shall prove that (ii) holds.

By (7.2.2), the left-hand side of (ii) is invariant under $g$. A straightforward computation shows that

$$\frac{|g(z) - g(w)|^2}{\operatorname{Im}[g(z)] \operatorname{Im}[g(w)]} = \frac{|z - w|^2}{\operatorname{Im}[z] \operatorname{Im}[w]},$$

thus the right-hand side of (ii) is also invariant under $g$. In fact, this is no more than the invariance of (3.3.3) established in Section 3.3.

Now select distinct $z$ and $w$ in $H^2$ and let $L$ be the unique Euclidean circle or line which contains $z$ and $w$ and which is orthogonal to the real axis. Now $L$ meets the real axis at some finite point $\alpha$ and by taking $g(z) = -(z - \alpha)^{-1} + \beta$ (for a suitable $\beta$) we may assume that $g$ in (7.2.1) maps $L$ onto the imaginary axis. It is only necessary, therefore, to verify (ii) when $z$ and $w$ lie on the imaginary axis.

We now assume that $z = ip$, $w = iq$ and also (as both sides of (ii) are symmetric in $z$ and $w$) that $0 < p < q$. If

$$\gamma(t) = x(t) + iy(t), \qquad 0 \le t \le 1,$$

is any curve joining $z$ to $w$, then

$$\|\gamma\| = \int_0^1 \frac{|x^{(1)}(t) + iy^{(1)}(t)|}{y(t)} \, dt$$

$$\ge \int_0^1 \frac{y^{(1)}(t)}{y(t)} \, dt$$

$$= \log(q/p)$$

as $y(1) = q$, $y(0) = p$. As equality holds when, for example,

$$\gamma(t) = i[p + t(q - p)],$$

we find that

$$\rho(ip, iq) = \log(q/p) \qquad (0 < p < q),$$

and it is easy to see that (ii) holds when $z = ip$ and $w = iq$. ☐

*Remark.* We have proved a little more than is stated in Theorem 7.2.1. First, we have obtained

$$\|\gamma\| = \rho(ip, iq),$$

(that is, $\|\gamma\|$ is minimal) if and only if $x(t) = 0$ and $y^{(1)}(t) > 0$ for all $t$ in [0, 1]. We shall return to this in the next section. Next, for future reference we record the formula

$$\rho(ip, iq) = |\log(p/q)|: \qquad (7.2.3)$$

in this form we do not need to assume that $p < q$.

We now consider the model $\Delta$. The map

$$f(z) = \frac{z - i}{z + i}$$

is a 1–1 map of $H^2$ onto $\Delta$, thus $\rho^*$ given by

$$\rho^*(z, w) = \rho(f^{-1}z, f^{-1}w) \qquad (z, w \in \Delta),$$

is a metric on $\Delta$. However, as

$$\frac{2|f^{(1)}(z)|}{1 - |f(z)|^2} = \frac{1}{\text{Im}[z]} \qquad (z \in H^2),$$

we can also identify $\rho^*$ with the metric derived from the differential (7.1.2). As we have already remarked, we prefer to use $\rho$ for $\rho^*$ and with this convention, $f$ is an isometry of $(H^2, \rho)$ onto $(\Delta, \rho)$.

We can derive formulae for the model $\Delta$ by simply rewriting Theorem 7.2.1 by means of $f$. It is more instructive, though, to work directly with $\Delta$: for example, corresponding to (7.2.3) we find that if $0 < r < 1$ then

$$\rho(0, r) = \int_0^r \frac{2dt}{1 - t^2} = \log\frac{1 + r}{1 - r}$$

(the reader should verify this).

Given distinct points $z$ and $w$ there is an isometry $g$ of $\Delta$ onto itself with $g(z) = 0$ and $g(w) = r, r > 0$. The invariance described by (3.4.3) yields

$$\frac{|z - w|^2}{(1 - |z|^2)(1 - |w|^2)} = \frac{r^2}{1 - r^2}$$

$$= \sinh^2[\tfrac{1}{2}\rho(0, r)]$$

$$= \sinh^2[\tfrac{1}{2}\rho(z, w)]. \qquad (7.2.4)$$

The identity (3.4.4) becomes

$$|1 - z\bar{w}|^2 = |z - w|^2 + (1 - |z|^2)(1 - |w|^2)$$

and this together with (7.2.4) yields

$$\cosh^2[\tfrac{1}{2}\rho(z, w)] = \frac{|1 - z\bar{w}|^2}{(1 - |z|^2)(1 - |w|^2)};$$

this is actually (3.4.5). Finally, we obtain

$$\tanh[\tfrac{1}{2}\rho(z, w)] = \left| \frac{z - w}{1 - z\bar{w}} \right|$$

and

$$\rho(z, w) = \log \frac{|1 - z\bar{w}| + |z - w|}{|1 - z\bar{w}| - |z - w|}. \tag{7.2.5}$$

As simple and useful examples of these ideas, we compute the length of a circle and the area of a disc (see (3.3.5)). Of course, length and area here are with respect to the hyperbolic metric and both remain invariant under isometries.

If $E$ is contained in $\Delta$, then the hyperbolic area of $E$ is

$$\text{h-area}(E) = \iint_E \left[ \frac{2}{1 - |z|^2} \right]^2 dx \, dy:$$

if $E$ is contained in $H^2$, the integrand is replaced by $1/y^2$. For any curve $C$ in $\Delta$, the hyperbolic length of $C$ is

$$\text{h-length}(C) = \int_C \frac{2|dz|}{1 - |z|^2}:$$

if $C$ is in $H^2$, the integrand is replaced by $1/y$.

**Theorem 7.2.2.** (i) *The area of a hyperbolic disc of radius $r$ is $4\pi \sinh^2(\tfrac{1}{2}r)$.*
(ii) *The length of a hyperbolic circle of radius $r$ is $2\pi \sinh r$.*

PROOF. We use the model $\Delta$ and let $C$ and $D$ be the circle and disc with centre $O$ and (hyperbolic) radius $r$. From (7.2.4) we see that

$$C = \{z : |z| = R\}, \qquad D = \{z : |z| \leq R\},$$

where

$$\sinh(\tfrac{1}{2}r) = \frac{R}{(1 - R^2)^{1/2}},$$

or, equivalently,

$$\tanh(\tfrac{1}{2}r) = R.$$

The stated results now follow by direct integration.                                            □

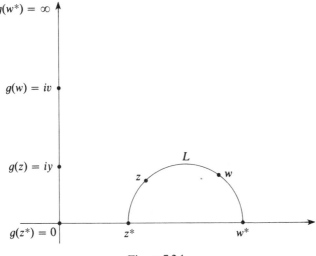

Figure 7.2.1

If we are prepared to use points on the circle at infinity we can also express $\rho(z, w)$ in terms of a cross-ratio. We recall from Section 4.4 that the cross-ratio is defined by

$$[z_1, z_2, z_3, z_4] = \frac{(z_1 - z_3)(z_2 - z_4)}{(z_1 - z_2)(z_3 - z_4)}.$$

Let $z$ and $w$ be distinct points in $H^2$ and let $g$ and $L$ be as in the proof of Theorem 7.2.1. Further, let $L$ meet the real axis at $z^*$ and $w^*$, these being labelled so that $z^*, z, w, w^*$ occur in this order along $L$ (see Figure 7.2.1). Now as $g(L)$ is the imaginary axis, $g(z^*) = 0$ or $g(z^*) = \infty$. If $g(z^*) = \infty$ we can apply the map $z \to -1/z$: thus we may assume $g$ to be chosen so that

$$g(z^*) = 0, \qquad g(z) = iy, \qquad g(w) = iv, \qquad g(w^*) = \infty,$$

where $y < v$. As the cross-ratio is invariant under Möbius transformations we obtain from (7.2.3),

$$\begin{aligned}
\rho(z, w) &= \rho(gz, gw) \\
&= \log(v/y) \\
&= \log[0, iy, iv, \infty] \\
&= \log[z^*, z, w, z^*].
\end{aligned} \tag{7.2.6}$$

Of course, this is equally valid in $\Delta$ for we can simply map $H^2$ isometrically onto $\Delta$ without changing the value of the cross-ratio.

We end this section with a few brief remarks about the metric topology of the hyperbolic plane. First, the Euclidean and hyperbolic metrics on $H^2$ (and $\Delta$) induce the same topologies. In particular, the closed hyperbolic plane is compact in the Euclidean topology and the subspace topology is the hyperbolic topology. It is convenient to introduce notation for the closure relative to the hyperbolic plane as well as the closed hyperbolic plane.

**Definition 7.2.3.** Let $E$ be a subset of the hyperbolic plane. Then

(i) $\tilde{E}$ denotes the *closure* of $E$ relative to the *hyperbolic plane*;
(ii) $\bar{E}$ denotes the *closure* of $E$ relative to the *closed hyperbolic plane*.

Of course, $\bar{E}$ is also the closure of $E$ in $\hat{\mathbb{C}}$.

EXERCISE 7.2

1. Let $L$ be the set of points $x + iy$ in $H^2$ where $x = y$. Find where

$$\inf\{\rho(z, w): z \in L\} \qquad (w \in H^2)$$

is attained and describe this point in geometric terms.

2. Suppose that $x_1 < x_2 < x_3 < x_4$. Let the semi-circle in $H^2$ with diameter $[x_1, x_3]$ meet the line $x = x_2$ at the point $z_3$. Similarly, let $z_4$ be the intersection of this line and the semi-circle with diameter $[x_1, x_4]$. Prove that

$$\rho(z_3, z_4) = \tfrac{1}{2} \log[x_2, x_3, x_4, \infty].$$

3. Show that if $\sigma$ is a metric on a set $X$ then $\tanh \sigma$ is also a metric on $X$. Deduce that

$$\rho_0(z, w) = \left| \frac{z - w}{z - \bar{w}} \right|$$

is a metric on $H^2$. Show that

$$\rho_0(u, v) = \rho_0(u, w) + \rho_0(w, v)$$

if and only if $w = u$ or $w = v$.

4. Show that $(H^2, \rho)$ is complete but not compact.

## §7.3. The Geodesics

We begin by defining a *hyperbolic line* or, more briefly, an *h-line* to be the intersection of the hyperbolic plane with a Euclidean circle or straight line which is orthogonal to the circle at infinity. With this definition, the following facts are easily established.

(1) *There is a unique h-line through any two distinct points of the hyperbolic plane.*
(2) *Two distinct h-lines intersect in at most one point in the hyperbolic plane.*
(3) *The reflection in an h-line is a $\rho$-isometry* (see Section 3.3).
(4) *Given any two h-lines $L_1$ and $L_2$, there is a $\rho$-isometry $g$ such that $g(L_1)$ $= L_2$* (see the proof of Theorem 7.2.1).

Given any $w$ in $H^2$, it is clear that

$$\{z \in H^2 : |z| = |w|\}$$

is the unique h-line which contains $w$ and which is orthogonal to the positive imaginary axis (an h-line). As the isometry in (4) can be taken to be a Möbius transformation we obtain:

(5) *given any h-line and any point $w$, there is a unique h-line through $w$ and orthogonal to L.*

Without going into the details, the reader should be aware that an essential feature of axiomatic geometry is the notion of "between" on a line. In our case, this notion can be described in terms of the metric.

Given two distinct points $z$ and $w$ on an h-line $L$, the set $L - \{z, w\}$ has three components exactly one of which has a compact closure (relative to the hyperbolic plane). This component is the *open segment* $(z, w)$ and $\zeta$ is *between* $z$ and $w$ if and only if $\zeta \in (z, w)$. The *closed segment* $[z, w]$ and segments $[z, w), (z, w]$ are defined in the obvious way.

The discussion preceding (7.2.3) shows that a curve $\gamma$ joining $ip$ to $iq$ satisfies

$$\|\gamma\| = \rho(ip, iq)$$

if and only if $\gamma$ is a parametrization of $[ip, iq]$ as a *simple* curve. Clearly, this can be phrased in an invariant form as follows.

**Theorem 7.3.1.** *Let $z$ and $w$ be any points in the hyperbolic plane. A curve $\gamma$ joining $z$ to $w$ satisfies*

$$\|\gamma\| = \rho(z, w)$$

*if and only if $\gamma$ is a parametrization of $[z, w]$ as a simple curve.*

It is for this reason that we refer to h-lines as geodesics (that is, curves of shortest length).

Now consider any three points $z$, $w$ and $\zeta$. It is clear from the special case (7.2.3) that if $\zeta$ is between $z$ and $w$, then

$$\rho(z, w) = \rho(z, \zeta) + \rho(\zeta, w).$$

Equally clearly, if $\zeta$ is not between $z$ and $w$ then the curve $\gamma$ comprising of the segments $[z, \zeta]$ and $[\zeta, w]$ satisfies (by Theorem 7.3.1)

$$\|\gamma\| > \rho(z, w).$$

Thus we obtain the next result.

**Theorem 7.3.2.** *Let $z$ and $w$ be distinct points in the hyperbolic plane. Then*

$$\rho(z, w) = \rho(z, \zeta) + \rho(\zeta, w)$$

*if and only if $\zeta \in [z, w]$.*

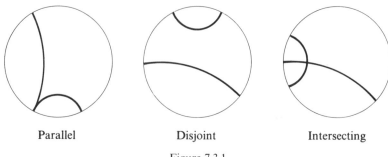

Parallel                          Disjoint                     Intersecting

Figure 7.3.1

We end this section with more terminology. First, the points $z_1, z_2, \ldots$ are *collinear* if they lie on a single geodesic. Each geodesic has two *end-points*, each on the circle at infinity. It is natural to extend the notation for a segment so as to include geodesics: thus $(\alpha, \beta)$ denotes the geodesic segment with end-points $\alpha$ and $\beta$ even if these are on the circle at infinity. A *ray from z* is a segment $[z, \alpha)$ where $\alpha$ lies on the circle at infinity: each geodesic $(\alpha, \beta)$ through $z$ determines exactly two rays from $z$, namely $[z, \alpha)$ and $[z, \beta)$.

**Definition 7.3.3.** Let $L_1$ and $L_2$ be distinct geodesics. We say that $L_1$ and $L_2$ are *parallel* if and only if they have exactly one end-point in common. If $L_1$ and $L_2$ have no end-points in common, then they are *intersecting* when $L_1 \cap L_2 \neq \varnothing$ and *disjoint* when $L_1 \cap L_2 = \varnothing$.

**Warning.** This terminology is not standard and the terms are illustrated in the model $\Delta$ in Figure 7.3.1. Much of the geometry is based on a discussion of these three mutually exclusive possibilities (parallel, intersecting and disjoint) and for this reason we prefer a particularly descriptive terminology.

EXERCISE 7.3

1. Let $w = u + iv$, $w' = iv$ and $z = ri$ be points in $H^2$. Prove that

$$\rho(w, z) \geq \rho(w', z)$$

with equality if and only if $w = w'$. Deduce Theorem 7.3.2.

## §7.4. The Isometries

The objective here is to identify all isometries of the hyperbolic plane. Let $z$, $w$ and $\zeta$ be distinct points in $H^2$ with $\zeta$ between $z$ and $w$. It is an immediate consequence of Theorem 7.3.2 that for any isometry $\phi$, the point $\phi(\zeta)$ is between $\phi(z)$ and $\phi(w)$. Thus $\phi$ maps the segment $[z, w]$ onto the segment $[\phi(z), \phi(w)]$: because of this, $\phi$ maps h-lines to h-lines.

Given any isometry $\phi$, there is an isometry

$$g(z) = \frac{az + b}{cz + d} \qquad (ad - bc > 0),$$

such that $g\phi$ leaves the positive imaginary axis $L$ invariant (simply choose $g$ to map $\phi(L)$ to $L$). By applying the isometries $z \to kz$ $(k > 0)$ and $z \to -1/z$ as necessary, we may assume that $g\phi$ fixes $i$ and leaves invariant the rays $(i, \infty)$, $(0, i)$. It is now an immediate consequence of (7.2.3) that $g\phi$ fixes each point of $L$.

Now select any $z$ in $H^2$ and write

$$z = x + iy, \qquad g\phi(z) = u + iv.$$

For all positive $t$,

$$\rho(z, it) = \rho(g\phi(z), g\phi(it))$$
$$= \rho(u + iv, it)$$

and so, by Theorem 7.2.1(iii),

$$[x^2 + (y - t)^2]v = [u^2 + (v - t)^2]y.$$

As this holds for all positive $t$ we have $y = v$ and $x^2 = u^2$: thus

$$g\phi(z) = z \text{ or } -\bar{z}.$$

A straightforward continuity argument (isometries are necessarily continuous) shows that one of these equations holds for all $z$ in $H^2$: for example, the set of $z$ in the open first quadrant with $g\phi(z) = z$ is both open and closed in that quadrant. This proves the next result.

**Theorem 7.4.1.** *The group of isometries of $(H^2, \rho)$ is precisely the group of maps of the form*

$$z \mapsto \frac{az + b}{cz + d}, \qquad z \mapsto \frac{a(-\bar{z}) + b}{c(-\bar{z}) + d},$$

*where $a$, $b$, $c$ and $d$ are real and $ad - bc > 0$. Further, the group of isometries is generated by reflections in h-lines.*

A similar development holds for the model $\Delta$: here, the isometries are

$$z \mapsto \frac{az + \bar{c}}{cz + \bar{a}}, \qquad z \mapsto \frac{a\bar{z} + \bar{c}}{c\bar{z} + \bar{a}},$$

where $|a|^2 - |c|^2 = 1$.

Note that if

$$g(z) = \frac{az + \bar{c}}{cz + \bar{a}} \qquad |a|^2 - |c|^2 = 1,$$

then from (7.2.4) we obtain the useful expressions

$$|c| = \sinh \tfrac{1}{2}\rho(0, g0), \qquad\qquad\qquad (7.4.1)$$

$$|a| = \cosh \tfrac{1}{2}\rho(0, g0) \qquad\qquad\qquad (7.4.2)$$

and so (see Section 5.2) we find again that

$$\|g\|^2 = 2 \cosh \rho(0, g0).$$

Of course, if $h$ is an isometry of $(H^2, \rho)$ then

$$\|h\|^2 = 2 \cosh \rho(i, hi):$$

the proof is by an elementary computation using Theorem 7.2.1(ii) (or by
Theorem 4.2.1).

### EXERCISE 7.4

1. Let $z_j, w_j$ ($j = 1, 2, 3$) be points in $H^2$. Show that there is an isometry $g$ with $g(z_j) = w_j$
for each $j$ if and only if for all $i$ and $j$,

$$\rho(z_i, z_j) = \rho(w_i, w_j).$$

## §7.5. Convex Sets

A subset $E$ of the hyperbolic plane is said to be *convex* if and only if for each
$z$ and $w$ in $E$, we have $[z, w] \subset E$. The following facts regarding convexity are
easily verified.

(1) *If $E$ is convex, then so is $g(E)$ for every isometry $g$.*
(2) *If $E$ is convex, then so are $E^0$ (the interior of $E$) and $\tilde{E}$.*
(3) *If $E_1, E_2, \ldots$ are convex and $E_1 \subset E_2 \subset \ldots$, then $\bigcup E_n$ is convex.*
(4) *If each $E_\alpha$ is convex, then so is $\bigcap_\alpha E_\alpha$.*

By definition, a geodesic is convex. The mapping $iy \to \log y$ is a homeomorphism of the hyperbolic geodesic $\{iy : y > 0\}$ onto the Euclidean geodesic
$\{x + iy : y = 0\}$ which preserves the relation "between". We deduce that
*the segments are the only convex subsets of a hyperbolic geodesic.*

An open *half-plane* is a component of the complement of a geodesic
and any open half-plane is convex. As an illustration of the use of the Klein
model, let $F : \Delta \to \Delta$ be the map described in Section 7.1. This maps the
geodesics of the Poincaré model $(\Delta, \rho)$ onto Euclidean segments in $\Delta$ and so
a subset $E$ of $\Delta$ is convex in the Poincaré model if and only if $F(E)$ is convex
in the Euclidean sense. In particular, a half-plane in the Poincaré model maps
onto the intersection of $\Delta$ with a Euclidean half-plane and this is indeed convex
in the Euclidean sense. In this way, the Klein model enables us to refer
hyperbolic convexity to the more familiar context of Euclidean convexity.

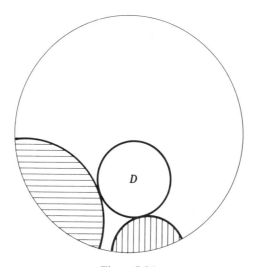

Figure 7.5.1

By (2), a closed half-plane is convex. If $E_\alpha$ $(\alpha \in A)$ is now any family of half-planes (open or closed), then the *complement* of $\bigcup E_\alpha$ is the intersection of half-planes and so is convex. For example, *a hyperbolic disc D is convex* for it is the complement of a union of (shaded) half-planes as in Figure 7.5.1.

There are two other examples of a similar nature which we shall use later. A *horocyclic region* is the interior of a Euclidean circle which is tangent to the circle at infinity. By taking the model $H^2$ and $\infty$ as the point of tangency, we may assume that the horocyclic region is $\{x + iy: y > t\}$. This region is convex for it is complement of the union of all half-planes of the form $\{z \in H^2: |z - x_0| \le t\}$ as $x_0$ varies over the real line. For future reference, a *horocycle* is the boundary of a horocyclic region.

A *hypercyclic region* is any region which is isometrically congruent to a region of the form

$$\{z \in H^2: |\arg(z) - \pi/2| < \theta\}$$

for some $\theta$ in $(0, \pi/2)$. The significance of this will appear later, however such a region in convex for it is the complement of the union of half-planes of the form

$$\{z \in H^2: |z - x_0| \le x_0 \cos \theta\} \qquad (x_0 \text{ real}).$$

The boundary of a hypercyclic region is called a *hypercycle*.

We end with a characterization of closed convex sets. A set $E$ is *locally convex* if and only if each $z$ in $E$ has an open neighbourhood $N$ such that $E \cap N$ is convex. The notions of convexity and local convexity are meaningful in both Euclidean and hyperbolic spaces and they extend in the obvious way to the closed hyperbolic plane.

**Theorem 7.5.1.** *Let P be the Euclidean plane or the closed hyperbolic plane. A closed subset E of P is convex if and only if it is connected and locally convex.*

PROOF. If the result is true when $P$ is the Euclidean plane, the relationship between the Poincaré and Klein models shows that the result is also true when $P$ is the closed hyperbolic plane. Thus it is only necessary to show that if $E$ is a closed, connected, locally convex subset of $\mathbb{R}^2$ then $E$ is convex (the reverse implication is trivial).

We say two points in $E$ are *polygonally connected* if they can be joined by a polygonal arc lying in $E$. This is an equivalence relation and the local convexity of $E$ implies that the equivalence classes are relatively open in $E$. As $E$ is connected, there is only one equivalence class so any two points of $E$ can be joined by a polygonal curve in $E$. Because of this it is sufficient to prove that if the Euclidean segments $[u, v]$, $[v, w]$ lie in $E$ then so does the segment $[u, v]$. If $u, v, w$ are collinear then this is trivial: thus we assume that these points are not collinear.

For each $a, b, c$ let $T(a, b, c)$ denote the closed triangle with vertices $a$, $b$, $c$ (by this, we mean the convex hull of the points $a, b, c$). Now let $K$ be the set of $x$ in $[v, u]$ with the property that for some $y$ in $(v, w)$ we have $T(v, x, y) \subset E$. As $E$ is locally convex at $v$, $K$ contains some interval of positive length. Clearly, $K$ is an interval of the form $[v, x_0)$ or $[v, x_0]$ where $x_0 \neq v$ and we shall now show that $K = [v, u]$.

Choose a neighbourhood $N$ of $x_0$ such that $E \cap N$ is convex and then choose $x_1$ in $[v, x_0) \cap N$ and $x_2$ in $[x_0, u] \cap N$: see Figure 7.5.2.

As $x_1 \in K$, there is some $y_1$ in $(v, w)$ with

$$T(v, x_1, y_1) \subset E.$$

Choose $z$ in $N \cap (x_1, y_1)$: as $E \cap N$ is convex we have

$$T(z, x_1, x_2) \subset E.$$

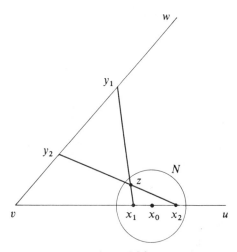

Figure 7.5.2

With $y_2$ as in Figure 7.5.2 we also have

$$T(v, x_2, y_2) \subset T(v, x_1, y_1) \cup T(x_1, x_2, z)$$
$$\subset E,$$

so $x_2 \in K$. This shows that $x_0 \in K$ and $x_0 = u$ so $K = [v, u]$. Note that as $u \in K$, there is some $y$ in $(v, w)$ with $T(v, u, y) \subset E$.

Now consider the set $K_1$ of $y$ in $[v, w]$ such that $T(v, u, y) \subset E$. Exactly as before, $K_1$ is some segment $[v, y_0)$ or $[v, y_0]$. As $E$ is closed, we see that $K_1 = [v, y_0]$. The argument in the preceding paragraph (with $u, v, w$ replaced by $u, y_0, w$) shows that $y_0 = w$ so $w \in K_1$ and

$$T(v, u, w) \subset E. \qquad \qquad \square$$

### Exercise 7.5

1. Let $z, z', w, w'$ be points in $H^2$. Prove that if $w \in [z, z']$ then

$$\rho(w, w') \leq \max\{\rho(w, z), \rho(w', z)\}.$$

Deduce (analytically) that a hyperbolic disc is convex.

2. Construct a subset $E$ of $H^2$ which is connected and locally convex but not convex (see Theorem 7.5.1).

3. Show that exactly one of the sets

$$\{x + iy \in H^2 : a < x < b, y < c\},$$
$$\{x + iy \in H^2 : a < x < b, y > c\}$$

is convex.

## §7.6. Angles

Our attitude to angles in the hyperbolic plane is consistent with the policy outlined in Section 7.1, namely we describe the angles of hyperbolic geometry in terms of Euclidean geometry. In hyperbolic geometry, *an angle at a point $z$* is an unordered pair of rays $(L, L')$ from $z$. Let $(L, L')$ be an angle at $z$ and suppose for the moment that $L$ and $L'$ are not on the same geodesic. The ray $L$ determines a geodesic, say $L^*$, and $L' - \{z\}$ does not meet $L^*$. It follows that $L' - \{z\}$ lies in one of the open half-planes say $\Sigma'$, complementary to $L^*$. Similarly, $L - \{z\}$ lies in one of the half-planes, say $\Sigma$, complementary to $L'$. We now define the *interior of the angle* $(L, L')$ to be $\Sigma \cap \Sigma'$. It is easy to see that the interior of $(L, L')$ is one component of the complement of $L \cup L'$: the other component is called *the exterior of* $(L, L')$.

If $L$ and $L'$ lie on the same geodesic then either $L \cup L'$ is a geodesic (and there is no canonical choice of interior or exterior) or $L = L'$ in which case we define the interior to be empty and the exterior to be the complement of $L$.

Given an angle $(L, L')$ at $z$ with $L$ and $L'$ defining different geodesics, *the interior of* $(L, L')$ *is convex* as it is the intersection of half-planes. To complement this, *the exterior cannot be convex* for otherwise a segment joining points on $L - \{z\}$ and $L' - \{z\}$ would lie in both the interior and the exterior of $(L, L')$. Of course, we can measure the interior and exterior angles at $z$ in the usual way and the measurements lie in $[0, \pi)$ and $(\pi, 2\pi]$, respectively.

## §7.7. Triangles

Let $z_1, z_2$ and $z_3$ be three non-collinear points in the hyperbolic plane and let $L_2$ and $L_3$ be the rays from $z_1$ through $z_2$ and $z_3$ respectively. Then $(L_2, L_3)$ is an angle at $z_1$: we denote its interior by $A_1$. In a similar way, $A_2$ and $A_3$ are the interiors of angles at $z_2$ and $z_3$. This notation will readily be absorbed by a glance at Figure 7.7.1. Note that by convexity, $(z_2, z_3) \subset A_1$ (see Section 7.6).

**Definition 7.7.1.** The *triangle* $T(z_1, z_2, z_3)$ is $A_1 \cap A_2 \cap A_3$.

The $z_j$ are the *vertices*, the $[z_i, z_j]$ are the *sides* and the $A_i$ are the *angles* of $T(z_1, z_2, z_3)$. Each angle of $T(z_1, z_2, z_3)$, being an interior angle, is less than $\pi$. For brevity, we write $T$ for $T(z_1, z_2, z_3)$. Observe that as each $A_j$ is convex, so is $T$. Moreover, the $A_j$ are also the angles of $T$ in the sense that for any sufficiently small open disc $D$ with centre, say, $z_1$, we have

$$D \cap T = D \cap A_1.$$

To see this, let $H_j$ be the half-plane containing $z_j$ and having the other two $z_i$ on its boundary. Then (if $D \subset H_1$)

$$\begin{aligned} D \cap A_1 &= (D \cap H_1) \cap (H_2 \cap H_3) \\ &= D \cap (H_2 \cap H_3) \cap (H_3 \cap H_1) \cap (H_1 \cap H_2) \\ &= D \cap T. \end{aligned}$$

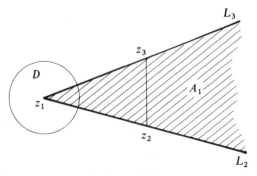

Figure 7.7.1

Next, define $\partial T$ by

$$\partial T = [z_1, z_2] \cup [z_2, z_3] \cup [z_3, z_1]:$$

this may be parametrized as a Jordan curve with interior, say $T_0$. As $\partial T \subset \tilde{H}_1$, so $T_0 \subset H_1$. The same argument holds for $H_2$ and $H_3$ so

$$T_0 \subset H_1 \cap H_2 \cap H_3 = T.$$

As $T$ is connected (in fact, convex) and does not meet $\partial T$, it lies inside or outside of $\partial T$. However, $T$ meets $T_0$ so $T \subset T_0$ and hence $T = T_0$.

In an axiomatic treatment, it is sometimes necessary to take as an axiom the fact that a ray $L$ from $z_1$ through a point $w$ in $T(z_1, z_2, z_3)$ necessarily meets the side $(z_2, z_3)$. In our case, we observe that the (connected) segment $L - \{z_1\}$ meets the interior of $\partial T$ (at $w$) and cannot meet the sides $[z_1, z_2]$ or $[z_1, z_3]$. As $L - \{z_1\}$ is unbounded, its closure meets the circle at infinity and so must meet $\partial T$.

The next result is used frequently in deriving trigonometric formulae (and so must be proved independently of these formulae).

**Theorem 7.7.2.** *Let $L$ be the geodesic containing the longest side, say $[z_2, z_3]$, of $T$. Then the geodesic $L_1$ through $z_1$ and orthogonal to $L$ meets $L$ at a point $w$ in $[z_2, z_3]$.*

PROOF. We may assume that $L$ is the positive imaginary axis so $w = i|z_1|$: see Figure 7.7.2.

It is easy to see that

$$\rho(z_1, z_2) \geq \rho(w, z_2)$$

and similarly

$$\rho(z_1, z_3) \geq \rho(w, z_3),$$

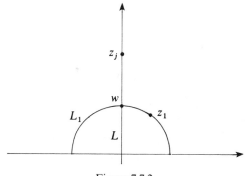

Figure 7.7.2

(see Section 7.20 which does not use trigonometry). As $[z_2, z_3]$ is the longest side we deduce that

$$\max\{\rho(z_2, w), \rho(z_3, w)\} \leq \rho(z_2, z_3).$$

The points $z_2$, $z_3$, $w$ are collinear and one must lie between the other two. Using Theorem 7.3.2, we see that $w$ must lie between $z_2$ and $z_3$ (or be equal to one of them).                                                                            □

Most of the material in this section extends without difficulty to the case when some (or all) of the vertices lie on the circle at infinity. The notable exception to this is Theorem 7.7.2 (consider $z_2$ but not $z_3$ on the circle at infinity).

EXERCISE 7.7

1. Show that in hyperbolic geometry, the vertices of a triangle may, but need not, lie on a circle.

2. Prove that the diameter of a triangle $T$, namely

$$\sup\{\rho(z, w): z, w \in T\},$$

is the length of the longest side (see Exercise 7.5.1).

## §7.8. Notation

In the next six sections we shall be concerned with hyperbolic triangles and it is convenient to adopt a standard notation which allows us to express trigonometric relations easily. A triangle $T$ will have vertices labelled $v_a$, $v_b$ and $v_c$: the sides opposite these vertices will have lengths $a$, $b$ and $c$ respectively and the interior angles at the vertices will be $\alpha$, $\beta$ and $\gamma$. This notation will readily be absorbed by a glance at Figure 7.8.1. As isometries preserve length and angles, trigonometric formulae remain invariant under isometries.

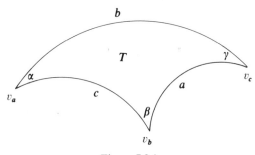

Figure 7.8.1

We shall allow some or all of the vertices of a triangle to be on the circle at infinity. If, for example, $v_a$ is at infinity, then

$$\alpha = 0, \qquad b = c = +\infty.$$

If two vertices are on the circle at infinity, then all three sides have infinite length.

EXERCISE 7.8

1. Let $T_1$ and $T_2$ be two triangles, each with all sides of infinite length. Show that there is an isometry mapping $T_1$ onto $T_2$.

## §7.9. The Angle of Parallelism

The Angle of Parallelism is the classical term for the trigonometric relation which holds for a triangle with angles $\alpha$, $0$, $\pi/2$: in this case, there are only two parameters, namely $\alpha$ and $b$.

**Theorem 7.9.1.** *Let $T$ be a triangle with angles $\alpha$, $0$, $\pi/2$ ($\alpha \neq 0$). Then*

(i) $\sinh b \tan \alpha = 1$;
(ii) $\cosh b \sin \alpha = 1$;
(iii) $\tanh b \sec \alpha = 1$.

PROOF. We work in $H^2$ and we may assume that

$$v_c = i, \qquad v_b = \infty, \qquad v_a = x + iy,$$

where $x^2 + y^2 = 1$: see Figure 7.9.1. As $y = \sin \alpha$, Theorem 7.2.1(ii) yields (ii). The remaining formulae are equivalent to (ii). $\qquad\square$

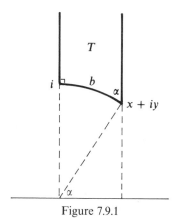

Figure 7.9.1

## §7.10. Triangles with a Vertex at Infinity

Consider a triangle with angles $\alpha$, $\beta$, 0 where $\alpha$ and $\beta$ are non-zero: then

$$a = b = +\infty, \qquad 0 < c < +\infty$$

and we shall determine the relationship between $\alpha$, $\beta$ and $c$.

**Theorem 7.10.1.** *For any triangle with angles $\alpha$, $\beta$, 0 we have*

(i)
$$\cosh c = \frac{1 + \cos \alpha \cos \beta}{\sin \alpha \sin \beta};$$

(ii)
$$\sinh c = \frac{\cos \alpha + \cos \beta}{\sin \alpha \sin \beta}.$$

PROOF. We work in $H^2$ with $v_c = \infty$. We may assume that $v_a$ and $v_b$ lie on the circle $|z| = 1$, say with

$$v_a = \exp(i\theta), \qquad v_b = \exp(i\phi),$$

where $0 < \theta < \phi < \pi$. Thus $\alpha = \theta$, $\beta = \pi - \phi$ and (i) follows from Theorem 7.2.1 as

$$\cosh c = \cosh \rho(v_a, v_b).$$

The verification of (ii) is left to the reader.                                       $\square$

## §7.11. Right-angled Triangles

We now consider a triangle with angles $\alpha$, $\beta$, $\pi/2$. By applying a suitable isometry we may assume that

$$v_c = i, \qquad v_b = ki, \qquad v_a = s + it,$$

where $k > 1$ and $s$ and $t$ are positive with $s^2 + t^2 = 1$: see Figure 7.11.1.
    We begin with the relationship between the *three sides*: this is the hyperbolic form of *Pythagoras' Theorem*.

**Theorem 7.11.1.** *For any triangle with angles $\alpha$, $\beta$, $\pi/2$ we have*

$$\cosh c = \cosh a \cosh b. \tag{7.11.1}$$

PROOF. Using Theorem 7.2.1(ii) we have

$$\cosh c = (1 + k^2)/2kt;$$
$$\cosh b = 1/t;$$
$$\cosh a = (1 + k^2)/2k. \qquad \square \tag{7.11.2}$$

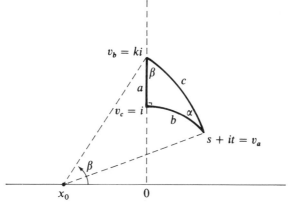

Figure 7.11.1

Note that we also obtain

$$\tanh b = s. \tag{7.11.3}$$

Next, we seek relations between two sides and one angle.

**Theorem 7.11.2.** *For any triangle with angles* $\alpha$, $\beta$, $\pi/2$ *we have*

(i) $\tanh b = \sinh a \tan \beta$;
(ii) $\sinh b = \sinh c \sin \beta$;
(iii) $\tanh a = \tanh c \cos \beta$.

PROOF. Let the geodesic through $v_a$ and $v_b$ have Euclidean centre $x_0$. Then by equating the distances from $v_a$ and $v_b$ to $x_0$ we see that

$$k^2 = 1 - 2x_0 s.$$

This shows that $x_0 < 0$. The Euclidean triangle with vertices $x_0$, $0$ and $ki$ has angle $\beta$ at $x_0$. Thus

$$\tan \beta = k/|x_0|$$
$$= 2sk/(k^2 - 1)$$

and this gives (i) because of (7.11.2) and (7.11.3).

Elimination of $a$ from (i) and (7.11.1) yields (ii): elimination of $b$ from (i) and (ii) yields (iii). $\qquad\square$

We end with the relations between one side and two angles.

**Theorem 7.11.3.** *For any triangle with angles* $\alpha$, $\beta$, $\pi/2$:

(i) $$\cosh a \sin \beta = \cos \alpha;$$

(ii) $$\cosh c = \cot \alpha \cot \beta.$$

PROOF. Theorem 7.11.2(i) gives

$$\sinh a \tan \beta = \tanh b,$$

$$\sinh b \tan \alpha = \tanh a$$

and elimination of $b$ gives (i).

To prove (ii), simply eliminate $\cosh a$ and $\cosh b$ from (7.11.1), Theorem 7.11.3(i) and the corresponding identity with $a$ and $\alpha$ interchanged with $b$ and $\beta$.                                                                □

## §7.12. The Sine and Cosine Rules

We now consider the general hyperbolic triangle with sides $a$, $b$ and $c$ and opposite angles $\alpha$, $\beta$ and $\gamma$. We assume that $\alpha$, $\beta$ and $\gamma$ are positive (so $a$, $b$ and $c$ are finite) and we prove the following results,

**The Sine Rule:**

$$\frac{\sinh a}{\sin \alpha} = \frac{\sinh b}{\sin \beta} = \frac{\sinh c}{\sin \gamma}.$$

**The Cosine Rule I:**

$$\cosh c = \cosh a \cosh b - \sinh a \sinh b \cos \gamma.$$

**The Cosine Rule II:**

$$\cosh c = \frac{\cos \alpha \cos \beta + \cos \gamma}{\sin \alpha \sin \beta}.$$

Note the existence of the second Cosine Rule. This has no analogue in Euclidean geometry: in hyperbolic geometry it implies that *if two triangles have the same angles, then there is an isometry mapping one triangle onto the other.*

PROOF OF THE COSINE RULE I. We shall use the model $\Delta$ and we may assume that $v_c = 0$ and $v_a > 0$: See Figure 7.12.1.

Note that

$$v_a = \tanh \tfrac{1}{2}\rho(0, v_a)$$

$$= \tanh(\tfrac{1}{2}b) \qquad\qquad (7.12.1)$$

and similarly,

$$v_b = e^{i\gamma} \tanh(\tfrac{1}{2}a). \qquad\qquad (7.12.2)$$

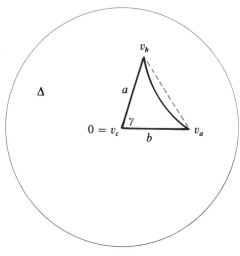

Figure 7.12.1

From (7.2.4) we have

$$\cosh c = 2 \sinh^2[\tfrac{1}{2}\rho(v_a, v_b)] + 1$$
$$= \frac{2|v_a - v_b|^2}{(1 - |v_a|^2)(1 - |v_b|^2)} + 1.$$

The Euclidean Cosine Rule gives $|v_a - v_b|^2$ in terms of $|v_a|$, $|v_b|$ and $\cos \gamma$ and using (7.12.1) and (7.12.2), the required result follows by straightforward simplification. □

PROOF OF THE SINE RULE. Using the Cosine Rule I we obtain

$$\left(\frac{\sinh c}{\sin \gamma}\right)^2 = \frac{\sinh^2 c}{1 - \left(\dfrac{\cosh a \cosh b - \cosh c}{\sinh a \sinh b}\right)^2}.$$

The Sine Rule will be valid provided that this is symmetric in $a$, $b$ and $c$ and this will be so if

$$(\sinh a \sinh b)^2 - (\cosh a \cosh b - \cosh c)^2$$

is symmetric. After writing $\sinh^2$ in terms of $\cosh^2$, we find that this is so. □

PROOF OF THE COSINE RULE II. For brevity, we shall write $A$ for $\cosh a$ and similarly for $B$ and $C$. The Cosine Rule I yields

$$\cos \gamma = \frac{(AB - C)}{(A^2 - 1)^{1/2}(B^2 - 1)^{1/2}}$$

and so

$$\sin^2 \gamma = \frac{D}{(A^2 - 1)(B^2 - 1)},$$

where

$$D = 1 + 2ABC - (A^2 + B^2 + C^2)$$

is symmetric in $A$, $B$ and $C$. The expression for $\sin^2 \gamma$ shows that $D \geq 0$.
Now observe that if we multiply both numerator and denominator of

$$\frac{\cos \alpha \cos \beta + \cos \gamma}{\sin \alpha \sin \beta}$$

by the positive value of

$$(A^2 - 1)^{1/2}(B^2 - 1)^{1/2}(C^2 - 1),$$

we obtain

$$\frac{\cos \alpha \cos \beta + \cos \gamma}{\sin \alpha \sin \beta} = \frac{[(BC - A)(CA - B) + (AB - C)(C^2 - 1)]}{D}$$

$$= C. \qquad \square$$

EXERCISE 7.12

1. For a general triangle, prove that $a \leq b \leq c$ if and only if $\alpha \leq \beta \leq \gamma$. [Use the Sine Rule and the Corollary of Theorem 7.13.1.]

2. Show that a triangle is an equilateral triangle if and only if $\alpha = \beta = \gamma$ and that in this case,

$$2 \cosh(\tfrac{1}{2}a) \sin(\tfrac{1}{2}\alpha) = 1.$$

3. Show that for a general triangle, the angle bisector at $v_a$ contains the mid-point of $[v_b, v_c]$ if and only if $b = c$ (Isosceles triangles).

4. Prove that there exists an isometry mapping a triangle $T_1$ onto a triangle $T_2$ if and only if $T_1$ and $T_2$ have the same angles (or sides of the same lengths).

## §7.13. The Area of a Triangle

**Theorem 7.13.1.** *For any triangle $T$ with angles $\alpha$, $\beta$ and $\gamma$,*

$$\text{h-area}(T) = \pi - (\alpha + \beta + \gamma).$$

**Corollary.** *The angle sum of a hyperbolic triangle is less than $\pi$.*

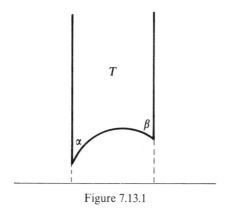

Figure 7.13.1

PROOF. Assume first that $\gamma = 0$. We may assume that $v_c = \infty$ and that $v_a$ and $v_b$ lie on $|z| = 1$. Referring to Figure 7.13.1 we find that

$$\text{h-area}(T) = \int_{\cos(\pi - \alpha)}^{\cos \beta} \left[ \int_{(1 - x^2)^{1/2}}^{\infty} \frac{dy}{y^2} \right] dx$$

$$= \pi - (\alpha + \beta),$$

which is the desired result when $\gamma = 0$. In general, any triangle is the difference of two such triangles (continue the ray from $v_a$ through $v_c$ to $w$ on the circle at infinity and consider $T(v_a, v_b, w)$) and the general case follows easily. $\square$

## §7.14. The Inscribed Circle

This is the last section on hyperbolic trigonometry and we leave the reader to provide most of the details.

**Theorem 7.14.1.** *The three angle bisectors of a triangle T meet at a point $\zeta$ in T.*

PROOF. We may assume that $\gamma$ is the smallest angle so $\gamma < \pi/2$. Now construct angle bisectors at $v_a$ and $v_b$: these must meet at a point $\zeta$ in $T$ (see Section 7.7). Next, define $\gamma_1$ and $\gamma_2$ as in Figure 7.14.1. As $\alpha/2$, $\beta/2$, $\gamma_1$ and $\gamma_2$ are each less than $\pi/2$, we can construct points $w_a$, $w_b$ and $w_c$ as in Figure 7.14.1 (and these points must lie on the open sides of $T$).

The Sine Rule applied to the two triangles with side $[\zeta, v_b]$ gives

$$\rho(\zeta, w_c) = \rho(\zeta, w_a).$$

The same result holds with $w_b$ instead of $w_a$ so the points $w_a$, $w_b$ and $w_c$ lie on a circle with centre $\zeta$. Moreover, elementary trigonometry now shows that $\gamma_1 = \gamma_2$. $\square$

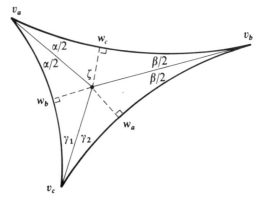

Figure 7.14.1

The circle centre $\zeta$ passing through $w_a$, $w_b$ and $w_c$ is called the *inscribed circle* of $T$.

**Theorem 7.14.2.** *The radius $R$ of the inscribed circle of $T$ is given by*

$$\tanh^2 R = \frac{\cos^2 \alpha + \cos^2 \beta + \cos^2 \alpha + 2 \cos \alpha \cos \beta \cos \gamma - 1}{2(1 + \cos \alpha)(1 + \cos \beta)(1 + \cos \gamma)}.$$

PROOF. Let $x = \rho(v_a, w_c)$ and $y = \rho(w_c, v_b)$. Then

$$\frac{\cos \alpha \cos \beta + \cos \gamma}{\sin \alpha \sin \beta} = \cosh x \cosh y + \sinh x \sinh y$$

so

$$[(\cos \alpha \cos \beta + \cos \gamma) - (\sin \alpha \sinh x)(\sin \beta \sinh y)]^2$$
$$= [(1 - \cos^2 \alpha) + \sin^2 \alpha \sinh^2 x][(1 - \cos^2 \beta) + \sin^2 \beta \sinh^2 y].$$

The identity

$$\sin \theta = (1 + \cos \theta) \tan(\theta/2)$$

together with the relation

$$\tanh R = \sinh x \tan(\alpha/2)$$

yields

$$\sin \alpha \sinh x = (1 + \cos \alpha) \tanh R.$$

A similar expression holds for $\beta$, $y$ and $R$ and substitution yields (after some simplification) the desired result.                                      $\square$

The next example is of interest.

**Example 7.14.3.** For each $\alpha$ in $(0, \pi)$ we can construct a triangle $T$ with angles $\alpha, 0, 0$. Then

$$4 \tanh^2 R = \tfrac{1}{2}(1 + \cos \alpha)$$
$$= \cos^2(\alpha/2)$$
$$= \sin^2[\tfrac{1}{2}\text{h-area}(T)].$$

In Euclidean geometry, a triangle may have a large area but a small inscribed circle. The next result shows that the situation in hyperbolic geometry is quite different: for a proof of this, see [10].

**Theorem 7.14.4.** *The radius $R$ of the inscribed circle of $T$ satisfies*

$$\tanh R \geq \tfrac{1}{2} \sin[\tfrac{1}{2}\text{h-area}(T)]$$

*and this lower bound is best possible for each value of $h\text{-area}(T)$.*

Example 7.14.3 shows that this lower bound is best possible.

## §7.15. The Area of a Polygon

A polygon $P$ is the interior of a closed Jordan curve

$$[z_1, z_2] \cup [z_2, z_3] \cup \cdots \cup [z_{n-1}, z_n] \cup [z_n, z_1].$$

The interior angle $\theta_j$ of the polygon at $z_j$ is the angle determined by $D \cap P$ for all sufficiently small discs $D$ centered at $z_j$. Note that this is not necessarily the interior of the angle determined by the two sides of $P$ leaving $z_j$; it is this interior angle if and only if $0 < \theta_j < \pi$. We allow the vertices to lie on the circle at infinity: if $z_j$ is such an infinite vertex, then $\theta_j = 0$.

**Theorem 7.15.1.** *If $P$ is any polygon with interior angles $\theta_1, \ldots, \theta_n$, then*

$$\text{h-area}(P) = (n - 2)\pi - (\theta_1 + \cdots + \theta_n).$$

PROOF. This has been proved for the case $n = 3$ (Section 7.13) and from this it follows for convex polygons by subdivision of $P$ into $n - 2$ triangles (the details are omitted). It is worth noting explicitly that Theorem 15.1 applies to all polygons whether convex or not.

The proof for non-convex polygons is also by subdivision into triangles: the subdivision is less tractable but we can compensate for this by using Euler's formula. We begin by extending each side of $P$ to a complete geodesic. This provides a subdivision of the entire hyperbolic plane into a finite number of non-overlapping convex polygons (convex as each is the intersection of half-planes).

We now consider only those polygons $P_j$ of the subdivision which lie in the original polygon $P$. By convexity, each $P_j$ can be subdivided into triangles. We have now subdivided $P$ into non-overlapping triangles $T_j$ such that each vertex of $P$ is a vertex of some $T_j$ and each side of a $T_j$ is either a side of some other $T_i$ or is part of a side of $P$ (and not of any other $T_i$).

Let this triangulation of $P$ have $N$ triangles, $E$ edges, $V$ vertices and let there be $E_0$ edges which lie in the sides of $P$. Euler's formula for the sphere yields

$$(N + 1) - E + V = 2.$$

As each of the $N$ triangles has three sides we count sides in different ways and obtain

$$3N = E_0 + 2(E - E_0).$$

Elimination of $E$ now gives

$$N - 2V + E_0 = -2. \tag{7.15.1}$$

We can now compute areas. Of the $V$ vertices in the subdivision, $n$ occur as vertices of $P$, $E_0 - n$ occur at points lying interior to a side of $P$ and $V - E_0$ occur inside $P$. Thus

$$\begin{aligned}
\text{area}(P) &= \sum_{j=1}^{N} \text{area}(T_j) \\
&= N\pi - (\theta_1 + \cdots + \theta_n) - (E_0 - n)\pi - (V - E_0)2\pi \\
&= (n - 2)\pi - (\theta_1 + \cdots + \theta_n)
\end{aligned}$$

by virtue of (7.15.1).                                                                    □

*Remark.* For a Euclidean polygon, of course, we have

$$(n - 2)\pi = \theta_1 + \cdots + \theta_n.$$

## §7.16. Convex Polygons

We establish two results concerning convex polygons. The first is a necessary and sufficient condition for a polygon to be convex: the second establishes the existence of convex polygons with prescribed angles.

**Theorem 7.16.1.** *Let $P$ be a polygon with interior angles $\theta_1, \ldots, \theta_n$. Then $P$ is convex if and only if each $\theta_j$ satisfies $0 \le \theta_j \le \pi$.*

This is an immediate consequence of Theorem 7.5.1. Observe that Theorem 7.15.1 shows that a necessary condition for the existence of a polygon with interior angles $\theta_1, \ldots, \theta_n$ is

$$\theta_1 + \cdots + \theta_n < (n - 2)\pi.$$

In fact, for convex polygons (and possibly for all polygons) this is also sufficient.

**Theorem 7.16.2.** *Let $\theta_1, \ldots, \theta_n$ be any ordered n-tuple with $0 \le \theta_j < \pi$, $j = 1, \ldots, n$. Then there exists a polygon P with interior angles $\theta_1, \ldots, \theta_n$, occurring in this order around $\partial P$, if and only if*

$$\theta_1 + \cdots + \theta_n < (n - 2)\pi. \tag{7.16.1}$$

In fact, we shall construct a polygon $P$ with these angles and with an inscribed disc touching all sides of $P$.

PROOF. Given $\theta_1, \ldots, \theta_n$ satisfying (7.16.1) and each lying in $[0, \pi)$, construct quadrilaterals $Q_1, \ldots, Q_n$ each with one vertex at the origin in $\Delta$ as in Figure 7.16.1. The length $d$ can take any positive value and is to be determined later: note that $Q_j$ is determined (to within a rotation about the origin) by $d$ and $\theta_j$. It is clear that we can construct the desired polygon $P$ as the union of non-overlapping $Q_j$ provided that

$$\sum_{j=1}^{n} \alpha_j = \pi. \tag{7.16.2}$$

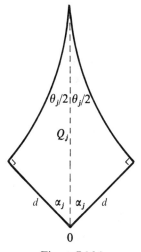

Figure 7.16.1

Now (Theorem 7.11.3)

$$\sin \alpha_j = \frac{\cos(\theta_j/2)}{\cosh d} \qquad (7.16.3)$$

and so it is appropriate to examine the function

$$g(t) = \sum_{j=1}^{n} \sin^{-1}\left(\frac{\cos(\theta_j/2)}{\cosh t}\right),$$

where $t \geq 0$ and where $\sin^{-1}$ takes values in $[0, \pi/2]$.

Clearly, $g$ is continuous and decreasing and $g(t) \to 0$ as $t \to +\infty$. Also,

$$g(0) = \sum_{j=1}^{n} \left(\frac{\pi - \theta_j}{2}\right)$$

$$= \tfrac{1}{2}[n\pi - (\theta_1 + \cdots + \theta_n)]$$

$$> \pi$$

because of (7.16.1). The Intermediate Value Theorem guarantees the existence of a positive $d$ with $g(d) = \pi$ and with $\alpha_j$ defined by (7.16.3), we see that (7.16.2) holds.                                                                                □

As an application of Theorem 7.16.2, observe that *there exists a polygon with n sides and all interior angles equal to $\pi/2$ if and only if $n \geq 5$.*

## §7.17. Quadrilaterals

It is a direct consequence of Theorem 7.16.2 that there exist quadrilaterals with angles $\pi/2$, $\pi/2$, $\pi/2$, $\phi$ if and only if $0 \leq \phi < \pi/2$: such a quadrilateral is illustrated in Figure 7.17.1. This quadrilateral is known as a *Lambert quadrilateral* (after J. H. Lambert, 1728–1777). If we reflect across one side we obtain a quadrilateral with angles $\pi/2$, $\pi/2$, $\phi$, $\phi$: this quadrilateral (illustrated in Figure 7.17.2) was used by G. Saccheri (1667–1733) in his study of the parallel postulate and is known as the *Saccheri quadrilateral*.

The next theorem refers to Figure 7.17.1.

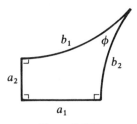

Figure 7.17.1

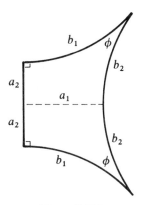

Figure 7.17.2

**Theorem 7.17.1.** (i) $\sinh a_1 \sinh a_2 = \cos \phi$;
(ii) $\cosh a_1 = \cosh b_1 \sin \phi$.

The proof depends on two useful preliminary results.

**Lemma 7.17.2.** *Let $L$ be a hyperbolic geodesic in $\Delta$ with Euclidean centre $\xi$ and radius $r$ and let $w$ be the point on $L$ which is nearest to the origin. Then*

$$\sinh \rho(0, w) = 1/r, \qquad \cosh \rho(0, w) = |\xi|/r.$$

PROOF. Clearly, $|\xi| = |w| + r$ and orthogonality gives $|\xi|^2 = 1 + r^2$. Using (7.2.4) we obtain $\sinh \frac{1}{2}\rho(0, w)$ and hence

$$\sinh \rho(0, w) = \frac{2|w|}{1 - |w|^2} = \frac{1}{r}.$$

The value for cosh follows immediately. ☐

**Lemma 7.17.3.** *Let $L$ and $L'$ be geodesics in the hyperbolic plane. Then the inversive product $(L, L')$ is*

$$\cosh \rho(L, L'), \quad 1, \quad \cos \phi$$

*according as $L$ and $L'$ are disjoint, parallel or intersecting at an angle $\phi$ where $0 \le \phi \le \pi/2$.*

PROOF. It is not difficult to see that disjoint geodesics have a common orthogonal geodesic (see Section 7.22) and (for the moment) $\rho(L, L')$ is defined to be the length of this orthogonal segment between $L$ and $L'$. By the usual invariance arguments we need only consider the cases

(i) $L, L'$ are in $H^2$ and are given by $|z| = r, |z| = R$;
(ii) $L, L'$ are in $H^2$ and are given by $x = 0, x = x_1$;
(iii) $L, L'$ are Euclidean diameters of $\Delta$.

In all these cases, the formula for $(L, L')$ given in Section 3.2 yields the desired result. □

PROOF OF THEOREM 7.17.1. We may suppose that the quadrilateral in Figure 7.17.1 has the sides $a_1$ and $a_2$ lying on the positive real and imaginary axes. Suppose that the sides labelled $b_1$ and $b_2$ lie on the circles

$$|z - iv| = R, \qquad |z - u| = r,$$

respectively, where $u, v, r$ and $R$ are positive. Then by Lemma 7.17.2,

$$\sinh a_1 \sinh a_2 = 1/rR.$$

Lemma 7.17.3 implies that

$$(L, L') = \cos \phi$$

and from Section 3.2 we have

$$(L, L') = \left| \frac{r^2 + R^2 - |u - iv|^2}{2rR} \right|$$

$$= \frac{|r^2 + R^2 - u^2 - v^2|}{2rR}$$

$$= 1/rR$$

because, for example, $u^2 = 1 + r^2$.

To prove (ii) we relocate the polygon so that the vertex with angle $\phi$ is at the origin and the side labelled $b_2$ is on the positive real axis: see Figure 7.17.3.

Now reflect the quadrilateral in the real axis: let $L$ be the geodesic containing the side labelled $a_2$ and let $L'$ be its reflection in the real axis. By Lemma 7.17.3 we have

$$(L, L') = \cosh(2a_1). \tag{7.17.1}$$

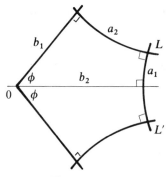

Figure 7.17.3

If $L$ (viewed) as a Euclidean circle) has centre $de^{i\phi}$ and radius $r$, then $L'$ has centre $de^{-i\phi}$ and radius $r$ and clearly,

$$|de^{i\phi} - de^{-i\phi}| > 2r.$$

Thus

$$(L, L') = \left| \frac{|de^{i\phi} - de^{-i\phi}|^2 - 2r^2}{2r^2} \right|$$

$$= \frac{2d^2 \sin^2 \phi - r^2}{r^2}$$

$$= 2 \cosh^2 b_1 \sin^2 \phi - 1$$

by virtue of Lemma 7.17.2. This with (7.17.1) yields (ii).                □

EXERCISE 7.17

1. Derive Lemma 7.17.2 directly from Theorem 7.9.1 (Lemma 7.17.2 is simply a restatement of the Angle of Parallelism formula).

## §7.18. Pentagons

We shall examine the metric relationships which exist for the pentagon illustrated in Figure 7.18.1 where $0 \le \phi < \pi$.

**Theorem 7.18.1.** (i) $\cosh a \cosh c + \cos \phi = \sinh a \cosh b \sinh c$.
(ii) *If* $\phi = \pi/2$ *then*

$$\tanh a \cosh b \tanh c = 1, \tag{7.18.1}$$

$$\sinh a \sinh b = \cosh d. \tag{7.18.2}$$

PROOF. It is easy to see that there is a geodesic through the vertex with angle $\phi$ which meets and is orthogonal to the side of length $b$. Let $b_1$ and $b_2$ be the

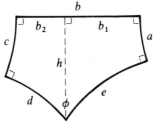

Figure 7.18.1

lengths as illustrated and let $\phi_1$, $\phi_2$ be the subdivision of $\phi$; $\phi_1$ being on the same side of this geodesic as the side of length $b_1$. By Theorem 7.17.1, we have

$$\cosh a = \cosh h \sin \phi_1;$$

$$\cosh c = \cosh h \sin \phi_2;$$

$$\sinh a \sinh b_1 = \cos \phi_1;$$

$$\sinh c \sinh b_2 = \cos \phi_2.$$

It follows that

$$(\cosh a \cosh c - \sin \phi_1 \sin \phi_2)^2$$
$$= (\cosh a \cosh c - \sin \phi_1 \sin \phi_2)^2 - (\cosh a \sin \phi_2 - \cosh c \sin \phi_1)^2$$
$$= (\cosh^2 a - \sin^2 \phi_1)(\cosh^2 c - \sin^2 \phi_2)$$
$$= (\sinh^2 a + \cos^2 \phi_1)(\sinh^2 c + \cos^2 \phi_2)$$
$$= (\sinh^2 a \cosh^2 b_1)(\sinh^2 c \cosh^2 b_2)$$

and so, taking positive square roots,

$$\cosh a \cosh c - \sin \phi_1 \sin \phi_2 = \sinh a \sinh c \cosh b_1 \cosh b_2.$$

This leads directly to (i) as

$$\cosh a \cosh c + \cos \phi = \cosh a \cosh c - \sin \phi_1 \sin \phi_2 + \cos \phi_1 \cos \phi_2$$
$$= \sinh a \sinh c (\cosh b_1 \cosh b_2 + \sinh b_1 \sinh b_2)$$
$$= \sinh a \sinh c \cosh b.$$

Putting $\phi = \pi/2$ in (i), we obtain (7.18.1). To prove (7.18.2), we apply (7.18.1) to the triple $b$, $c$, $d$ and eliminate $c$ from the resulting expression and (7.18.1). □

## §7.19. Hexagons

We shall only consider the right-angled hexagon illustrated in Figure 7.19.1. If we join the end-points of the sides labelled $a_1$ and $b_1$ to form a quadrilateral $Q$, we find that each interior angle of $Q$ is less than $\pi/2$. This implies that the sides labelled $a_1$ and $b_1$ have a common orthogonal of length, say $t$, as illustrated.

**Theorem 7.19.1.**

$$\frac{\sinh a_1}{\sinh b_1} = \frac{\sinh a_2}{\sinh b_2} = \frac{\sinh a_3}{\sinh b_3}.$$

PROOF. From Theorem 7.18.1 we obtain

$$\sinh b_2 \sinh a_3 = \cosh t = \sinh a_2 \sinh b_3$$

and the result follows by symmetry considerations. □

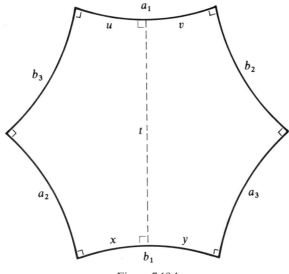

Figure 7.19.1

**Theorem 7.19.2.**

$$\cosh b_1 \sinh a_2 \sinh a_3 = \cosh a_1 + \cosh a_2 \cosh a_3$$

PROOF. From Theorem 7.18.1 we obtain the identities

$$\sinh x \sinh a_2 = \cosh u;$$
$$\sinh y \sinh a_3 = \cosh v;$$
$$\sinh u \sinh t = \cosh a_2;$$
$$\sinh v \sinh t = \cosh a_3.$$

Next, we obtain the identity

$$(\cosh^2 a_2 + \sinh^2 u)(\cosh^2 a_3 + \sinh^2 v)$$
$$= (\cosh a_2 \cosh a_3 + \sinh u \sinh v)^2$$

by expressing both sides as functions of $u$, $v$ and $t$. Thus

$$\cosh b_1 \sinh a_2 \sinh a_3$$
$$= (\cosh x \cosh y + \sinh x \sinh y)\sinh a_2 \sinh a_3$$
$$= \cosh x \cosh y \sinh a_2 \sinh a_3 + \cosh u \cosh v$$
$$= (\cosh x \sinh a_2)(\cosh y \sinh a_3) + \cosh u \cosh v$$
$$= (\sinh^2 a_2 + \cosh^2 u)^{1/2}(\sinh^2 a_3 + \cosh^2 v)^{1/2} + \cosh u \cosh v$$
$$= (\cosh^2 a_2 + \sinh^2 u)^{1/2}(\cosh^2 a_3 + \sinh^2 v)^{1/2} + \cosh u \cosh v$$
$$= \cosh a_2 \cosh a_3 + \sinh u \sinh v + \cosh u \cosh v. \qquad \square$$

*Remark.* Theorem 7.19.2 shows that the lengths of all sides of the hexagon are determined by the lengths $a_1$, $a_2$ and $a_3$.

## §7.20. The Distance of a Point from a Line

For each point $z$ and each geodesic $L$, define

$$\rho(z, L) = \inf\{\rho(z, w) : w \in L\}.$$

There is a unique geodesic $L_1$ through $z$ and orthogonal to $L$ and $\rho(z, L)$ *is the distance from $z$ to $L$ measured along $L_1$.*

We work in $H^2$ and we may assume that $L$ is the positive imaginary axis. Then

$$L_1 = \{\zeta \in H^2 : |\zeta| = |z|\}$$

and we are asserting that

$$\rho(z, L) = \rho(z, i|z|). \tag{7.20.1}$$

Each point on $L$ is of the form $it$ ($t > 0$) and from Theorem 7.2.1,

$$\cosh \rho(z, it) = \frac{x^2 + y^2 + t^2}{2yt} \qquad (z = x + iy)$$

$$= \frac{|z|}{2y}\left(\frac{|z|}{t} + \frac{t}{|z|}\right)$$

$$\geq \frac{|z|}{y}. \tag{7.20.2}$$

As equality holds here if and only if $t = |z|$, this verifies (7.20.1).
    With $\theta$ as in Figure 7.20.1, we can use (7.20.2) and

$$\cosh \rho(z, L) = 1/\cos\theta;$$

$$\sinh \rho(z, L) = \tan\theta; \tag{7.20.3}$$

$$\tanh \rho(z, L) = \sin\theta.$$

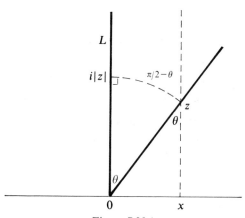

Figure 7.20.1

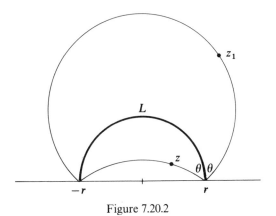

Figure 7.20.2

As an application of these formulae, note that the regions

$$\{z \in H^2 : \rho(z, L) < k\} \qquad (k > 0)$$

are precisely the hypercyclic regions described in Section 7.5.

We can also obtain a formula for $\rho(z, L)$ when $L$ is the Euclidean semi-circle $\{w : |w| = r\}$ in $H^2$: see Figure 7.20.2.

Suppose first that $|z| < r$. With $\theta$ as in Figure 7.20.2, the Euclidean circle through $z$, $r$ and $-r$ has centre $-ir(\tan \theta)$ and radius $r/\cos \theta$. Thus

$$|z + ir(\tan \theta)|^2 = r^2/\cos^2 \theta$$

and so

$$\tan \theta = \frac{r^2 - |z|^2}{2yr} \qquad (z = x + iy).$$

A similar formula holds for $z_1$ when $|z_1| > r$ with $|z_1|^2 - r^2$ replacing $r^2 - |z|^2$. Thus if $L$ is given by $|w| = r$ we obtain from (7.20.3),

$$\sinh \rho(z, L) = \left| \frac{|z|^2 - r^2}{2yr} \right|. \qquad (7.20.4)$$

We shall also need a formula for the model $\Delta$ when $L$ is the real diameter $(-1, 1)$ of $\Delta$. In this case we show that for all $w$ in $\Delta$,

$$\sinh \rho(w, L) = \frac{2|\operatorname{Im}(w)|}{1 - |w|^2}. \qquad (7.20.5)$$

First, there is a unique geodesic $L'$ through $w$ and orthogonal to $L$. Let $L$ and $L'$ meet at $\zeta$: then there is an isometry $g$ of $\Delta$ which fixes $-1$ and $1$, which maps $\zeta$ to $0$ and which leaves $L$ invariant. Now $g$ maps $L'$ to the segment $(-i, i)$ and so $g(w) = it$ for some real $t$. The relationship between $w$ and $t$ is best found by noting that as $g$ preserves both differentials $|dz|/y$ and $2|dz|/(1 - |z|^2)$ we have

$$\frac{t}{1 - t^2} = \frac{\text{Im}(w)}{1 - |w|^2}.$$

On the other hand,

$$\rho(w, L) = \rho(w, \zeta)$$
$$= \rho(it, 0)$$
$$= \log\left(\frac{1 + |t|}{1 - |t|}\right)$$

and this gives (7.20.5).

EXERCISE 7.20

1. Let $L$ be the geodesic $(-e^{i\theta}, e^{i\theta})$ in $\Delta$. Find an explicit formula for $\sinh \rho(z, L)$, $z \in \Delta$.

## §7.21. The Perpendicular Bisector of a Segment

Let $z_1$ and $z_2$ be distinct points and let $w$ be the mid-point of $[z_1, z_2]$. We shall prove that

$$\{z : \rho(z, z_1) = \rho(z, z_2)\}$$

is the unique geodesic through $w$ and orthogonal to $[z_1, z_2]$: this is the *perpendicular bisector* of $[z_1, z_2]$.

We work in $H^2$ and assume that $z_1 = i$ and $z_2 = r^2 i$ where $r > 1$: thus $w = ri$. From Theorem 7.2.1,

$$\cosh \rho(z, z_1) = \cosh \rho(z, z_2)$$

if and only if

$$\frac{|z - z_1|^2}{y} = \frac{|z - z_2|^2}{r^2 y}$$

and this simplifies to $|z| = r$.

In the model $\Delta$, the direct isometries are of the form

$$g(z) = \frac{az + \bar{c}}{cz + \bar{a}}, \qquad |a|^2 - |c|^2 = 1.$$

Using (7.2.4) we find that $z$ is on the perpendicular bisector of $[0, g0]$ if and only if

$$\frac{|z|^2}{(1 - |z|^2)(1 - |0|^2)} = \frac{|z - g0|^2}{(1 - |z|^2)(1 - |g0|^2)},$$

or, equivalently,

$$|\bar{a}z - \bar{c}|^2 = |z|^2.$$

As

$$|\bar{a}z - \bar{c}|^2 - |cz - a|^2 = |z|^2 - 1,$$

we see that *the perpendicular bisector of $[0, g0]$ is the isometric circle of $g^{-1}$*.

EXERCISE 7.21

1. Show that the perpendicular bisector of the two points $z_j = x_j + iy_j$ $(j = 1, 2)$ in $H^2$ is

$$L = \{z : y_1 | z - z_2 |^2 = y_2 | z - z_1 |^2\}.$$

Deduce that for any $z_1$ and any compact subset $K$ of $R^2$, $L \cap K = \varnothing$ when $|z_2|$ is sufficiently large.

# §7.22. The Common Orthogonal of Disjoint Geodesics

*If $L_1$ and $L_2$ are disjoint geodesics then there exists a unique geodesic which is orthogonal to both $L_1$ and $L_2$.*

The assertion remains invariant under isometries so we may assume that $L_1$ and $L_2$ are in $H^2$ with equations

$$x = 0, \qquad (x - a)^2 + y^2 = r^2,$$

respectively, where $a > r > 0$. The only geodesics orthogonal to $L_1$ are those with equations $|z| = t$ and such a geodesic is orthogonal to $L_2$ if and only if $a^2 = r^2 + t^2$. As $a > r$ there is a unique positive $t$ satisfying this equation.

EXERCISE 7.22

1. Prove that if two distinct geodesics have a common orthogonal then they are disjoint.

## §7.23. The Distance Between Disjoint Geodesics

For disjoint geodesics $L_1$ and $L_2$ define

$$\rho(L_1, L_2) = \inf\{\rho(z, w) : z \in L_1, w \in L_2\}.$$

*The distance $\rho(L_1, L_2)$ between $L_1$ and $L_2$ is the distance measured along their common orthogonal.*

We work in $H^2$ and assume that the common orthogonal is the positive imaginary axis. Then $L_1$ and $L_2$ are given by $|z| = r$, $|z| = R$, say and the result follows from (7.20.4) (see also Section 5.4).

There are other convenient expressions for $\rho(L_1, L_2)$, for example Lemma 7.17.2. Also, $\rho(L_1, L_2)$ can be expressed as a cross-ratio: if $L_1$ has end-points $z_1$ and $z_2$ and if $L_2$ has end-points $w_1$ and $w_2$, these occurring in the order $z_1, w_1, w_2, z_2$ around the circle at infinity, then

$$[z_1, w_1, w_2, z_2] \cdot \tanh^2[\tfrac{1}{2}\rho(L_1, L_2)] = 1. \qquad (7.23.1)$$

EXERCISE 7.23

1. Verify (7.23.1) by working in $H^2$ and taking $z_1 = 0$, $w_1 = 1$ and $z_2 = \infty$.

## §7.24. The Angle Between Intersecting Geodesics

The angle $\theta$, $0 < \theta < \pi$, between intersecting geodesics can be expressed both in terms of the inversive product (Lemma 7.17.2) and the cross-ratio. If $L_1 = (z_1, z_2)$ and $L_2 = (w_1, w_2)$ with the end-points occurring in the order $z_1, w_1, z_2, w_2$ around the circle at infinity, then

$$[z_1, w_1, z_2, w_2] \sin^2(\theta/2) = 1.$$

For the proof, use $\Delta$ and $L_1 = (-1, 1)$, $L_2 = (e^{i\theta}, -e^{i\theta})$.

## §7.25. The Bisector of Two Geodesics

Let $L_1$ and $L_2$ be distinct geodesics: the *bisector of $L_1$ and $L_2$* is

$$L = \{z : \rho(z, L_1) = \rho(z, L_2)\}.$$

We show that $L$ is one or two geodesics.

*Case 1: $L_1$ and $L_2$ are parallel.*
In this case, take $L_1$ and $L_2$ to be $x = a$ and $x = -a$ in $H^2$. From (7.20.3) we see that $z$ is on $L$ if and only if $|x - a| = |x + a|$, equivalently, $x = 0$.

*Case 2: $L_1$ and $L_2$ are disjoint.*
Take $L_1$ and $L_2$ to be $|z| = 1$ and $|z| = r^2$ in $H^2$. By (7.20.4), $z$ is on $L$ if and only if

$$(|z|^2 - r^4)^2 = r^4(|z|^2 - 1)^2$$

and this reduces to $|z| = r$.

*Case 3: $L_1$ and $L_2$ are intersecting.*
Take

$$L_1 = (e^{-i\theta}, -e^{-i\theta}), \qquad L_2 = (e^{i\theta}, -e^{i\theta})$$

in $\Delta$ where $0 < \theta < \pi/2$. Let $L' = (-1, 1)$: then $z$ is on $L$ if and only if

$$\rho(e^{i\theta}z, L') = \rho(z, L_1)$$
$$= \rho(z, L_2)$$
$$= \rho(e^{-i\theta}z, L').$$

Using (7.20.5) with $z = re^{it}$ this becomes

$$[\sin(\theta + t)]^2 = [\sin(\theta - t)]^2,$$

which gives $L$ as the union of the two geodesics $(-1, 1)$ and $(-i, i)$.


## §7.26. Transversals


Let $L_1$ and $L_2$ be disjoint geodesics. A geodesic $L$ is a *$\theta$-transversal* $(0 < \theta < \pi/2)$ of $L_1$ and $L_2$ if and only if $L$ meets both $L_1$ and $L_2$ at an angle $\theta$. As an example of how $\theta$-transversals arise naturally consider the isometry $g(z) = kz\,(k > 0)$ of $H^2$ and the geodesic $L$ given by $x = 0$. If $L_1$ is any geodesic meeting $L$, the $L$ is a $\theta$-transversal of $L_1$ and $g(L_1)$. We need to investigate the metric relations which exist for $\theta$-transversals.

The common orthogonal of $L_1$ and $L_2$ is $h$ the unique $\pi/2$-transversal of $L_1$ and $L_2$. We shall see that for all other values of $\theta$ there are exactly four $\theta$-transversals. Let $L_0$ be the common orthogonal of $L_1$ and $L_2$ and let $L^*$ be the bisector of $L_1$ and $L_2$. We shall work in $\Delta$ and we assume that

$$L_0 = (-1, 1), \qquad L^* = (-i, i).$$

The situation is then as illustrated in Figure 7.26.1 where the four transversals are shown, two in each case. We omit the proofs (which are not difficult) that any $\theta$ in $(0, \pi/2)$ can be attained in this way and that there are no other $\theta$-transversals.

With an obvious reference to Euclidean geometry, we call the $\theta$-transversals in Case (i) the *alternate transversals*: those in Case (ii) are the *complementary transversals*. Let $t_\theta$ denote the length of the segment of a $\theta$-transversal which

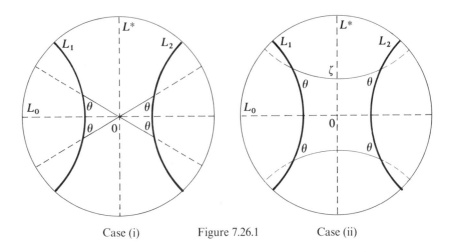

Case (i)            Figure 7.26.1            Case (ii)

lies between $L_1$ and $L_2$. For alternate transversals we have from Theorem 7.11.2,

$$\sinh \tfrac{1}{2}\rho(L_1, L_2) = \sinh(\tfrac{1}{2}t_\theta) \sin \theta:$$

for complementary transversals, Theorem 7.17.1 yields

$$\cosh \tfrac{1}{2}\rho(L_1, L_2) = \cosh(\tfrac{1}{2}t_\theta)\sin \theta.$$

EXERCISE 7.26

1. For a given $\theta$, are the alternate $\theta$-transversals longer or shorter than the complementary $\theta$-transversals?

2. Let the two alternate $\theta$-transversals meet $L_1$ at $z_1$ and $z_2$: let the complementary $\theta$-transversals meet $L_1$ at $w_1$ and $w_2$. Which of $\rho(z_1, z_2)$ and $\rho(w_1, w_2)$ is the greater?

## §7.27. The General Theory of Pencils

Much of the hyperbolic geometry required for a detailed discussion of Fuchsian groups is best described in terms of pencils of geodesics. For example, we see that circles, horocycles and hypercycles are simply variations of the same idea and this brings a greater unity into the subject. We shall also see that the classification of pencils leads naturally to the classification of isometries which is more illuminating than that given in Section 4.3. In this section we merely describe the notion of a pencil and list its main properties: the details occur in the next three sections.

Each pair of geodesics, say $L$ and $L'$, lie in a geometrically defined one-parameter family $\mathcal{P}$ of geodesics called the pencil determined by $L$ and $L'$.

Associated with each pencil $\mathscr{P}$ there is an orthogonal family $\mathscr{C}$ of curves: the curves in $\mathscr{C}$ are not in general geodesics and $\mathscr{C}$ is called the complementary family of $\mathscr{P}$. The interest centres on the following joint properties of $\mathscr{P}$ and $\mathscr{C}$.

P1: *each point in the hyperbolic plane lies on exactly one curve in $\mathscr{C}$.*
P2: *with possibly one exception, each point in the plane lies on exactly one geodesic in $\mathscr{P}$.*
P3: *every geodesic in $\mathscr{P}$ is orthogonal to every curve in $\mathscr{C}$.*
P4: *Every curve in $\mathscr{C}$ is invariant under the reflection in any geodesic in $\mathscr{P}$.*
P5: *any two curves $C_1$ and $C_2$ in $\mathscr{C}$ are equidistant:* that is, for each $z_1$ on $C_1$ there is some $z_2$ on $C_2$ such that

$$\rho(z_1, z_2) = \rho(C_1, C_2),$$

*moreover, $z_1$ and $z_2$ lie on the same geodesic in $\mathscr{P}$.*
P6: *two points $z$ and $w$ lie on the same curve in $\mathscr{C}$ if and only if the perpendicular bisector of $[z, w]$ is in $\mathscr{P}$.*
P7: *the set $\mathscr{P}$ is precisely the set of geodesics of the form*

$$\{z : a \sinh \rho(z, L) = b \sinh \rho(z, L')\}$$

*for some positive constants $a$ and $b$.*

The pencil determined by $L$ and $L'$ is

(i) *parabolic* if $L$ and $L'$ are parallel;
(ii) *elliptic* if $L$ and $L'$ are intersecting;
(iii) *hyperbolic* if $L$ and $L'$ are disjoint.

We shall examine these pencils in detail in the next three sections.

## §7.28. Parabolic Pencils

Let $L$ and $L'$ be parallel geodesics with common end-point $w$. We define $\mathscr{P}$ to be the family of all geodesics with end-point $w$ and $\mathscr{C}$ to be the family of all horocycles tangent to the circle at infinity at $w$ (see Section 7.5). We use the model $H^2$ with $w = \infty$: in this case the geodesics in $\mathscr{P}$ are the lines $x =$ constant, the curves in $\mathscr{C}$ are given by $y =$ constant and P1, P2, P3 and P4 are obvious.

Now consider two horocycles, say $y = k$ and $y = K$. From Theorem 7.2.1 we obtain

$$\cosh \rho(x + ik, s + iK) = 1 + \frac{(x - s)^2 + k^2 + K^2}{2kK}$$

$$\geq \cosh \rho(x + ik, x + iK)$$

and this established P5.

The bisector of $[z_1, z_2]$ ($z_j = x_j + iy_j$) is given by

$$\frac{|z - z_1|^2}{y_1} = \frac{|z - z_2|^2}{y_2}$$

and this geodesic ends at $\infty$ if and only if $y_1 = y_2$: this proves P6.

EXERCISE 7.28

1. Verify P7 by showing that if $L$ and $L'$ are given by $x = a$ and $x = a'$ then

$$\frac{\sinh \rho(u + iv, L)}{\sinh \rho(u + iv, L')} = \left| \frac{v - a}{v - a'} \right|.$$

## §7.29. Elliptic Pencils

Let $L$ and $L'$ be geodesics which intersect at the point $w$ in the hyperbolic plane. We define $\mathscr{P}$ to be the family of all geodesics through $w$ and $\mathscr{C}$ to be the family of all circles

$$C_r = \{z: \rho(z, w) = r\}.$$

If we use the model $\Delta$ with $w = 0$, the geodesics in $\mathscr{P}$ are the Euclidean diameters of $\Delta$ and the circles in $\mathscr{C}$ are the Euclidean circles with centre at the origin. It is now clear that P1, P2, P3 and P4 hold, the exceptional point in P2 being $w$.

To prove that P5 holds, we assume that $z$ is on $C_r$ and that $z'$ is on $C_t$. Using (7.2.4) we see that the minimum of $\rho(z, z')$ is attained precisely when $|z - z'|$ attains its minimum: this establishes P5 for this occurs precisely when $z$ and $z'$ lie on a geodesic in $\mathscr{P}$. The proof of P6 is trivial as

$$\rho(z, w) = \rho(z', w)$$

expresses both the fact that $z$ and $z'$ lie on the same $C_r$ as well as the fact that $w$ lies on the perpendicular bisector of $[z, z']$.

EXERCISE 7.29

1. Verify P7 (but see Section 7.25, Case 3).

## §7.30. Hyperbolic Pencils

Let $L$ and $L'$ be disjoint geodesics with $L_0$ as the common orthogonal geodesic. We define $\mathscr{P}$ to be the family of all geodesics which are orthogonal to $L_0$ and $\mathscr{C}$ to be the family of all hypercycles (defined in Section 7.5) which have the same end-points as $L_0$. As a standard model of this situation, we use the half-plane $H^2$ and take $L_0$ to be the positive imaginary axis. Then $\mathcal{P}$ consists of all geodesics given by $|z|$ = constant and $\mathscr{C}$ consists of all curves given by $\arg(z)$ = constant. It is immediate that P1, P2, P3 and P4 hold.

To verify P5, consider the two curves

$$C_1 = \{z : \arg(z) = \theta\}, \qquad C_2 = \{z : \arg(z) = \phi\}$$

in $\mathscr{C}$. From Theorem 7.2.1 we obtain

$$\sinh^2[\tfrac{1}{2}\rho(te^{i\theta}, re^{i\phi})] = \frac{|te^{i\theta} - re^{i\phi}|^2}{4tr \sin \theta \sin \phi}$$

$$= \frac{1}{4 \sin \theta \sin \phi}\left[\frac{t}{r} + \frac{r}{t} - 2\cos(\theta - \phi)\right]$$

and this is minimal precisely when $t = r$. This proves P5.

Now consider two points $w_j = u_j + iv_j$: the perpendicular bisector of $[w_1, w_2]$ has equation

$$\frac{|z - w_1|^2}{v_1} = \frac{|z - w_2|^2}{v_2},$$

or, equivalently,

$$(v_2 - v_1)|z|^2 - 2x(u_1v_2 - u_2v_1) = \text{constant}.$$

This geodesic is in $\mathcal{P}$ if and only if $u_1v_2 = u_2v_1$, that is, if and only if $w_1$ and $w_2$ lie on the same curve in $\mathscr{C}$: this proves P6.

EXERCISE 7.30

1. Verify P7.

2. Let $\mathcal{P}$ be any pencil (not necessarily hyperbolic). Show that no three distinct points on any curve in $\mathscr{C}$ are collinear.

3. Prove that the three perpendicular bisectors of the sides of a hyperbolic triangle lie in one pencil.

## §7.31. The Classification of Isometries

If we recall the classification of Möbius transformations given in Definition 4.3.2 and take account of Theorem 5.2.1, we see that every conformal isometry of the hyperbolic plane is either parabolic, elliptic or hyperbolic. These can be recognized by the location of their fixed points or by the function trace$^2$. In fact, each such isometry can be expressed as a product of two involutions and the geometric action of the isometry is intimately connected with the theory of pencils. We explore this idea in the next three sections.

## §7.32. Parabolic Isometries

An isometry $g$ is *parabolic* if and only if it can be represented as $g = \sigma_2 \sigma_1$ where $\sigma_j$ is a reflection in the geodesic $L_j$ and where $L_1$ and $L_2$ determine a parabolic pencil (and so are parallel geodesics). This is clear when $g(z) = z + 1$ acting on $H^2$ and so is true in general by invariance.

Given a parabolic isometry $g$, the associated parabolic pencil is the pencil containing all geodesics which end at the fixed point of $g$ and (and this is most important) either $L_1$ or $L_2$ may be chosen arbitrarily from this pencil. Also, $L_2$ is the bisector of $L_1$ and $g(L_1)$.

EXERCISE 7.32

1. Let $g$ be parabolic with fixed point $w$, let $L$ be a geodesic ending at $w$. For any $z$, let $z'$ be the point on $L$ where $[z, z']$ is orthogonal to $L$. Prove that

$$\rho(z, gz) \geq \rho(z', gz').$$

2. Let $g$ be a parabolic isometry acting on $H^2$. Show that there is a conformal isometry $h$ of $H^2$ such that $hgh^{-1}$ is $z \mapsto z + t$ for some real non-zero $t$. Let $T_g$ be the set of $t$ obtainable in this way (for varying $h$ but fixed $g$). Prove that $T_g$ is either $(-\infty, 0)$ or $(0, +\infty)$ and call $g$ negative or positive respectively. Find a necessary and sufficient condition for

$$g(z) = \frac{az + b}{cz + d} \qquad (ad - bc = 1)$$

to be positive in terms of $a$, $b$, $c$ and $d$.

## §7.33. Elliptic Isometries

An isometry $g$ is elliptic if and only if it can be represented as $g = \sigma_2 \sigma_1$ where $\sigma_j$ is the reflection in $L_j$ and $L_1$ and $L_2$ lie in an elliptic pencil. This is true when $g(z) = e^{i\theta} z$ and hence in general by invariance.

Given an elliptic isometry $g$, the associated elliptic pencil is the pencil containing all geodesics passing through the fixed point $v$ of $g$ in the hyperbolic plane. Moreover, $L_1$ (or $L_2$) can be chosen arbitrarily from this pencil and the other $L_j$ is then uniquely determined by $g$.

An elliptic isometry $g$ is completely determined by and completely determines its fixed point $v$ in the hyperbolic plane and a real number

$\theta$ in $[0, 2\pi)$. Indeed, $g$ also fixes $v_1$ (the reflection of $v$ in the circle at infinity) and we can write

$$\frac{g(z) - v}{g(z) - v_1} = e^{i\theta}\left(\frac{z - v}{z - v_1}\right):$$

this shows that $g^{(1)}(v) = e^{i\theta}$. We call $\theta$ the *angle of rotation* of $g$. As $g$ is conjugate (in $\mathcal{M}$) to $z \mapsto e^{i\theta}z$, we have

$$\text{trace}^2(g) = 4\cos^2(\theta/2).$$

### EXERCISE 7.33

1. Show that the elliptic elements $g$ and $h$ with angles of rotation $\theta$ and $\phi$ in $(0, 2\pi)$ are conjugate in the group of conformal hyperbolic isometries if and only if $\theta = \phi$.

## §7.34. Hyperbolic Isometries

An isometry $g$ is hyperbolic if and only if it can be represented as $g = \sigma_2\sigma_1$ where $\sigma_j$ is the reflection in $L_j$ and where $L_1$ and $L_2$ determine a hyperbolic pencil. The *axis* of $g$ (in the hyperbolic plane) is the axis of the pencil, that is the unique geodesic orthogonal to all lines in the pencil and ending at the fixed points of $g$. Of course, the axis of $g$ is the unique $g$-invariant geodesic. We can choose $L_1$ (or $L_2$) arbitrarily and the other $L_j$ is determined by $g$. These facts are easily verified when $g(z) = kz$, $k > 0$, and they are true in general by invariance.

Observe that if $g(z) = kz$, then by Theorem 7.2.1,

$$\sinh \tfrac{1}{2}\rho(z, gz) = \frac{|z||1 - k|}{2y\sqrt{k}}$$

and this attains its minimum (over all $z$ in $H^2$) at and only at each $z$ on the axis of $g$ (the line $x = 0$). As $\inf_z \rho(z, gz)$ remains invariant under conjugation we can define, for a general hyperbolic $g$, the *translation length* $T$ of $g$ by

$$T = \inf_z \rho(z, gz).$$

Observe that $T$ is positive and (again by invariance)

$$\cosh^2(\tfrac{1}{2}T) = 1 + \frac{(1 - k)^2}{4k}$$

$$= \text{trace}^2(g)/4$$

so

$$\tfrac{1}{2}|\text{trace}(g)| = \cosh(\tfrac{1}{2}T).$$

There is another representation of a hyperbolic $g$ as two involutions. An isometry $g$ is hyperbolic if and only if it can be represented as $g = \varepsilon_2 \varepsilon_1$ where $\varepsilon_j$ is a rotation of order two about some point $v_j$ lying on the axis of $g$. Here, $v_1$ (or $v_2$) can be chosen arbitrarily and the other $v_j$ is determined by $g$. The proof is only needed in the special case $g(z) = kz$ and this is straightforward. Observe that

$$T = 2\rho(v_1, v_2)$$

and that $\varepsilon_2(v_1) = g(v_1)$ (so the ray from $v_1$ through $v_2$ ends at the attractive fixed point of $g$).

## §7.35. The Displacement Function

Let $g$ be an isometry of the hyperbolic plane. It is easy to see that the *displacement function*

$$z \mapsto \rho(z, gz) = \rho(z, g^{-1}z)$$

determines and is determined by the pair $\{g, g^{-1}\}$. This is a particularly attractive way of discussing isometries; however, for technical reasons, it is preferable to use the function

$$z \mapsto \sinh \tfrac{1}{2}\rho(z, gz).$$

We shall evaluate this function in purely geometric terms.

**Theorem 7.35.1.** (i) *If $g$ is hyperbolic with axis $A$ and translation length $T$ then*

$$\sinh \tfrac{1}{2}\rho(z, gz) = \cosh \rho(z, A) \sinh(\tfrac{1}{2}T).$$

(ii) *If $g$ is elliptic with fixed point $v$ and angle of rotation $\theta$, then*

$$\sinh \tfrac{1}{2}\rho(z, gz) = \sinh \rho(z, v)|\sin(\theta/2)|,$$

*where here we take $\theta$ in the range $[-\pi, \pi]$.*
(iii) *If $g$ is parabolic with fixed point $v$ then*

$$P(z, v) \sinh \tfrac{1}{2}\rho(z, gz)$$

*is constant (which depends on $g$) where $P(z, v)$ is the Poisson kernel of the hyperbolic plane.*

*Remark.* The Poisson kernel is discussed in Section 1.6.

PROOF. By conjugation, we may assume in (i) that $g$ acts on $H^2$ and that $g(z) = kz, k > 1$. By Theorem 7.2.1 we have

$$\sinh \tfrac{1}{2}\rho(z, gz) = \frac{|z - kz|}{2y\sqrt{k}}$$

$$= \left(\frac{|z|}{y}\right)\frac{1}{2}\left(\sqrt{k} - \frac{1}{\sqrt{k}}\right)$$

$$= \left(\frac{|z|}{y}\right)\sinh(\tfrac{1}{2}T)$$

as $\rho(z, gz) = T$ when $|z| = y$ (i.e. $x = 0$): see Section 7.3.4. Finally, as $A$ is the positive imaginary axis, we can use Section 7.20 and obtain

$$\cosh \rho(z, A) = |z|/y.$$

To prove (ii), we may assume that $g$ acts on $\Delta$ and that $g(z) = e^{i\theta}z$. As

$$\rho(z, e^{i\theta}z) = \rho(z, e^{2\pi i - i\theta}z)$$

and

$$|\sin(\theta/2)| = |\sin(\pi - \theta/2)|,$$

we may assume that $0 < \theta < \pi$ (the cases $\theta = 0$ and $\theta = \pi$ are trivial).

Now construct the triangle with vertices $0, z, gz$ and corresponding angles $\theta, \phi, \phi$, say. Bisecting the angle at the origin yields a right-angled triangle with angles $\theta/2, \phi, \pi/2$ and opposite sides of lengths $\tfrac{1}{2}\rho(z, gz)$, $s$ (irrelevant), $\rho(z, 0)$. From Section 7.11 we obtain

$$\sinh \tfrac{1}{2}\rho(z, gz) = \sinh \rho(z, 0) \sin(\theta/2).$$

To prove (iii), we need only consider the case when $g(z) = z + 1$ acting on $H^2$: the general case follows by the usual invariance argument and the discussion of the Poisson kernel given in Chapter 1. The significance of the Poisson kernel here is that its level curves coincide with the level curves of the displacement function: indeed, this is all that (iii) says.

If $g(z) = z + 1$, then $v = \infty$ and

$$P(z, v) \sinh \tfrac{1}{2}\rho(z, gz) = y\left(\frac{1}{2y}\right) = \tfrac{1}{2}. \qquad \square$$

In conclusion, note that in all cases, *the level curves of the displacement function are precisely the curves in the family $\mathscr{C}$ orthogonal to the pencil $\mathscr{P}$ associated with g.*

EXERCISE 7.35

1. For any isometry $g$ let $m$ be the infimum of $\rho(z, gz)$. Show that $g$ is hyperbolic if and only if $m > 0$. If $m = 0$, show that $g$ is elliptic when $m$ is attained and parabolic when $m$ is not attained.

   Let $w$ be any point such that $\rho(w, gw) > m$. Show that the value of $\rho(w, gw)$ together with the set $\{z: \rho(z, gz) = m\}$ determines the pair $\{g, g^{-1}\}$.

## §7.36. Isometric Circles

Recall from Section 4.1 that for any Möbius transformation $g$, the isometric circle $I_g$ of $g$ is the set of points on which $g$ acts as a Euclidean isometry.

If $g$ is an isometry of the hyperbolic plane $\Delta$, then (see Section 7.21)

$$I_g = \{z: \rho(z, 0) = \rho(z, g^{-1}0)\}$$

and it is instructive to give an alternative proof of this.

PROOF. According to Sections 7.32–7.34 we can write $g = \sigma_2\sigma_1$ where $\sigma_j$ denotes reflection in $L_j$. Choose $L_2$ to pass through the origin so $\sigma_2$ is a Euclidean isometry. We deduce that $z$ is on $I_g$ if and only if the Euclidean distortion of $\sigma_1$ at $z$ is unity: hence $I_g = L_1$. With this available, we see that

$$\sigma_1(0) = \sigma_1\sigma_2(0) = g^{-1}(0)$$

so $I_g (= L_1)$ is the bisector of $0$ and $g^{-1}(0)$.                                            □

It is this geometric proof which reveals the true nature of the isometric circle in plane hyperbolic geometry. Given any point $w$ in the hyperbolic plane or on the circle at infinity, we suppose that $g(w) \neq w$ and we write $g = \sigma_2\sigma_1$ where $L_2$ is chosen to pass through $w$. We call $L_1$ the $w$-isometric circle of $g$ and write it as $I_g(w)$. In this form, there is a useful invariance property, namely,

$$I_{hgh^{-1}}(hw) = I_g(w)$$

and, of course, the isometric circle is the case $w = 0$. Now note that $g$ acts symmetrically about $\partial\Delta$ so we can allow $w$ to be any point in the extended plane and then

$$I_g(w) = I_g(1/\bar{w}).$$

In particular,

$$I_g(0) = I_g(\infty)$$

and this is simply the dependence of the classical isometric circle $I_g$ on the special point $\infty$. For more details, see Section 9.5.

EXERCISE 7.36

1. Prove that $g$ is elliptic, parabolic or hyperbolic according as $I_g$ and $I_{g^{-1}}$ are intersecting, parallel or disjoint respectively.


## §7.37. Canonical Regions

To each conformal isometry $g$ of the hyperbolic plane we shall associate a "canonical" region $\Sigma_g$ which is intimately connected with the geometric action of $g$ and which uniquely determines the pair $\{g, g^{-1}\}$.

**Definition 7.37.1.** Let $g$ be a conformal isometry which is not the identity nor elliptic of order two. The *canonical region* $\Sigma_g$ of $g$ is defined by

$$\Sigma_g = \{z: \sinh \tfrac{1}{2}\rho(z, gz) < \tfrac{1}{2}|\text{trace}(g)|\}.$$

If $g$ is of order two with fixed point $v$, then $\Sigma_g$ is $\{v\}$.

The properties of canonical regions are described in the next theorem.

**Theorem 7.37.2.** (i) $\Sigma_g$ *is conjugation invariant: explicitly*

$$\Sigma_{hgh^{-1}} = h(\Sigma_g).$$

(ii) $\Sigma_g$ *determines the pair* $\{g, g^{-1}\}$: *explicitly,* $\Sigma_g = \Sigma_h$ *if and only if* $h = g$ *or* $h = g^{-1}$.

Before proving this we give a geometric construction of $\Sigma_g$.

**The geometric construction of $\Sigma_g$.** If $g$ is not of order two, then $\Sigma_g$ may be constructed as follows. For each $z$ on the circle at infinity let $L_z$ be the geodesic joining $z$ to $gz$. Then if $P$ denotes the hyperbolic plane, we have

$$\Sigma_g = P - \bigcup_z L_z.$$

Suppose first that $g$ is parabolic: it is only necessary to consider the case when $g$ acts on $H^2$ and is given by $g(z) = z + 1$. In this case,

$$P - \bigcup_z L_z = \{x + iy: y > \tfrac{1}{2}\}.$$

On the other hand,

$$|\text{trace}(g)| = 2$$

so by Theorem 7.2.1, $z$ is in $\Sigma_g$ if and only if

$$1 > \sinh \tfrac{1}{2}\rho(z, gz)$$
$$= 1/2y.$$

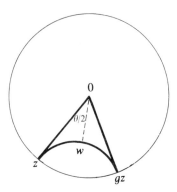

Figure 7.37.1

Next, suppose that $g$ is elliptic. We may suppose that $g$ acts on $\Delta$ and is given by $g(z) = e^{i\theta}z$, $0 < |\theta| < \pi$. In this case the family of lines $L_z$ contains the geodesics that subtend an angle $\theta$ at the origin: see Figure 7.37.1.

From Section 7.9 we obtain

$$\sinh \rho(0, w) \tan(\theta/2) = 1,$$

thus

$$P - \bigcup_z L_z = \{z : \sinh \rho(z, 0) \tan(\theta/2) < 1\}.$$

However, by Theorem 7.35.1,

$$\sinh \tfrac{1}{2}\rho(z, gz) = \sinh \rho(z, 0)|\sin(\theta/2)|$$
$$= \sinh \rho(z, 0)|\tan(\theta/2)|(\tfrac{1}{2}|\operatorname{trace}(g)|)$$

and this is the desired result.

Finally, we suppose that $g$ is hyperbolic: without loss of generality, $g(z) = kz$, $k > 1$, and $g$ acts on $H^2$. In this case, $P - \bigcup_z L_z$ is the hypercyclic region shaded in Figure 7.37.2

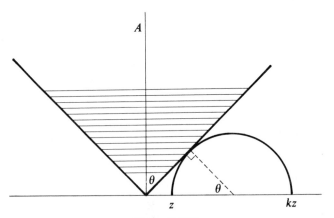

Figure 7.37.2

However, from Section 7.20, $z$ lies in this region if and only if

$$\cosh \rho(z, A) < 1/\cos \theta$$

$$= \frac{(k + 1)/2}{(k - 1)/2}$$

$$= \frac{e^T + 1}{e^T - 1}$$

$$= \frac{\cosh(\frac{1}{2}T)}{\sinh(\frac{1}{2}T)}.$$

Using Theorem 7.35.1, we see that $\Sigma_g$ is indeed this shaded region.

PROOF OF THEOREM 7.37.2. First, (i) is trivially true. Next, observe from the geometric construction of $\Sigma_g$ that $\Sigma_g$ determines the fixed points of $g$ and also the pairs $\{z, gz\}$ on the circle at infinity. It follows that $\Sigma_g$ determines the pair $\{g, g^{-1}\}$.                                                   □

Observe that $\Sigma_g$ can be constructed from the fixed points of $g$ and *one* pair $\{z, gz\}$ on the circle at infinity. Also, *the boundary of $\Sigma_g$ consists of one or two curves from the family $\mathscr{C}$ of curves orthogonal to the pencil $\mathscr{P}$ associated with $g$*.

# §7.38. The Geometry of Products of Isometries

We know that any conformal isometry of the hyperbolic plane can be expressed as a product $f = \sigma_1\sigma_2$ of reflections $\sigma_j$ in geodesics $L_j$. The relative geometric positions of $L_1$ and $L_2$ determine the nature of $f$: for example, if $L_1$ and $L_2$ cross, then $f$ is elliptic. The relative *metric* positions of $L_1$ and $L_2$ determine the geometric parameters of $f$ (for example, the angle of rotation of $f$) in a particularly simple way.

**Theorem 7.38.1.** *Let $L_1$ and $L_2$ be distinct geodesics, let $\sigma_j$ denote reflection in $L_j$ and let $f = \sigma_1\sigma_2$. Then the inversive product $(L_1, L_2)$ satisfies*

$$(L_1, L_2) = \tfrac{1}{2}|\mathrm{trace}(f)|.$$

PROOF. If $L_1$ and $L_2$ are disjoint, then their common orthogonal geodesic $L$ is invariant under $\sigma_1$ and $\sigma_2$. It follows that $f$ is hyperbolic, that $L$ is the axis of $f$ and consequently, the translation length $T$ of $f$ satisfies

$$\tfrac{1}{2}T = \rho(L_1, L_2).$$

We also know that the inversive product $(L_1, L_2)$ is given by

$$(L_1, L_2) = \cosh \rho(L_1, L_2)$$

(Lemma 7.17.2) and the result follows in this case for (see Section 7.34)

$$|\text{trace}(f)| = 2 \cosh(\tfrac{1}{2}T).$$

If $L_1$ amd $L_2$ are parallel, then the inversive product equals one and, as $f$ is then parabolic, $|\text{trace}(f)| = 2$.

Finally, suppose that $L_1$ and $L_2$ intersect in an angle $\theta, 0 < \theta \le \pi/2$, then

$$(L_1, L_2) = \cos \theta.$$

However, in this case, $f$ is a rotation of angle $2\theta$ about the point of intersection of $L_1$ and $L_2$ and

$$|\text{trace}(f)| = 2 \cos \theta. \qquad \Box$$

Given two isometries $g$ and $h$ we can write

$$g = \sigma_1 \sigma_2, \qquad h = \sigma_3 \sigma_4,$$

where $\sigma_j$ represent reflections in the geodesics $L_j$ chosen from certain pencils $\mathscr{P}_1$ and $\mathscr{P}_2$. Suppose now that $\mathscr{P}_1$ and $\mathscr{P}_2$ have a common geodesic $L$: then we can take $L_2 = L = L_3$ so $\sigma_2 = \sigma_3$ and

$$gh = (\sigma_1 \sigma_2)(\sigma_3 \sigma_4) = \sigma_1 \sigma_4.$$

Thus we have obtained a simple representation of the product $gh$ from which we can study the geometric action of $gh$. In particular,

$$|\text{trace}(gh)| = 2(L_1, L_4),$$

*thus the geometry of the relative positions of $L_1$ and $L_4$ enables us to predict the nature of $gh$.* The results in this section are examples of this technique: other results are available and the choice of the material given here has been dictated by later use.

**Theorem 7.38.2.** *Let $g$ and $h$ be elliptic isometries with $g$ a rotation of $2\theta$ about $u$ and $h$ a rotation of $2\phi$ about $v$. We suppose that $g$ and $h$ are rotations in the same sense with $u \ne v$ and $\theta, \phi$ in $(0, \pi)$. Then*

$$\tfrac{1}{2}|\text{trace}(gh)| = \cosh \rho(u, v) \sin \theta \sin \phi - \cos \theta \cos \phi.$$

PROOF. We may assume that $g$ and $h$ act on $H^2$, that $u$ and $v$ lie on the positive imaginary axis $L$ and that

$$g = \sigma_1 \sigma_2, \qquad h = \sigma_3 \sigma_4,$$

where $L_2 = L = L_3$. This is illustrated in Figure 7.38.1 and by Theorem 7.38.1, it is simply a matter of computing $(L_1, L_4)$.

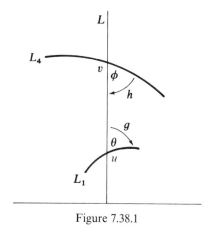

Figure 7.38.1

Now $L_1$ has Euclidean equation

$$x^2 + y^2 - 2x|u| \cot \theta - |u|^2 = 0,$$

while $L_4$ has equation

$$x^2 + y^2 + 2x|v| \cot \phi - |v|^2 = 0.$$

The definition of the inversive product gives

$$(L_1, L_4) = \frac{1}{2}\left(\frac{|v|}{|u|} + \frac{|u|}{|v|}\right) \sin \theta \sin \phi - \cos \theta \cos \phi$$

and this is the required result as

$$\cosh \rho(u, v) = \cosh\left(\log \frac{|v|}{|u|}\right)$$

$$= \frac{1}{2}\left(\frac{|v|}{|u|} + \frac{|u|}{|v|}\right). \qquad \square$$

*Remark.* As an explicit example of Theorem 7.38.2, observe that $gh$ is parabolic if and only if

$$\cosh \rho(u, v) = \frac{1 + \cos \theta \cos \phi}{\sin \theta \sin \phi}.$$

Of course, $gh$ is parabolic if and only if $L_1$ and $L_4$ are parallel and this formula is seen to be in agreement with that given in Section 7.10.

Next, we examine $gh$ when both $g$ and $h$ are hyperbolic.

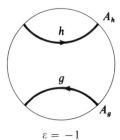 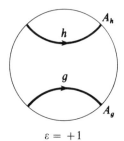

$$\varepsilon = -1 \qquad\qquad\qquad \varepsilon = +1$$

Figure 7.38.2

**Theorem 7.38.3.** *Let $g$ and $h$ be hyperbolic with translation lengths $T_g$, $T_h$ and disjoint axes $A_g$, $A_h$. Then*

$$\tfrac{1}{2}|\mathrm{trace}(gh)| = |\cosh \rho(A_g, A_h) \sinh(\tfrac{1}{2}T_g) \sinh(\tfrac{1}{2}T_h)$$
$$+ \, \varepsilon \cosh(\tfrac{1}{2}T_g) \cosh(\tfrac{1}{2}T_h)|,$$

*where $\varepsilon$ is $+1$ or $-1$ according to the relative directions of $g$ and $h$ as given in Figure 7.38.2.*

**Corollary 7.38.4.** *If $g$ and $h$ are directed so that $\varepsilon = +1$, then $gh$ is hyperbolic.*

PROOF OF THEOREM 7.38.3. We refer to Figure 7.38.3 (which is the case $\varepsilon = -1$) where we have assumed (as we may) that the positive imaginary axis $L_2$ is the common orthogonal of $A_g$ and $A_h$. In this case

$$gh = (\sigma_3\sigma_2)(\sigma_2\sigma_1) = \sigma_3\sigma_1$$

so

$$\tfrac{1}{2}|\mathrm{trace}(gh)| = (L_3, L_1).$$

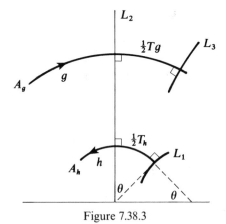

Figure 7.38.3

In order to compute $(L_3, L_1)$ suppose that $A_h$ is given by $|z| = t$. Then $L_1$ has equation

$$\left| z - \frac{t}{\sin \theta} \right| = \frac{t}{\tan \theta},$$

or, equivalently,

$$(x^2 + y^2) \sin \theta - 2xt + t^2 \sin \theta = 0.$$

Now from Section 7.9,

$$\sin \theta = \tanh(\tfrac{1}{2} T_h)$$

so $L_1$ has coefficient vector (to within a scalar multiple)

$$(\tanh(\tfrac{1}{2} T_h), t, 0, -\tfrac{1}{2}t^2 \tanh(\tfrac{1}{2} T_h)).$$

A similar result holds for $A_g$, given by $|z| = s$ say, and

$$\cosh \rho(A_g, A_h) = \frac{1}{2} \left( \frac{s}{t} + \frac{t}{s} \right).$$

The result when $\varepsilon = -1$ now follows by a direct computation of the inversive product. To establish the result when $\varepsilon = +1$, we simply modify Figure 7.38.3 so that $L_1$ and $L_3$ occur on opposite sides of $L_2$.                $\square$

**Corollary 7.38.5.** *Suppose that $g$ and $h$ are hyperbolic with disjoint axes and the same translation length $T$. If $gh$ and $gh^{-1}$ are not elliptic, then*

$$\sinh \tfrac{1}{2}\rho(A_g, A_h) \sinh(\tfrac{1}{2} T) \geq 1.$$

PROOF. With these assumptions we have

$$\tfrac{1}{2}|\operatorname{trace}(gh)| \geq 1$$

and similarly for $gh^{-1}$. By using $h$ or $h^{-1}$ we may assume that $\varepsilon = -1$ in Theorem 7.38.3 and the result follows as

$$\cosh \rho(A_g, A_h) \sinh(\tfrac{1}{2} T_g) \sinh(\tfrac{1}{2} T_h) - \cosh(\tfrac{1}{2} T_g) \cosh(\tfrac{1}{2} T_h)$$
$$= [1 + 2 \sinh^2 \tfrac{1}{2}\rho(A_g, A_h)] \sinh^2(\tfrac{1}{2} T) - [1 + \sinh^2(\tfrac{1}{2} T)]$$
$$= 2 \sinh^2 \tfrac{1}{2}\rho(A_g, A_h) \sinh^2(\tfrac{1}{2} T) - 1.                \square$$

Finally, we consider the case when $A_g$ and $A_h$ cross.

**Theorem 7.38.6.** *Let $g$ and $h$ be hyperbolic and suppose that $A_g$ and $A_h$ intersect at a point $v_2$ in an angle $\theta, 0 < \theta < \pi$, this being the angle between the half-rays from $v_2$ to the attractive fixed points of $g$ and $h$. Then $gh$ is hyperbolic and*

$$\tfrac{1}{2}|\operatorname{trace}(gh)| = \cosh(\tfrac{1}{2} T_g) \cosh(\tfrac{1}{2} T_h) + \sinh(\tfrac{1}{2} T_g) \sinh(\tfrac{1}{2} T_h) \cos \theta.$$

PROOF. This proof uses the alternative expression of a hyperbolic element as a product of two rotations of order two (see Section 7.34).

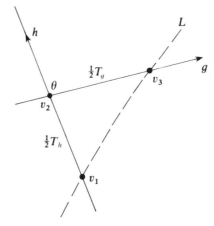

Figure 7.38.4

We refer to Figure 7.38.4: then

$$gh = (\varepsilon_3 \varepsilon_2)(\varepsilon_2 \varepsilon_1) = \varepsilon_3 \varepsilon_1$$

and it is immediate that $gh$ is hyperbolic with axis $L$ and translation length $2\rho(v_1, v_3)$. Thus

$$\tfrac{1}{2}|\text{trace}(gh)| = \cosh \rho(v_1, v_3)$$

and the result follows from the Cosine Rule (Section 7.12). ☐

EXERCISE 7.38

1. Derive Corollary 7.38.5 by constructing the following geodesics in $\Delta$. The common orthogonal $L$ of $A_g$ and $A_h$ may be taken as the real segment $(-1, 1)$: we may also take the origin to be the mid-point of the segment of $L$ between $A_g$ and $A_h$. By selecting $g$ of $g^{-1}$ and $h$ or $h^{-1}$ as appropriate, we may write $g = \sigma_1 \sigma, h = \sigma \sigma_2$ where $\sigma$ is reflection in $L$, $\sigma_j$ is reflection in $L_j$ ($L_j$ lying in the *lower* half of $\Delta$): in addition,

$$\rho(L, L_1) = \rho(L, L_2) = \tfrac{1}{2}T.$$

Now apply the results of Sections 7.18 and 7.19 to the polygon whose sides lie on $L, L_1, L_2, A_g$ and $A_h$.

## §7.39. The Geometry of Commutators

Recall that the commutator $[g, h]$ is $ghg^{-1}h^{-1}$. Our aim here is to discuss the geometry of $[g, h]$ and we shall do this by regarding $[g, h]$ as the product of $g$ and the conjugate $hg^{-1}h^{-1}$ of $g^{-1}$ and then considering, in turn, the various possibilities for $g$. Note that if, say, $g$ is a rotation of angle $\theta$ then $hg^{-1}h^{-1}$ is also a rotation of angle $\theta$ but in the opposite sense. We can restrict our attention to the possibilities for $g$ (rather than $h$) because

$$[h, g] = [g, h]^{-1} \tag{7.39.1}$$

and we need only consider conjugates of $g$ and $h$ because

$$[fgf^{-1}, fhf^{-1}] = f[g, h]f^{-1}.$$

**Theorem 7.39.1.** *Let $g$ be parabolic and suppose that $g$ and $h$ have no common fixed point. Then $[g, h]$ is hyperbolic.*

PROOF. A matrix proof (with $g(z) = z + 1$) is easy enough but the geometry is more revealing. Let $g$ fix the point $v$ and let $L_2$ be the geodesic from $v$ to $h(v)$. For a suitable $L_1$ and $L_3$ we can write

$$g = \sigma_1\sigma_2, \qquad hg^{-1}h^{-1} = \sigma_2\sigma_3, \qquad [g, h] = \sigma_1\sigma_3,$$

where $L_1$ and $L_3$ end at $v$ and $h(v)$ respectively. As $g$ and $hg^{-1}h^{-1}$ act in opposite directions, it is clear that $L_1$ and $L_3$ lie on different sides of $L_2$ and so are disjoint. Thus $\sigma_1\sigma_3$ is hyperbolic with translation length $2\rho(L_1, L_3)$. □

**Theorem 7.39.2.** *Let $g$ be elliptic with fixed point $v$ and angle of rotation $2\theta$, $0 < \theta \leq \pi$. Let $h$ be any isometry not fixing $v$: then $[g, h]$ is hyperbolic with translation length $T$ and*

$$\sinh(T/4) = \sinh \tfrac{1}{2}\rho(v, hv) \sin \theta.$$

PROOF. We write $g = \sigma_1\sigma_2$ where $L_2$ joins $v$ to $h(v)$. Now construct $L_3$ as in Figure 7.39.1 so $hg^{-1}h^{-1} = \sigma_2\sigma_3$ and $[g, h] = \sigma_1\sigma_3$.

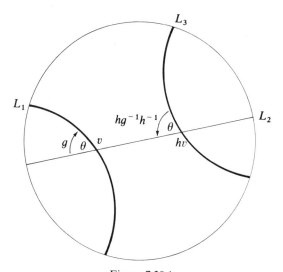

Figure 7.39.1

As $L_1$ and $L_3$ make equal angles with $L_2$ they are disjoint so $[g, h]$ is hyperbolic with

$$T = 2\rho(L_1, L_3).$$

From Section 7.26, we see that

$$\sinh \tfrac{1}{2}\rho(L_1, L_3) = \sinh \tfrac{1}{2}\rho(v, vh) \sin \theta. \qquad \Box$$

Finally, we consider $[g, h]$ when $g$ is hyperbolic. If $h$ is elliptic or parabolic, the previous cases apply by virtue of (7.39.1): thus we may assume that *both* $g$ and $h$ are hyperbolic. Note that $hg^{-1}h^{-1}$ has translation length $T_g$ and axis $h(A_g)$.

**Theorem 7.39.3.** *Let $g$ and $h$ be hyperbolic and suppose that $h(A_g)$ and $A_g$ cross at an angle $\theta$ (between the positive directions of $g$ and $hg^{-1}h^{-1}$). Then $[g, h]$ is hyperbolic with translation length $T$ where*

$$\cosh(\tfrac{1}{2}T) = 1 + 2 \sinh^2(\tfrac{1}{2}T_g) \cos^2(\theta/2).$$

PROOF. Apply Theorem 7.38.6 with $h$ in that theorem replaced by $hg^{-1}h^{-1}$: thus

$$\cosh(\tfrac{1}{2}T) = \cosh^2(\tfrac{1}{2}T_g) + \sinh^2(\tfrac{1}{2}T_g) \cos \theta. \qquad \Box$$

It is possible to consider many other situations with $g$ and $h$ hyperbolic and thereby construct an "animated film" of the behaviour of $[g, h]$ as the three parameters $T_g$, $T_h$ and $(A_g, A_h)$ (the inversive product) vary. It is extremely instructive to do this but the reader will benefit most if he does this for himself: we simply give three "frames" of the film in Figure 7.39.2 in which $[g, h]$ ($= \sigma_3 \sigma_1$) is respectively elliptic, parabolic and hyperbolic.

We end with two results concerning crossing axes.

**Theorem 7.39.4.** *Let $g$ and $h$ be hyperbolic with their axes $A_g$ and $A_h$ crossing at an angle $\theta$, $0 < \theta < \pi$. If $[g, h]$ is not elliptic then*

$$\sinh(\tfrac{1}{2}T_g) \sinh(\tfrac{1}{2}T_h) \sin \theta \geq 1.$$

PROOF. The situation is that described in one of the last two diagrams in Figure 7.39.2. We may apply Theorem 7.38.3 with $h$ in that theorem replaced

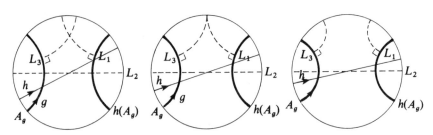

Figure 7.39.2

by $hg^{-1}h^{-1}$ and with $\varepsilon = -1$. Thus

$$1 \le \tfrac{1}{2}|\text{trace}[g, h]|$$
$$= |\cosh \rho(A_g, hA_g) \sinh^2(\tfrac{1}{2}T_g) - \cosh^2(\tfrac{1}{2}T_g)|$$
$$= |\sinh^2(\tfrac{1}{2}T_g)[1 + 2 \sinh^2 \tfrac{1}{2}\rho(A_g, hA_g)] - [1 + \sinh^2(\tfrac{1}{2}T_g)]|$$
$$= |2 \sinh^2(\tfrac{1}{2}T_g) \sinh^2 \tfrac{1}{2}\rho(A_g, hA_g) - 1|$$

so

$$\sinh(\tfrac{1}{2}T_g) \sinh \tfrac{1}{2}\rho(A_g, hA_g) \ge 1.$$

However, from Section 7.26 we obtain

$$\sinh \tfrac{1}{2}\rho(A_g, hA_g) = \sinh(\tfrac{1}{2}T_h) \sin \theta. \qquad \square$$

**Corollary 7.39.5.** *Let $g_1, \ldots, g_n$ be conjugate hyperbolic elements in a group $G$ with no elliptic elements, let $T$ be the common translation length and suppose that the axes $A_j$ of $g_j$ are concurrent. Then*

$$\sinh^2(\tfrac{1}{2}T) \sin(\pi/n) \ge 1.$$

PROOF. Two axes $A_i$ and $A_j$ must cross at an angle $\theta$ where $\theta \le \pi/n$: now apply Theorem 7.39.4. $\qquad \square$

EXERCISE 7.39

1. Derive Theorem 7.39.4 from the last two diagrams in Figure 7.39.2 by using the results of Sections 7.18 and 7.19 on the polygons with sides lying on $A_g, A_h, L_1, L_2$ and $L_3$.

2. Let $L$ be the positive imaginary axis in $H^2$ and let

$$h(z) = \frac{az + b}{cz + d} \qquad (ad - bc = 1)$$

preserve $H^2$. Show that the inversive product $(L, hL)$ and the cross-ratio $[0, \infty, h0, h\infty]$ can be expressed in terms of each other and in terms of the coefficients $a, b, c$ and $d$: for example, show that $(L, hL) = 2ad - 1$. Show also that $L$ and $hL$ cross each other if and only if $ad \in (0, 1)$. These ideas will be found useful when $L$ is the axis of some $g$ so $hL$ is the axis of $hgh^{-1}$.

# §7.40. Notes

For a general introduction to hyperbolic geometry we mention [21], [32], [66], [68] and [112]: for a discussion of hyperbolic isometries, see, for example, [55], [56], [57] and [98]. Convexity is discussed in [102]; convex hyperbolic polygons are considered in [10]. The metric relations for polygons (Sections 7.17, 7.18 and 7.19) are used in [29] for a discussion of plane geometry (and Riemann surfaces) and in the account [101] of recent developments in the theory of 3-manifolds.

# CHAPTER 8

# Fuchsian Groups

## §8.1. Fuchsian Groups

We recall Definition 6.2.2: a *Fuchsian group G* is a discrete subgroup of $\mathcal{M}$ with an invariant disc $D$ (so $G$ acts discontinuously in $D$). We may assume that the unit disc $\Delta$ (or the half-plane $H^2$) is $G$-invariant and so we may regard $G$ as a discrete group of isometries of the hyperbolic plane. We shall see in Chapter 9 that this induces a tesselation, or "tiling," of the plane by hyperbolic polygons and it is the geometry of this action of $G$ which, from now on, is our only concern.

If $G$ is non-elementary then (Theorem 5.3.7) the limit set $\Lambda$ of $G$ lies on the unit circle $\partial \Delta$ (this is also true for elementary Fuchsian groups) and it is important to distinguish between the cases in which $\Lambda$ is or is not the entire circle $\partial \Delta$.

**Definition 8.1.1.** Let $G$ be a Fuchsian group with an invariant disc $D$. We say that $G$ is of the *first kind* if $\Lambda = \partial D$ and of the *second kind* if $\Lambda$ is a proper subset of $\partial D$.

The elementary discrete groups are given in Section 5.1 and it is worthwhile to describe explicitly all elementary Fuchsian groups. Note that these are all of the second kind.

First, consider a Fuchsian group $G$ consisting only of elliptic elements and $I$. By Theorem 4.3.7 the elements of $G$ have a common fixed point $\zeta$ in $H^3$. We may suppose that $H^2$ is $G$-invariant so each elliptic $g$ in $G$ has fixed points, say, $w$ and $\bar{w}$ in $\hat{\mathbb{C}}$ (see Section 5.2). As the axis of $g$ is a geodesic in $H^3$ which contains $\zeta$ and ends at $w$ and $\bar{w}$, we see that $w$ independent of $g$.

Thus all elements of $G$ have the same fixed points and it is now easy to see that $G$ is a finite cyclic group.

An algebraic (but less illuminating) proof can be given. We may suppose that $\Delta$ is $G$-invariant and that

$$
g = \begin{pmatrix} u & 0 \\ 0 & \bar{u} \end{pmatrix}, \qquad h = \begin{pmatrix} a & \bar{c} \\ c & \bar{a} \end{pmatrix}, \qquad |u| = 1,
$$

are elliptic elements in $G$. As

$$
\text{trace}[g, h] = 2 + 4|c|^2 (\text{Im}[u])^2,
$$

we find that $c = 0$ or $\text{Im}[u] = 0$ (else $[g, h]$ is hyperbolic). As $|u| = 1$ and $u^2 \neq 1$, we see that $c = 0$ so $h$ also fixes $0$ and $\infty$.

In order to find all elementary Fuchsian groups we first consider an arbitrary Fuchsian group $G$ which leaves $\Delta$ invariant and which fixes a single point $w$. The fixed points of elliptic elements cannot occur on $\partial \Delta$: the fixed points of parabolic and hyperbolic elements of $G$ must occur on $\partial \Delta$. Moreover, by Theorem 5.1.2, parabolic and hyperbolic elements of $G$ cannot have a common fixed point. We deduce that $G$ can only contain elements of one type and the next result follows easily from the discreteness of $G$.

**Theorem 8.1.2.** *Let $G$ be any Fuchsian group. Then for each $w$, the stabilizer*

$$
G_w = \{g \in G : g(w) = w\}
$$

*is cyclic.*

More generally, it is easy to see that *any elementary Fuchsian group is either cyclic or is conjugate to some group $\langle g, h \rangle$ where $g(z) = kz$ $(k > 1)$ and $h(z) = -1/z$.*

**Definition 8.1.3.** A parabolic or hyperbolic element $g$ of a Fuchsian group $G$ is said to be *primitive* if and only if $g$ generates the stabilizer of each of its fixed points. If $g$ is elliptic, it is *primitive* when it generates the stabilizer and has an angle of rotation of the form $2\pi/n$.

**Remark 8.1.4.** Let $G_0$ be the stabilizer of each of the fixed points of $g$. Then $g$ is primitive if and only if $\Sigma_g \supset \Sigma_h$ for all $h$ in $G_0$ where $\Sigma_g$ denotes the canonical region associated with $g$ (see Section 7.37). In some, but not all, cases this can be described in terms of the trace function.

Finally, we discuss the classification of hyperbolic elements in a Fuchsian group into the *simple* and *non-simple* hyperbolics. This classification depends on the way in which the hyperbolic element lies in the entire group and it is not an "absolute" classification of hyperbolic elements.

**Definition 8.1.5.** Let $h$ be a hyperbolic element of a Fuchsian group $G$ and let $A$ be the axis of $h$. We say that $h$ is a *simple* element of $G$ if and only if for all $g$ in $G$, either $g(A) = A$ or $g(A) \cap A = \emptyset$. Otherwise, we say that $h$ is *non-simple*.

This situation has been described in Section 6.3 and in the terminology introduced there, $h$ is simple if and only if the axis $A$ is $G$-stable.

Let us assume that $G$ acts on $\Delta$ and that $G$ has no elliptic elements. If $h$ is simple, then the projection $\pi(A)$ of $A$ into $\Delta/G$ is the same as $A/\langle g \rangle$ where $g$ generates the cyclic stabilizer of $A$. Thus $\pi(A)$ is a simple closed curve on $\Delta/G$. If $h$ is non-simple there is an image $f(A)$ crossing $A$ at, say, $w$. As $G$ has no elliptic elements, the projection $\pi$ is a homeomorphism near $w$ and so $\pi(A)$ is a closed curve which intersects itself.

EXERCISE 8.1

1. Let $G$ be a Fuchsian group acting on $H^2$ and suppose that $g: z \mapsto kz$ $(k > 1)$ is in $G$. Show that $g$ is simple if and only if for all

$$h(z) = \frac{az + b}{cz + d} \qquad (ad - bc = 1)$$

in $G$, we have $abcd \geq 0$ (equivalently, $|ad - \frac{1}{2}| \geq \frac{1}{2}$).

# §8.2. Purely Hyperbolic Groups

In this section we study those groups which contain only hyperbolic elements and $I$: in Section 8.3 we allow parabolic, but not elliptic, elements. These are an important class of groups from the point of view of Riemann surfaces (see Chapter 6): in particular, they represent compact surfaces of genus at least two.

A group of Möbius transformations is a *purely hyperbolic group* if every non-trivial element of $G$ is hyperbolic. By Theorem 5.2.1, a non-elementary purely hyperbolic group has an invariant disc: in fact, *it is also necessarily discrete* and so is a Fuchsian group. A purely algebraic proof of this will be given (together with a geometric interpretation of the proof) but a stronger quantitative result will be established by geometry alone. It is worth noting that this stronger result (Theorem 8.2.1) contains much information yet requires no further development of the theory for its proof.

**Theorem 8.2.1.** *Let $G$ be a purely hyperbolic group with $\Delta$ as its invariant disc. Then $G$ is either discrete or elementary. Further, if $g$, $h \in G$ and $\langle g, h \rangle$ is non-elementary, then for all $z$ in $\Delta$,*

$$\sinh \tfrac{1}{2}\rho(z, gz) \sinh \tfrac{1}{2}\rho(z, hz) \geq 1. \tag{8.2.1}$$

*The lower bound is best possible.*

We mention three corollaries.

**Corollary 8.2.2.** *If $\langle g, h \rangle$ is non-elementary and purely hyperbolic, then for all $z$,*

$$\max\{\rho(z, gz), \rho(z, hz)\} \geq 2 \sinh^{-1}(1) > 1 \cdot 76$$

*and this is best possible.*

Example 8.2.5 (to follow) shows that this lower bound is best possible. As $G$ preserves $\Delta$, (7.2.4) yields

$$\sinh^2 \tfrac{1}{2}\rho(0, g0) = \frac{|g(0)|^2}{1 - |g(0)|^2}.$$

For $z = 0$, the inequality in Theorem 8.2.1 is

$$|g(0)|^2 \cdot |h(0)|^2 \geq (1 - |g(0)|^2)(1 - |h(0)|^2)$$

and this is equivalent to the next inequality (which is a Euclidean version of Theorem 8.2.1).

**Corollary 8.2.3.** *If $\langle g, h \rangle$ is non-elementary and purely hyperbolic, then*

$$|g(0)|^2 + |h(0)|^2 \geq 1.$$

Another inequality (which relates more directly to the concept of discreteness in $SL(2, \mathbb{C})$) can be obtained by observing that if

$$g = \begin{pmatrix} a & \bar{c} \\ c & \bar{a} \end{pmatrix}, \qquad |a|^2 - |c|^2 = 1,$$

then

$$\|g - I\|^2 \geq 2|c|^2$$
$$= 2 \sinh^2 \tfrac{1}{2}\rho(0, g0)$$

Thus we also have the following result.

**Corollary 8.2.4.** *Let $\langle g, h \rangle$ be a purely hyperbolic non-elementary group preserving $\Delta$. If $A$ and $B$ are matrices in $SL(2, \mathbb{C})$ representing $g$ and $h$, then*

$$\|A - I\| \cdot \|B - I\| \geq 2.$$

Theorem 8.2.1 and its consequences are similar in character to Jørgensen's inequality (Theorem 5.4.1) in that both imply that $g$ and $h$ cannot both be near to $I$. However, the latter inequality, namely

$$|\text{trace}^2(g) - 4| + |\text{trace}[g, h] - 2| \geq 1,$$

gives no information unless $\text{trace}^2(g)$ lies between 3 and 5 whereas Theorem 8.2.1 (involving a product instead of a sum) and the corollaries give useful information in all cases.

Now let $R$ be any Riemann surface of the form $\Delta/G$ where $G$ is non-elementary and purely hyperbolic. From any point on $R$, construct two closed curves $\mathscr{L}_1$ and $\mathscr{L}_2$ of lengths $\ell_1$ and $\ell_2$ respectively. By Theorem 8.2.1 (and Section 6.2),

$$\sinh(\tfrac{1}{2}\ell_1)\sinh(\tfrac{1}{2}\ell_2) \geq 1$$

unless the corresponding group $\langle g, h \rangle$ obtained by lifting $\mathscr{L}_1$ and $\mathscr{L}_2$ to $\Delta$ is elementary (this only arises when $\mathscr{L}_1$ or $\mathscr{L}_2$ is homotopic to its initial point or when $\mathscr{L}_1$ and $\mathscr{L}_2$ are both homotopic to some power of a single closed curve in which case $\langle g, h \rangle$ is cyclic).

The next example shows that the lower bound in Theorem 8.2.1 is best possible.

**Example 8.2.5.** Construct four disjoint geodesics $L_j$ in $\Delta$ as in Figure 8.2.1. Let $g$ be the hyperbolic element which fixes $1, -1$ and which maps $L_1$ to $L_2$: let $h$ be the hyperbolic element which fixes $i, -i$ and which maps $L_3$ to $L_4$ and let $G = \langle g, h \rangle$. Obviously, $G$ is non-elementary.

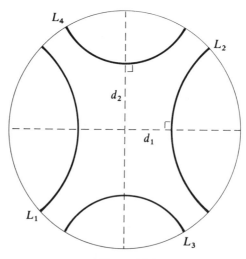

Figure 8.2.1

Using Corollary 5.3.15 (with $G_1 = \langle g \rangle$, $G_2 = \langle h \rangle$ and $D$ the region bounded by the $L_j$), we see that $G$ acts discontinuously in $\Delta$. It will be apparent from later considerations (Chapter 9) that $G$ is purely hyperbolic ($D$ is a fundamental region for $G$ and no elliptic or parabolic fixed points occur on $\tilde{D}$) so the hypotheses of Theorem 8.2.1 are satisfied.

In this example the origin lies on the axis of both $g$ and $h$ so from Theorem 7.18.1, we have

$$\sinh \tfrac{1}{2}\rho(0, g0) \sinh \tfrac{1}{2}\rho(0, h0) = \sinh(\tfrac{1}{2}T_g) \sinh(\tfrac{1}{2}T_h)$$
$$= \sinh(d_1) \sinh(d_2)$$
$$= \cosh \rho(L_2, L_4).$$

As the construction can be achieved with $\rho(L_2, L_4)$ arbitrarily small, the lower bound in Theorem 8.2.1 is best possible.

PROOF OF THEOREM 8.2.1. We begin by showing that if $\langle g, h \rangle$ is non-elementary and purely hyperbolic, then (8.2.1) holds. We are not assuming that $\langle g, h \rangle$ is discrete: indeed *discreteness will be derived from* (8.2.1).

Let $A_g$ and $A_h$ be the axes of $g$ and $h$. As $\langle g, h \rangle$ is non-elementary, these axes either cross or are disjoint. Recalling Definition 8.1.5, we now see that one of the following cases must arise.

*Case* 1: $A_g$ and $A_h$ cross.
*Case* 2: Both $g$ and $h$ are non-simple.
*Case* 3: $A_g$ and $A_h$ are disjoint and (without loss of generality) $g$ is simple.

In Case 2 we can apply Corollary 7.39.5 (with $n = 2$) and obtain (as an image of $A_g$ meets $A_g$)

$$\sinh(\tfrac{1}{2}T_g) \geq 1.$$

A similar inequality holds for $h$ and so

$$\sinh(\tfrac{1}{2}T_g) \sinh(\tfrac{1}{2}T_h) \geq 1.$$

Observe that by Theorem 7.39.4, this also holds in Case 1. Applying Theorem 7.35.1, we find that in Cases 1 and 2,

$$\sinh \tfrac{1}{2}\rho(z, gz) \sinh \tfrac{1}{2}\rho(z, hz)$$
$$= \cosh \rho(z, A_g) \cosh \rho(z, A_h) \sinh(\tfrac{1}{2}T_g) \sinh(\tfrac{1}{2}T_h)$$
$$\geq 1.$$

and this is (8.2.1).

The proof of (8.2.1) in Case 3 is more difficult. As $g$ is simple and $\langle g, h \rangle$ is non-elementary, the geodesics $A_g$, $h(A_g)$ are disjoint. Thus the three geodesics $A_g$, $A_h$, $h(A_g)$ are pairwise disjoint and by applying a suitable isometry, the situation is as illustrated in Figure 8.2.2 (construct $L_0$ first, then $L$ so that $h$ is the reflection in $L_0$ followed by reflection in $L$).

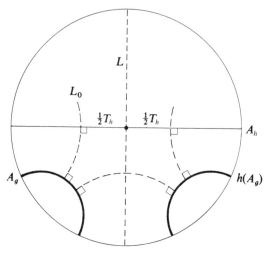

Figure 8.2.2

Applying Theorem 7.19.2 we have

$$\cosh T_h \sinh^2 \rho(A_g, A_h) = \cosh^2 \rho(A_g, A_h) + \cosh \rho(A_g, hA_g).$$

Thus

$$\cosh^2 \rho(A_g, A_h)[\cosh T_h - 1] \geq 2 \sinh^2 \tfrac{1}{2}\rho(A_g, hA_g)$$
$$= 2 \sinh^2 \rho(A_g, L)$$

and this yields

$$\cosh \rho(A_g, A_h) \sinh(\tfrac{1}{2}T_h) \geq \sinh \rho(A_g, L). \qquad (8.2.2)$$

Now construct lines $L_n$ ($n \in \mathbb{Z}$) orthogonal to $A_g$ so that if $\sigma_j$ denotes reflection in $L_j$, then $\sigma_n\sigma_0 = g^n$ (or $g^{-n}$): thus $\rho(L_0, L_n) = nT_g/2$. Now no $L_n$ can meet $L$ as if it does, then

$$\sigma_n\sigma = (\sigma_n\sigma_0)(\sigma_0\sigma) \in G$$

($\sigma$ denotes reflection in $L$ so $\sigma_0\sigma$ is $h$ or $h^{-1}$) and this is elliptic fixing the point of intersection of $L_n$ and $L$. It follows that for some value, say $m$, of $n$, the lines $L_m, L_{m+1}$ as are illustrated in Figure 8.2.3. In order to focus attention on the relevant features, this situation is illustrated again (after applying an isometry) in Figure 8.2.4.

We may assume (without loss of generality) that $d_1 \leq d_2$ so

$$d_1 \leq \tfrac{1}{2}(\tfrac{1}{2}T_g) = \tfrac{1}{4}T_g$$

and applying Theorem 7.18.1 we obtain

$$\sinh(T_g/4) \sinh \rho(A_g, L) \geq \sinh(d_1) \sinh \rho(A_g, L)$$
$$= \cosh \rho(L_{m+1}, L)$$
$$\geq 1.$$

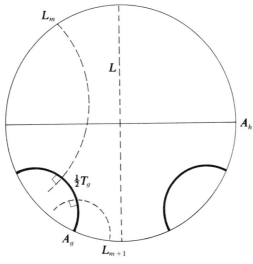

Figure 8.2.3

Using this we obtain

$$\sinh(\tfrac{1}{2}T_g) \sinh \rho(A_g, L) = 2 \sinh(T_g/4) \cosh(T_g/4) \sinh \rho(A_g, L)$$
$$\geq 2$$

and this with (8.2.2) yields

$$\cosh \rho(A_g, A_h) \sinh(\tfrac{1}{2}T_h) \sinh(\tfrac{1}{2}T_g) \geq 2. \qquad (8.2.3)$$

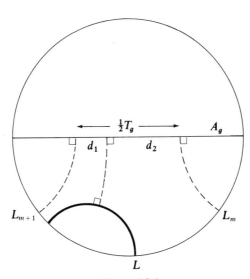

Figure 8.2.4

Now observe that

$$2 \cosh \rho(z, A_g) \cosh \rho(z, A_h) \geq \cosh[\rho(z, A_g) + \rho(z, A_h)]$$
$$\geq \cosh \rho(A_g, A_h)$$

so by (8.2.3),

$$\cosh \rho(z, A_g) \sinh(\tfrac{1}{2}T_g) \cosh \rho(z, A_h) \sinh(\tfrac{1}{2}T_h) \geq 1$$

and, by virtue of Theorem 7.35.1, this is (8.2.1).

To complete the proof of Theorem 8.2.1, we must show that any purely hyperbolic group $G$ is either discrete or elementary. We assume, then, that $G$ is purely hyperbolic but not discrete so there are distinct hyperbolic elements $g_n$ in $G$ with $g_n \to I$. It follows that

$$\rho(0, g_n 0) \to 0$$

and, by discarding some of the $g_n$, we may assume that for all $n$,

$$\sinh \tfrac{1}{2}\rho(0, g_n 0) < 1.$$

From the first part of the proof we see that for all $m$ and $n$, the group $\langle g_m, g_n \rangle$ is elementary. As $G$ has no parabolic elements, $g_n$ and $g_m$ cannot have a single common fixed point (Theorem 4.3.5) so there are distinct points $u$ and $v$ fixed by every $g_n$.

Finally, for every $h$ in $G$ ($h \neq I$),

$$\sinh \tfrac{1}{2}\rho(0, h0) \sinh \tfrac{1}{2}\rho(0, g_n 0) \to 0$$

and so for large $n$, $\langle g_n, h \rangle$ is elementary. We deduce (as above) that $h$ fixes $u$ and $v$ and as $h$ is any element of $G$ we see that $G$ is elementary. $\quad\square$

AN ALGEBRAIC PROOF OF THEOREM 8.2.1. We prove only that $G$ is discrete although a more thorough investigation may also yield (8.2.1).

Assume that $G$ is non-elementary and acts on $H^2$. Thus from Theorem 5.1.3, $G$ contains a hyperbolic element which we may assume is

$$h = \begin{pmatrix} u & 0 \\ 0 & 1/u \end{pmatrix}, \qquad u > 0.$$

Now select any sequence

$$g_n = \begin{pmatrix} a_n & b_n \\ c_n & d_n \end{pmatrix}, \qquad a_n d_n - b_n c_n = 1,$$

in $G$ with $g_n \to I$. In order to prove that $G$ is discrete we must show that $g_n = I$ for all sufficiently large $n$. A computation shows that

$$\text{trace}[h, g_n] = 2 - b_n c_n \left( u - \frac{1}{u} \right)^2$$

$$\to 2$$

as $n \to \infty$, because $b_n c_n \to 0$ (as $g_n \to I$). Because $G$ is purely hyperbolic, the traces of elements in $G$ cannot lie in the interval $(-2, 2)$ so for all sufficiently large $n$, we have $b_n c_n \leq 0$.

Now write

$$f_n = [h, g_n] = \begin{pmatrix} A_n & B_n \\ C_n & D_n \end{pmatrix},$$

with $A_n D_n - B_n C_n = 1$. Exactly the same reasoning (note that $f_n \to I$ because $g_n \to I$) shows that for all sufficiently large $n$,

$$B_n C_n \leq 0.$$

However, a computation shows that

$$\text{trace}[h, f_n] = 2 - B_n C_n \left( u - \frac{1}{u} \right)^2$$

$$= 2 + b_n c_n (1 + b_n c_n) \left( u - \frac{1}{u} \right)^4$$

so for all sufficiently large $n$,

$$b_n c_n \geq 0.$$

We deduce that for all sufficiently large $n$,

$$b_n c_n = 0.$$

This means that for these $n$, the hyperbolic elements $h$ and $g_n$ have a common fixed point. By Theorem 5.1.3, $G$ contains three hyperbolic elements $h_1, h_2$ and $h_3$, no two of which have a common fixed point. It follows that for sufficiently large $n$, each $g_n$ has three fixed points (one in common with each $h_j$) so $g_n = I$.

*The Geometric Interpretation.* The method of proof is simply to extract information from the fact that a commutator is not elliptic. Now the axes $A$ (of $h$) and $g_n(A)$ (of $g_n h g_n^{-1}$) cannot be close and disjoint else $[g_n, h]$ is elliptic (Corollary 7.38.5): this is the condition $b_n c_n \leq 0$. Indeed, $A$ is the positive imaginary axis, $g_n(A)$ is the geodesic with end-points $b_n/d_n$ and $a_n/c_n$ and the inversive product of $A$ and $g_n(A)$ is

$$(A, g_n(A)) = \frac{\frac{1}{2}|(b_n/d_n) + (a_n/c_n)|}{\frac{1}{2}|(b_n/d_n) - (a_n/c_n)|}$$

$$= |1 + 2b_n c_n|.$$

This shows that if $|b_n c_n|$ is small, then $b_n c_n \leq 0$ as otherwise, $A$ and $g_n(A)$ are close and disjoint.

As $b_n c_n \to 0$, we see that for large $n$, the axes $A$ and $g_n(A)$ cross or are parallel and $b_n c_n \leq 0$. If they cross, then they do so at a small angle (as $b_n c_n \to 0$) and Theorem 7.38.6 shows that the commutator

$$f_n = [h, g_n]$$

has a small translation length and an axis which crosses $A$. It follows that the axes $A$ (of $h$) and $f_n(A)$ (of $f_n h f_n^{-1}$) are close and disjoint so the second commutator $[h, f_n]$ is elliptic. As this cannot happen we see that the axes $A$ and $g_n(A)$ have a common end-point and this is $b_n c_n = 0$.

For an alternative interpretation, note that $b_n c_n \to 0$ and $b_n c_n \neq 0$ implies that there is a sequence of axes $g_n(A)$ of elements conjugate to $h$ which converge to (but are distinct from) the axis $A$ of $h$ and this clearly violates discreteness.                                                    $\square$

It is worth noting explicitly that the algebraic proof of Theorem 8.2.1. actually proves that $G$ is discrete providing only that $G$ has no elliptic elements. We state this as our next result: a geometric proof of this is given in the next section.

**Theorem 8.2.6.** *Let $G$ be a non-elementary group of isometries of the hyperbolic plane. If $G$ has no elliptic elements, then $G$ is discrete.*

EXERCISE 8.2

1. Verify the details given in the geometric interpretation of the algebraic proof of Theorem 8.2.1.

2. Show that if $G$ is a group of isometries acting on $H^2$ without elliptic elements and if $g: z \mapsto z + 1$ is in $G$, then for all

$$h(z) = \frac{az + b}{cz + d} \qquad (ad - bc = 1)$$

in $G$, either $c = 0$ or $|c| \geq 4$. [Consider the trace of the matrix representing $g^n h$.]

## §8.3. Groups Without Elliptic Elements

We now obtain a direct extension of Theorem 8.2.1 to allow groups with parabolic (but not elliptic) elements. The conclusion is the same as for Theorem 8.2.1 and the conclusions of Corollaries 8.2.2, 8.2.3 and 8.2.4 remain valid: however, the reader will benefit from reading the proof of Theorem 8.2.1 first. More general results (which allow elliptic elements) are considered in Section 8.4 and Chapter 11.

**Theorem 8.3.1.** *Let $G$ be a group of isometries of the hyperbolic plane and suppose that $G$ has no elliptic elements. Then $G$ is either elementary or discrete. Further, if $g, h \in G$ and $\langle g, h \rangle$ is non-elementary, then for all $z$ in $\Delta$,*

$$\sinh \tfrac{1}{2}\rho(z, gz) \sinh \tfrac{1}{2}\rho(z, hz) \geq 1 \qquad (8.3.1)$$

*and this is best possible.*

PROOF. Example 8.2.5 shows that the lower bound is best possible: indeed, as $G$ may now contain parabolic elements, we can construct the four geodesics in that example with each consecutive pair being tangent and so the lower bound in (8.3.1) can actually be attained.

Now let $G$ be any non-elementary group without elliptic elements. Theorem 8.2.6 shows that $G$ is discrete but we prefer to ignore this and keep to the spirit of the geometric proof of Theorem 8.2.1. If $G$ has no parabolic elements, this result is Theorem 8.2.1, thus we may assume that $G$ has some parabolic elements.

We shall suppose that $G$ acts on $H^2$ and that $\infty$ is fixed by some parabolic element, say $h(z) = z + 1$, in $G$. If $G$ contains a hyperbolic element $f$ fixing $\infty$, we may assume that $f$ also fixes the origin, say $f(z) = kz$, and $G$ then contains translations $z \mapsto z + t$ for arbitrarily small $t$: see Figure 8.3.1. Thus $G$ contains $z \mapsto z + t$ for a set $T$ of $t$ which is dense in $\mathbb{R}$.

As $G$ is non-elementary, it contains a hyperbolic element $g$ which does not fix $\infty$. Thus there are geodesics $L_0$ (ending at $\infty$) and $L$ (the isometric circle of $g$) with $g = \sigma_0\sigma$ ($\sigma$ being the reflection in $L$). As $T$ is dense in $\mathbb{R}$, there is a vertical geodesic $L^*$ (with reflection $\sigma^*$) crossing $L$ and with $\sigma^*\sigma_0$ a Euclidean translation in $G$. Thus $\sigma^*\sigma$ is an elliptic element of $G$, a contradiction. We deduce that *a parabolic fixed point is not fixed by any hyperbolic element of $G$* (compare Theorem 5.1.2 in which discreteness is assumed).

Exactly the same argument shows that *the stabilizer of any parabolic fixed point of $G$ is a discrete (hence cyclic) subgroup of parabolic elements of $G$.*

Now consider any $g$ and $h$ in $G$ with $\langle g, h \rangle$ non-elementary. If $g$ and $h$ are hyperbolic, then they cannot have a single common fixed point (else $[g, h]$ is parabolic and this has been excluded above). In all other cases, the proof of (8.3.1), which is the same as (8.2.1), as given in the proof of Theorem

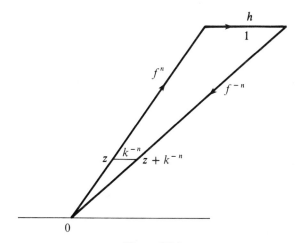

Figure 8.3.1

8.2.1 remains valid (with weak inequalities) and so it only remains to consider the following case:

*Case 4: h is parabolic, g is parabolic or hyperbolic.*

We may suppose that $h(z) = z + 1$ and that $h$ generates the stabilizer of $\infty$ because for all non-zero integers $n$,

$$\sinh \tfrac{1}{2}\rho(z, hz) \le \sinh \tfrac{1}{2}\rho(z, h^n z).$$

Now take

$$g(z) = \frac{az + b}{cz + d}, \qquad ad - bc = 1,$$

with $g(\infty) \ne \infty$ so $c \ne 0$. With $L_0$ as above, let $L_n$ be the vertical geodesic given (in the obvious sense) by $L_0 + n/2$: thus $\sigma_n \sigma_0 = h^n$. The reasoning given above shows that none of the lines $L_n$ can meet the isometric circle of $g$ so necessarily, $2/|c| \le \tfrac{1}{2}$: thus

$$|c| \ge 4.$$

Now suppose that $g$ has fixed points $u$ and $v$ (possibly coincident but not $\infty$). Then, as $u$ and $v$ are real, we have

$$
\begin{aligned}
|z - g(z)| \cdot |cz + d| &= |z(cz + d) - (az + b)| \\
&= |c| \cdot |z - u| \cdot |z - v| \\
&\ge |c| y^2 \\
&\ge 4y^2.
\end{aligned}
$$

Using Theorem 7.2.1, we have

$$
\begin{aligned}
\sinh \tfrac{1}{2}\rho(z, gz) \sinh \tfrac{1}{2}\rho(z, hz) &= \frac{|z - g(z)|}{2(y \operatorname{Im}[gz])^{1/2}} \cdot \frac{1}{2y} \\
&= |z - g(z)| \cdot |cz + d|/4y^2 \\
&\ge 1
\end{aligned}
$$

and this completes the proof in Case 4.

The discreteness of $G$ follows as in the proof of Theorem 8.2.1.  $\square$

## §8.4. Criteria for Discreteness

The following result is the culmination of several earlier results.

**Theorem 8.4.1.** *Let G be a non-elementary group of isometries of the hyperbolic plane: the following statements are equivalent.*

(1) *G is discrete*;
(2) *G acts discontinuously in* $\Delta$;
(3) *the fixed points of elliptic elements of G do not accumulate in* $\Delta$;
(4) *the elliptic elements of G do not accumulate at I*;
(5) *each elliptic element of G has finite order*;
(6) *every cyclic subgroup of G is discrete*.

The structure of the proof is illustrated below: the solid arrows ($A \to B$ means $A$ implies $B$) denote implications which are trivial or already known; the implications given in dotted arrows are proved below.

*Remark.* If $G$ has no elliptic elements then all six conditions are known to be true thus we assume that $G$ has elliptic elements.

PROOF THAT (2) IMPLIES (3). Select any $z$ in $\Delta$ and any compact neighbourhood $N$ of $z$. By (2), $g(N)$ meets $N$ for only a finite set of $g$ in $G$ so only finitely many fixed points lie in $N$. □

PROOF THAT (3) IMPLIES (5). If (5) fails, then $G$ contains an elliptic element $g$ of infinite order. If $g$ fixes $v$ say, then the points $g^n(z)$, $n \in \mathbb{Z}$, are dense on the hyperbolic circle centre $v$ and radius $\rho(z, v)$. As $G$ is non-elementary, there is some $f$ with $f(v) \neq v$ and so the points $g^n f(v)$ are elliptic fixed points which accumulate in $\Delta$. □

PROOF THAT (4) IMPLIES (5). If (5) fails we may assume that $G$ contains $g(z) = \exp(2\pi i\theta)z$ where $\theta$ is irrational. The numbers $\exp(2n\pi i\theta)$, $n \in \mathbb{Z}$, are dense on the unit circle so on a suitable subsequence we have $g^n \to I$. □

PROOF THAT (5) IMPLIES (1). We view $G$ as a group of matrices and let $G_0$ be any finitely generated subgroup of $G$. By a result of Selberg (see Section 2.2), $G_0$ contains a subgroup $G_1$ of finite index which has no elements of finite order.

Because (5) holds, we see that $G_1$ has no elliptic elements and so by Theorem 8.3.1, $G_1$ is discrete. It is easy to see that as $G_1$ is of finite index in $G_0$, the subgroup $G_0$ is also discrete. Finally, by Theorem 5.4.2, $G$ itself is discrete. □

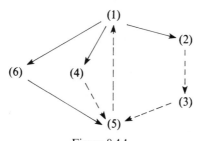

Figure 8.4.1

## §8.5. The Nielsen Region

Let $G$ be a Fuchsian group acting in the hyperbolic plane $\Delta$: we shall be concerned here with non-empty $G$-invariant convex sets.

First, suppose that $G$ *is of the first kind*. In this case, the orbit of every point accumulates at every point of $\partial\Delta$ and so any non-empty $G$-invariant convex set is necessarily the entire hyperbolic plane.

Now suppose that $G$ *is of the second kind*. Then $\partial\Delta$ is the disjoint union of the limit set $\Lambda$ of $G$ and a countable union of mutually disjoint open arcs $\sigma_j$. Let $L_j$ be the geodesic with the same end-points as $\sigma_j$ and let $H_j$ be the open half-plane bounded by $L_j$ and separated from $\sigma_j$ by $L_j$. As the collection $\{\sigma_j\}$ is $G$-invariant, so is the collection $\{H_j\}$ and so

$$N = \bigcap_j H_j \tag{8.5.1}$$

is a $G$-invariant convex subset of $\Delta$. If $G$ is non-elementary, then $\Lambda$ is infinite and so $N$ is non-empty. Also, in this case, there are infinitely many arcs $\sigma_j$ and so as $j \to +\infty$, the Euclidean length of $\sigma_j$ tends to zero. This means that each open disc $\{|z| < r\}, r < 1$, lies in all but a finite number of the $H_j$ and this in turn implies that $N$ is open. To summarize: $N$ is a non-empty $G$-invariant open convex subset of $\Delta$.

**Definition 8.5.1.** Let $G$ be a non-elementary Fuchsian group acting in $\Delta$. Let $N$ be defined by (8.5.1) if $G$ is of the second kind and let $N = \Delta$ if $G$ is of the first kind. Then $N$ is called the *Nielsen region* of $G$.

The next result shows that $N$ may be defined without reference to the circle at infinity.

**Theorem 8.5.2.** *$N$ is the smallest non-empty $G$-invariant open convex subset of $\Delta$.*

PROOF. As $N$ has these properties except possibly of being the smallest such set, we must show that any non-empty $G$-invariant open convex set $E$ contains $N$. As $E$ is non-empty and $G$-invariant, it contains some $G$-orbit which necessarily accumulates at each point of $\Lambda$. It follows that $\tilde{E} \supset \tilde{N}$. Now for any open convex set $A$, we have $(\tilde{A})^0 = A$ and so $E \supset N$.

EXERCISE 8.5

1. Prove carefully that for each $z$, $C(z) \supset N$ where $C(z)$ is the convex hull of the $G$-orbit of $z$.

# §8.6. Notes

For a general account of Fuchsian groups, we refer the reader to [30], [52], [57], [103] and [114]. The geometric ideas explored in this chapter have their origins in the work of Fenchel and Nielsen (see, for example, [29], [99]). The algebraic proof of Theorem 8.2.1 is given in [95]: to the best of my knowledge, Theorem 8.3.1 is new. The ideas in Section 8.4 originate in [42].

# CHAPTER 9

# Fundamental Domains

## §9.1. Fundamental Domains

Let $G$ be a Fuchsian group acting on the hyperbolic plane $\Delta$ (or $H^2$). A *fundamental set* for $G$ is a subset $F$ of $\Delta$ which contains exactly one point from each orbit in $\Delta$. Thus no two distinct points in $F$ are $G$-equivalent and

$$\bigcup_{f \in G} f(F) = \Delta.$$

The Axiom of Choice guarantees the existence (but little else) of a fundamental set for $G$. A fundamental domain is a domain which, with part of its boundary, forms a fundamental set for $G$.

**Definition 9.1.1.** A subset $D$ of the hyperbolic plane is a *fundamental domain* for a Fuchsian group $G$ if and only if

(1) $D$ is a domain;
(2) there is some fundamental set $F$ with $D \subset F \subset \tilde{D}$;
(3) h-area($\partial D$) = 0.

The existence of a fundamental domain will be established in Section 9.4. If $D$ is a fundamental domain, then for all $g$ in $G$ ($g \neq I$)

$$g(D) \cap D = \varnothing, \qquad \bigcup_{f \in G} f(\tilde{D}) = \Delta$$

and, with a slight abuse of terminology, we say that $D$ and its images *tesselate* $\Delta$.

**Remark 9.1.2.** It is not sufficient (as is sometimes suggested) to replace (2) by the requirement that each point of $\partial D$ is the image of some other point of $\partial D$. For example, the group generated by $z \mapsto 2z$ acts discontinuously on $H^2$ but the set $\{x + iy: y > 0, 1 < x < 2\}$ (which has this property) is not a fundamental domain for $G$.

The properties (2) and (3) of Definition 9.1.1 imply that $F$ is measurable and

$$\text{h-area}(D) = \text{h-area}(F).$$

In fact, the next result shows that h-area($D$) depends only on $G$ and not on the choice of $D$. Later (Section 10.4) we shall see that in all cases

$$\text{h-area}(D) \geq \pi/21.$$

**Theorem 9.1.3.** *Let $F_1$ and $F_2$ be measurable fundamental sets for $G$. Then*

$$\text{h-area}(F_1) = \text{h-area}(F_2).$$

*Let $F_0$ be a measurable fundamental set for a subgroup $G_0$ of index $k$ in $G$. Then*

$$\text{h-area}(F_0) = k \cdot \text{h-area}(F_1),$$

PROOF. Denote h-area by $\mu$. As $\mu$ is invariant under each isometry we have

$$\mu(F_1) = \mu\left(F_1 \cap \left[\bigcup_g gF_2\right]\right)$$

$$= \sum_g \mu(F_1 \cap gF_2)$$

$$= \sum_g \mu(F_2 \cap g^{-1}F_1)$$

$$= \mu(F_2).$$

Next, write $G$ as a disjoint union of cosets, say

$$G = \bigcup_n G_0 g_n$$

and let

$$F^* = \bigcup_n g_n(F_1).$$

If $w \in \Delta$, then $g(w) \in F_1$ for some $g$ in $G$ and $g^{-1} = h^{-1}g_n$ for some $n$ and some $h$ in $G_0$. Thus $h(w) \in g_n(F_1)$ and so $F^*$ contains at least one point from each orbit.

Now suppose that $z$ and $f(z)$ are in $F^*$ where $f \in G_0$ and $z$ is not fixed by any non-trivial element of $G$. For some $m$ and $n$, the points $g_n^{-1}(z), g_m^{-1}(fz)$ lie in $F_1$ and so $g_n g_m^{-1}f$ fixes $z$. We deduce that

$$g_m g_n^{-1} = f \in G_0:$$

so $G_0 g_m = G_0 g_n$ and therefore $n = m$. This shows that $f$ fixes $z$ so $f = I$.

These facts show that $F^*$ contains exactly one point from each orbit not containing fixed points and at least one point from each orbit of fixed points. If we now delete a suitable (countable) set of fixed points from $F^*$, the resulting set is a fundamental set for $G_0$ and by the first part,

$$\mu(F^*) = \mu(F_0).$$

Clearly, $F_1$ intersects an image of itself in at most a countable set (of fixed points) so

$$\mu(F^*) = \sum_n \mu(g_n F_1)$$

$$= k\mu(F_1). \qquad \square$$

In terms of quotient spaces, Theorem 9.1.3 is to be expected. As discussed in Section 6.2, the differential $ds$ for the hyperbolic metric projects to a metric on the quotient surface $\Delta/G$ and Theorem 9.1.3 merely states that for any measurable fundamental set $F$, we have

$$\text{h-area}(F) = \text{h-area}(\Delta/G).$$

EXERCISE 9.1

1. Let $D$ be a fundamental domain for $G$. Show that if $w \in D$, then $D - \{w\}$ is also a fundamental domain (so a fundamental domain need not be simply connected). Now let $E = (\tilde{D})^\circ$ (the interior of the closure of $D$ relative to $\Delta$). Show that $E$ is a simply connected fundamental domain which contains $D$.

2. Let $D$ be a fundamental domain for $G$ and suppose that $D_1$ and $D_2$ are open subsets of $D$ with

$$(\tilde{D}_1 \cup \tilde{D}_2)^\circ = D.$$

Under which circumstances is

$$(g(\tilde{D}_1) \cup \tilde{D}_2)^\circ$$

a fundamental domain for $G$?

## §9.2. Locally Finite Fundamental Domains

There is another condition that is required before we can develop any reasonably interesting theory of fundamental domains. We motivate this in the next example: the fact that this is not a Fuchsian group is of no consequence for we have merely selected the simplest example to illustrate the condition.

**Example 9.2.1.** Let $\mathbb{C}^*$ be the set of non-zero complex numbers and let $G$ be the cyclic group generated by $g: z \to 2z$. The quotient space $\mathbb{C}^*/G$ is a

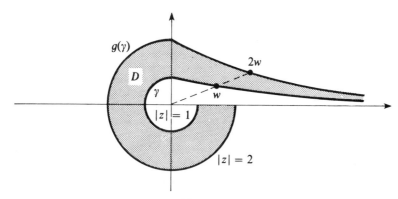

Figure 9.2.1

torus. Now let $\gamma$ be the curve illustrated in Figure 9.2.1: in the first open quadrant, $\gamma$ is the curve $y = e^{-x}$; elsewhere $\gamma$ is given by $|z| = 1$. The region $D$ lying between $\gamma$ and $g(\gamma)$ is a fundamental domain for $G$ in the sense that each point of $\mathbb{C}^*$ is equivalent to at least one point of $\bar{D}$ and at most one point of $D$. Nevertheless, if we identify equivalent points on $\partial D$, we find that the quotient space $\bar{D}/G$ is not compact; thus $\bar{D}/G$ and $\mathbb{C}^*/G$ are *not* homeomorphic.

The same situation can arise for a Fuchsian group, *even when D is a convex polygon with only finitely many sides* (Example 9.2.5) and we wish to impose a condition which prevents this unpleasant possibility.

Let $G$ be a Fuchsian group acting in $\Delta$ and let $D$ be a fundamental domain for $G$ in $\Delta$. The group $G$ induces the natural, continuous, open projection $\pi$: $\Delta \to \Delta/G$. We can also use $G$ to induce an equivalence relation on $\tilde{D}$ by identifying equivalent points (necessarily on $\partial D$) and so with $\tilde{D}/G$ inheriting the quotient topology, there is another continuous projection $\tilde{\pi}: \tilde{D} \to \tilde{D}/G$. The elements of $\Delta/G$ are the orbits $G(z)$: the elements of $\tilde{D}/G$ are the sets $\tilde{D} \cap G(z)$ and

$$\pi(z) = G(z), \qquad \tilde{\pi}(z) = \tilde{D} \cap G(z).$$

Next, let $\tau: \tilde{D} \to \Delta$ denote the inclusion map (the identity restricted to $\tilde{D}$). We now construct a map $\theta: \tilde{D}/G \to \Delta/G$ by the rule

$$\theta: \tilde{D} \cap G(z) \to G(z).$$

The map $\theta$ is properly defined because for each $z$, $\tilde{D} \cap G(z) \neq \varnothing$ and, of course.

$$\theta\tilde{\pi} = \pi\tau: \qquad (9.2.1)$$

these maps are illustrated in Figure 9.2.2.

We study now the relationship between $\tilde{D}/G$ and $\Delta/G$.

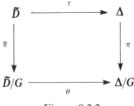

Figure 9.2.2

**Proposition 9.2.2.** (i) $\theta$ and $\tau$ are injective;
(ii) $\pi$, $\tilde{\pi}$ and $\theta$ are surjective;
(iii) $\pi$, $\tilde{\pi}$, $\theta$ and $\tau$ are continuous;
(iv) $\pi$ is an open map.

PROOF. The only assertion which is not completely trivial is that $\theta$ is continuous. If $A$ is any open subset of $\Delta/G$, we may apply (9.2.1) to obtain

$$\tilde{\pi}^{-1}(\theta^{-1}A) = \tilde{D} \cap \pi^{-1}(A)$$

and this is open in $\tilde{D}$ as $\pi$ is continuous. For any $B$, $\tilde{\pi}^{-1}(B)$ is open in $\tilde{D}$ if and only if $B$ is open in $\tilde{D}/G$: thus $\theta^{-1}(A)$ is open in $\tilde{D}/G$ and $\theta$ is continuous (in fact, this is Proposition 1.4.2). ☐

We come now to the property which, if satisfied, guarantees that $\theta$ is a homeomorphism and hence that $\Delta/G$ and $\tilde{D}/G$ are topologically equivalent.

**Definition 9.2.3.** A fundamental domain $D$ for $G$ is said to be *locally finite* if and only if each compact subset of $\Delta$ meets only finitely many $G$-images of $\tilde{D}$.

In order to appreciate the implications of Definition 9.2.3, suppose that $D$ is locally finite. Each $z$ in $\Delta$ has a compact neighbourhood $N$ and this meets only finitely many $G$-images, say $g_i(\tilde{D})$, of $\tilde{D}$. By decreasing $N$ if necessary, we may assume that all these images actually contain $z$. Finally, if $h(D)$ meets $N$, then $h(D)$ meets the union of the $g_i(\tilde{D})$ and so (as $\partial D$ has measure zero) $h = g_i$ for some $i$. To summarize, if $D$ is locally finite, each $z$ has a compact neighbourhood $N$ and an associated finite subset $g_1, \ldots, g_n$ of $G$ with

(1) $z \in g_1(\tilde{D}) \cap \cdots \cap g_n(\tilde{D})$;
(2) $N \subset g_1(\tilde{D}) \cup \cdots \cup g_n(\tilde{D})$;
(3) $h(D) \cap N = \varnothing$ unless $h$ is some $g_j$.

We shall use these facts consistently throughout the following discussion.

**Theorem 9.2.4.** $D$ is locally finite if and only if $\theta$ is a homeomorphism of $\tilde{D}/G$ onto $\Delta/G$.

PROOF. First, we suppose that $\theta$ is a homeomorphism and that $D$ is not locally finite and we seek a contradiction. As $D$ is not locally finite there exists some $w$ in $\Delta$, points $z_1, z_2, \ldots$ in $D$ and distinct $g_1, g_2, \ldots$ in $G$ with

$$g_n(z_n) \to w \quad \text{as } n \to \infty. \tag{9.2.2}$$

Now write

$$K = \{z_1, z_2, \ldots\}.$$

First, $K \subset D$. Next, every neighbourhood of $w$ meets infinitely many of the distinct images $g_n(D)$, thus $w \notin h(D)$ for any $h$ in $G$. We deduce that

$$\pi(w) \notin \pi(K).$$

The contradiction we seek is obtained by proving that

$$\pi(w) \in \pi(K). \tag{9.2.3}$$

The points $g_n^{-1}(w)$ cannot accumulate in $\Delta$ as $G$ is discrete. Because of (9.2.2), the points $z_n$ cannot accumulate in $\Delta$ and this shows that $K$ is closed in $D$. As $K \subset D$, we have

$$\tilde{\pi}^{-1}(\tilde{\pi} K) = K$$

and the definition of the quotient topology on $\tilde{D}/G$ may be invoked to deduce that $\tilde{\pi}(K)$ is closed in $\tilde{D}/G$. By (9.2.1),

$$\pi(K) = \pi\tau(K) = \theta(\tilde{\pi} K)$$

and as $\theta$ is a homeomorphism, this is closed in $\Delta/G$. We conclude that

$$\pi(w) = \lim \pi(g_n z_n) = \lim \pi(z_n) \in \pi(K)$$

and this is (9.2.3).

To complete the proof, we must show that if $D$ is locally finite, then $\theta$ is a homeomorphism. We assume, then, that $D$ is locally finite: by Proposition 9.2.1, we need only prove that $\theta$ maps open sets to open sets.

Accordingly, we select any non-empty open subset $A$ of $\tilde{D}/G$. As $\tilde{\pi}$ is both surjective and continuous, there exists an open subset $B$ of $\Delta$ with

$$\tilde{\pi}^{-1}(A) = \tilde{D} \cap B, \qquad \tilde{\pi}(\tilde{D} \cap B) = A.$$

Now put

$$V = \bigcup_{g \in G} g(\tilde{D} \cap B).$$

Then

$$\begin{aligned}
\pi(V) &= \pi(\tilde{D} \cap B) \\
&= \pi\tau(\tilde{D} \cap B) \\
&= \theta\tilde{\pi}(\tilde{D} \cap B) \\
&= \theta(A).
\end{aligned}$$

We need to prove that $\theta(A)$ is open but as $\pi$ is an open map, it is sufficient to prove that $V$ is an open subset of $\Delta$. This has nothing to do with quotient spaces and depends only on the assumption that $D$ is locally finite.

Consider any $z$ in $V$: we must show that $V$ contains an open set $N$ which contains $z$. As $V$ is $G$-invariant, we may assume that

$$z \in \tilde{D} \cap B.$$

As $D$ is locally finite there exists an open hyperbolic disc $N$ with centre $z$ which meets only the images

$$g_0(\tilde{D}), g_1(\tilde{D}), \ldots, g_m(\tilde{D})$$

of $\tilde{D}$ where $g_0 = I$: also, we may suppose that each of these sets contains $z$. Then

$$g_j^{-1}(z) \in \tilde{D}, \qquad j = 0, \ldots, m,$$

and this means that $\tilde{\pi}$ is defined at $g_j^{-1}(z)$. Clearly $\tilde{\pi}$ maps this point to $\tilde{\pi}(z)$ in $A$ so

$$g_j^{-1}(z) \in \tilde{\pi}^{-1}(A) = \tilde{D} \cap B.$$

It follows that $z \in g_j(B)$ and by decreasing the radius of $N$ still further, we may assume that

$$N \subset g_0(B) \cap \cdots \cap g_m(B).$$

It is now clear that $N \subset V$. Indeed, if $w \in N$, then for some $j$, $w$ is in both $g_j(\tilde{D})$ and $g_j(B)$:

$$w \in g_j(\tilde{D} \cap B) \subset V.$$

The proof is now complete.                                                    $\square$

Next, we give an example to show that convexity is *not* sufficient to ensure local finiteness.

**Example 9.2.5.** We shall exhibit a convex five-sided polygon which is a fundamental domain for a Fuchsian group $G$ but which is *not* locally finite. The group $G$ is the group acting on $H^2$ and generated by

$$f(z) = 2z, \qquad g(z) = \frac{3z + 4}{2z + 3}.$$

Our first task is to show that $G$ is discrete and to identify a fundamental domain for $G$. To do this, consider Figure 9.2.3.

A computation shows that $f(\gamma_1) = \gamma_2$ and $g(\sigma_1) = \sigma_2$ and a straightforward application of Theorem 5.3.15 (with $G_1 = \langle f \rangle$, $D_1$ the region between $\gamma_1$ and $\gamma_2$ and similarly for $g$) shows that $G$ is discrete and $h(D) \cap D = \varnothing$ whenever $h \in G$, $h \neq I$ ($D$ being the region bounded by $\gamma_1$, $\gamma_2$, $\sigma_1$ and $\sigma_2$).

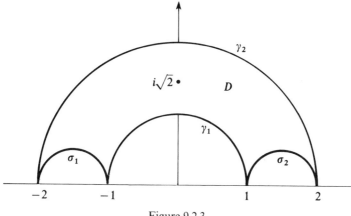

Figure 9.2.3

In fact, $D$ is a (locally finite) fundamental domain for $G$. To see this, take any $z$ in $H^2$ and select an image of $z$ which is closest to $i\sqrt{2}$ (this is possible as $G$ is discrete). By relabelling, we may assume that $z$ itself has this property. It is now easy to see that

$$\rho(z, i\sqrt{2}) \leq \rho(z, f(i\sqrt{2})) = \rho(f^{-1}z, i\sqrt{2})$$

if and only if $|z| \leq 2$. Similarly, $z$ is closer to $i\sqrt{2}$ than to $f^{-1}(i\sqrt{2})$ if and only if $|z| \geq 1$. With a little more computation (Theorem 7.2.1) or geometry we find that $z$ lies outside or on $\sigma_1$ and $\sigma_2$ because

$$\rho(z, i\sqrt{2}) \leq \rho(gz, i\sqrt{2}) = \rho(z, g^{-1}(i\sqrt{2}))$$

and similarly for $g^{-1}$. We deduce that $z \in \tilde{D}$ and this proves that $D$ is a fundamental domain for $G$.

We proceed by modifying $D$ to obtain a new fundamental domain $\Sigma$. The essential feature of this process is to replace parts of $D$ by various images of these parts in such a way that the modified domain is still a fundamental domain. First, we replace

$$D_1 = D \cap \{z: \operatorname{Re}[z] < 0\}$$

by $g(D_1)$: the new domain is illustrated in Figure 9.2.4 and this is still a fundamental domain for $G$.

Next, construct the vertical geodesics $x = 1$ and $x = 2$ and let $w$, $\zeta$ and $\zeta'$ be as in Figure 9.2.5. We now replace the closed triangle $T(w, 1, 2w)$ with vertices $w, 1, 2w$ by the triangle $T(2w, 2, 4w)$ ($= f(T)$). Each *Euclidean* segment $[\zeta, 2\zeta]$, where $\zeta$ lies on $|z| = 1$ and is *strictly* between $w$ and $i$, is replaced by the equivalent segment $[\zeta', 2\zeta']$. Finally, the segment $[i, 2i]$ is *deleted*: note, however, that $[i, 2i]$ is equivalent to the hyperbolic segment $[g(i), g(2i)]$ on the boundary of $g(D_1)$ and, as this segment is retained, the new domain $\Sigma$ still contains in its closure at least one point from every orbit.

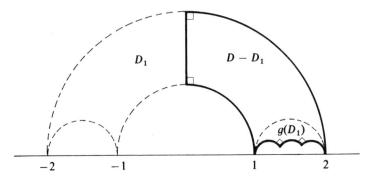

Figure 9.2.4

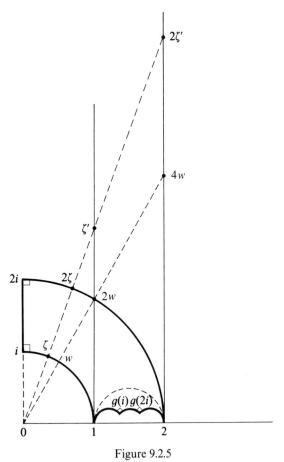

Figure 9.2.5

The construction given above replaces the quadrilateral $D$ by the pentagon $\Sigma$ with vertices $1, g(i), g(2i), 2, \infty$. By construction, $\Sigma$ is a fundamental domain for $G$ and by Theorem 7.16.1, $\Sigma$ is convex. Observe that the points on the boundary segment $[g(i), g(2i)]$ have no equivalent points on $\partial\Sigma$: the only possibility, then, is that the images of $\Sigma$ accumulate (from below) along the segment $[g(i), g(2i)]$. For a more explicit proof that $\Sigma$ is *not* locally finite, we need only observe that the points $z_n = 1 + 2^n i, n = 1, 2, \ldots$, are in $\tilde{\Sigma}$ and

$$f^{-n}(z_n) = i + 2^{-n} \to i \quad \text{as } n \to \infty.$$

We conclude that $\Sigma$ is a convex non-locally finite fundamental domain. As the original domain $D$ is locally finite, the quotient space $H^2/G$ is homeomorphic to $\tilde{D}/G$ and this is a torus with one point removed. The reader should now examine $\tilde{\Sigma}/G$ and also the projection of $\Sigma$ into $H^2/G$. □

In view of Theorem 9.2.4 and Example 9.2.5 it is of interest to record the following criterion for a fundamental domain to be locally finite.

**Theorem 9.2.6.** *Let $D$ be a fundamental domain for a Fuchsian group $G$ and suppose that for each $z$ in $\partial D$ we have*

(1) *there is some $g$ in $G$ with $g \neq I$ and $g(z) \in \partial D$;*
(2) *$z$ can be joined to a point in $D$ by a curve lying entirely in $D \cup \{z\}$.*

*Then $D$ is locally finite.*

PROOF. Neither (1) nor (2) is sufficient to ensure that $D$ is locally finite. We shall restrict ourselves here to a brief sketch of the proof in the most interesting case, namely when $D$ is convex (convexity being stronger than (2)).

It is convenient to say that $z$ in $\Delta$ is *regular* if there is a neighbourhood of $z$ which meets only finitely many copies of $\tilde{D}$: if $z$ is not regular, we say that $z$ is *exceptional*. Now $D$ is locally finite if there are no exceptional points and we shall show that this is so by proving:

(a) the set of exceptional points is countable; and
(b) if there is one exceptional point, then there are uncountably many such points.

By (1), each exceptional $z$ lies in some set $g(\tilde{D}) \cap h(\tilde{D}), g \neq h$. By convexity, the interior points of the intersection

$$\sigma(g, h) = g(\tilde{D}) \cap h(\tilde{D}),$$

(which, by convexity, is a hyperbolic segment) are regular; thus there are at most two exceptional points in $\sigma(g, h)$. As $G$ is countable, (a) follows.

To prove (b), assume that $w$ is exceptional so there exist points $z_1, z_2, \ldots$ in $D$ and distinct $g_1, g_2, \ldots$ in $G$ with $g_n(z_n) \to w$. Now we may assume that $D$ is unbounded in the hyperbolic metric (as clearly if $\tilde{D}$ is compact, then $D$ is locally finite) so there is some $\zeta$ in $\bar{D}$ with $|\zeta| = 1$. Let $L_n$ be the ray $[z_n, \zeta)$.

The $g_n(L_n)$ accumulate at some ray $[w, \zeta^*)$, $|\zeta^*| = 1$, and by construction, every point of $[w, \zeta^*)$ is exceptional.                                              □

It is clear from Theorem 9.2.4 that the concept of local finiteness is important. We end this section with some of the properties shared by all locally finite fundamental domains: we stress that these properties are derived from local finiteness *without any additional assumptions on D*.

**Theorem 9.2.7.** *Let D be any locally finite fundamental domain for a Fuchsian group G. Then*

$$G_0 = \{g \in G : g(\tilde{D}) \cap \tilde{D} \neq \varnothing\}$$

*generates G.*

PROOF. Let $G^*$ be the group generated by $G_0$. We may suppose that $G$ acts in $\Delta$ so for any $z$ in $\Delta$ there is some $g$ in $G$ with $g(z) \in \tilde{D}$. Suppose also that $h(z) \in \tilde{D}$. Then $h(z)$ is in both $\tilde{D}$ and $hg^{-1}(\tilde{D})$ so $hg^{-1} \in G_0$: thus we have equality of cosets, namely

$$G^*h = G^*g.$$

This fact means that there is a properly defined map

$$\phi : \Delta \to G^* \backslash G$$

given by

$$\phi(z) = G^*g,$$

where $g(z) \in \tilde{D}$: our proof is based on a discussion of this map.

Consider any $z$ in $\Delta$. As $D$ is locally finite, there exist a finite number of images

$$g_1(\tilde{D}), \ldots, g_m(\tilde{D})$$

each containing $z$ and such that their union covers an open neighbourhood $N$ of $z$. If $w \in N$, then $w \in g_j(\tilde{D})$ for some $j$ and

$$\phi(w) = G^*(g_j)^{-1} = \phi(z).$$

We deduce that each $z$ has an open neighbourhood $N$ on which $\phi$ is constant.

Now any function $\phi$ with this property is constant on $\Delta$ (give $\phi(\Delta)$ the discrete topology: $\phi$ is continuous and $\phi(\Delta)$ is connected, thus $\phi(\Delta)$ contains only one point). This shows that

$$\phi(z) = \phi(w)$$

for all $z$ and $w$ in $\Delta$. Given any $g$ in $G$ we select $z$ in $D$ and $w$ in $g^{-1}(D)$. Then as $\phi$ is constant,

$$G^* = \phi(z) = \phi(w) = G^*g$$

and so $g \in G^*$. This proves that $G \subset G^*$. Clearly $G^* \subset G$ so $G^* = G$ and $G_0$ generates $G$. $\qquad\qquad\square$

The next result relates local finiteness to invariant regions: for the definitions of horocyclic and hypercyclic regions, see Section 7.5.

**Theorem 9.2.8.** *Let D by any locally finite fundamental domain for a Fuchsian group G.*

(i) *Let g be an elliptic element in G and let K be a compact disc with $g(K) = K$. Then $\tilde{D}$ meets a positive but only finite number of distinct images of K.*
(ii) *Let g be a parabolic element in G and let K be a horocyclic region with $g(K) = K$. Then $\tilde{D}$ meets a positive but only finite number of distinct images of K.*
(iii) *Let g be a hyperbolic element in G and let K be a hypercyclic region with $g(K) = K$. Then $\tilde{D}$ meets a positive but only finite number of distinct images of K.*

PROOF. In all cases, choose $w$ in $K$. For some $h$ in $G$, $h(w) \in \tilde{D}$ so $\tilde{D}$ meets some image of $K$. Now (i) is trivial for $K$ is compact since if $\tilde{D}$ meets $h(K)$, then $h^{-1}(\tilde{D})$ meets $K$ and this can only happen for a finite set of $h$.

To prove (ii) it is convenient to suppose that $G$ acts on $H^2$ and that $g(z) = z + 1$. It follows that $K$ must be of the form

$$K = \{x + iy : y > k\}.$$

Now write

$$K_0 = \{x + iy : y \geq k_0\}$$

and

$$K_1 = \{x + iy : k \leq y \leq k_0\},$$

where $k_0$ is chosen so that

$$\bigcup_{f \in G} f(K_0) \neq H^2.$$

This last condition implies that $K_0$ cannot contain an image of $\tilde{D}$ so that if $f(\tilde{D})$ meets $K$ then necessarily, it also meets $K_1$. Observe that this choice of $K_0$ is made possible by Jørgensen's inequality, namely if

$$f(z) = \frac{az + b}{cz + d}$$

is in $G$ and does not fix $\infty$, then $|c| \geq 1$. Thus

$$\mathrm{Im}[fz] \leq 1/y$$

so with $k_0 > 1$ we find that $K_0$ does not meet the orbit $G(i)$.

Now suppose that $\tilde{D}$ meets $h(K)$: then $h^{-1}(\tilde{D})$ meets $K$ and hence it meets $K_1$. If

$$E = \{x + iy : 0 \le x \le 1; k \le y \le k_0\},$$

then

$$\bigcup_n g^n(E) = K_1$$

so for some $n$, $g^n h^{-1}(\tilde{D})$ meets $E$. Now as $E$ is compact and $D$ is locally finite, only a finite number of images of $\tilde{D}$, say

$$g_1(\tilde{D}), \ldots, g_s(\tilde{D})$$

meet $E$. Thus $g^n h^{-1} = g_j$ for some $j$ and $n$ and so $h(K) = g_j^{-1}(K)$.

The proof of (iii) is similar. We may assume that $G$ acts on $H^2$ and that $g(z) = kz$, $k > 1$. The hypercyclic region is necessarily of the form

$$K = \{re^{i\theta} : r > 0, |\theta - \pi/2| < \varepsilon\}:$$

we write

$$E = \{z \in K; 1 \le |z| \le k\}$$

so $\bigcup g^n(E) = K$. Only finitely many images of $\tilde{D}$ meet the compact set $E$: let these be $g_1(\tilde{D}), \ldots, g_s(\tilde{D})$. Suppose now that $h(K)$ meets $\tilde{D}$: then for some $n$, $g^n h^{-1}(\tilde{D})$ meets $E$ and so for some $j$, $h(K) = g_j^{-1}(K)$.  $\square$

We mention just one consequence of Theorem 9.2.8.

**Corollary 9.2.9.** *Let $G$ be a Fuchsian group, $D$ any locally finite fundamental domain for $G$ and let $\zeta$ be fixed by some parabolic element of $G$. Then for some $g$ in $G$, $g(\zeta)$ lies in the Euclidean closure of $D$.*

PROOF. We may suppose that $G$ acts on $H^2$, that $\zeta = \infty$ and that the stabilizer of $\zeta$ is generated by $p : z \mapsto z + 1$.

Now let $K$ be a horocyclic region invariant under $p$. Choose any sequence of points $z_1, z_2, \ldots$ in $K$ with $\text{Im}[z_n] \to +\infty$. There are elements $h_1, h_2, \ldots$ in $G$ with $h_n(z_n) \in \tilde{D}$ so $\tilde{D}$ meets each image $h_n(K)$. An application of Theorem 9.2.8 (after taking a subsequence and relabelling) shows that

$$h_1(K) = h_2(K) = \cdots.$$

It follows that there are integers $t_2, t_3, \ldots$ such that $h_n = h_1 p^{t_n}$: hence $h_1(w_n) \in \tilde{D}$ where $w_n = p^{t_n}(z_n)$. As

$$\text{Im}[w_n] = \text{Im}[z_n] \to +\infty,$$

we see that $w_n \to \infty$ and so $h_1(\infty)$ lies in the Euclidean closure of $D$.  $\square$

EXERCISE 9.2

1. Modify Definition 9.2.3 to apply to Example 9.2.1 and show that $D$ is not locally finite.

2. Construct a Fuchsian group $G(= \{g_1, g_2, \ldots\})$ acting on $H^2$ and a locally finite fundamental domain $D$ for $G$ with

$$\text{Euclidean diameter } g_n(D) = +\infty$$

   for every $n$.

3. Let $G$ be a Fuchsian group acting on $\Delta$ with a fundamental domain $D$. Show that $D$ is bounded in the hyperbolic metric if and only if (i) $\Delta/G$ is compact and (ii) $D$ is locally finite.

4. Let $G$ be generated by $g: z \mapsto z + 1$ and $h: z \mapsto z + i$. Despite the fact that $\mathbb{C}/G$ is compact, construct a fundamental domain $D$ for $G$ which is not locally finite in $\mathbb{C}$.

5. Show that the convex fundamental domain $\Sigma$ in Example 9.2.2 contains a hyperbolic fixed point on its (Euclidean) boundary. By contrast, show that a fixed point of a hyperbolic $g$ in $G$ cannot be on the Euclidean boundary of any convex locally finite fundamental domain.

# §9.3. Convex Fundamental Polygons

It is natural to pay special attention to fundamental domains that are polygons. Non-convex polygons are rarely used but on the other hand, convexity is not enough to guarantee satisfactory results (Example 9.2.5). With these preliminary remarks, we embark on a discussion of convex, locally finite fundamental polygons. It is a striking fact that the polygonal nature actually follows from the convexity and local finiteness: accordingly we begin with a rather stark definition which does not explicitly mention the polygonal structure.

**Definition 9.3.1.** Let $G$ be a Fuchsian group. Then $P$ is a *convex fundamental polygon* for $G$ if and only if $P$ is a convex, locally finite fundamental domain for $G$.

We emphasize that this is a definition of the phrase "convex fundamental polygon" and it does not presuppose any particular structure of the boundary of $P$. We now add a little flesh to this skeletal definition: the discussion is elementary but, as might be guessed, it is important to derive results in the optimal order. Throughout, $P$ is taken to be a convex fundamental polygon for $G$. It is perhaps worth mentioning now that $P$ is a hyperbolic polygon in a more general sense than is usually allowed. For example, $P$ may have vertices on the circle at infinity (possibly infinitely many) and the boundary of $P$ can even contain arcs of the circle at infinity. Explicitly, it will be shown that

$$P = \bigcap H_i,$$

where the $H_i$ are a countable number of open half-planes with the property that any compact subset of the hyperbolic plane is contained in all but a finite number of the $H_i$.

As $P$ is locally finite, for any $z$ in $\Delta$ there is an open hyperbolic disc $N$ centre $z$ and distinct elements $g_1, \ldots, g_t$ in $G$ such that

$$z \in g_1(\tilde{P}) \cap \cdots \cap g_t(\tilde{P}),$$
$$N \subset g_1(\tilde{P}) \cup \cdots \cup g_t(\tilde{P}),$$

and, if $g(\tilde{P})$ meets $N$, then necessarily $g = g_j$ for some $j$. If $z \in \partial P$, then $g_1 = I$, say and $t \geq 2$ (else $z \in N$, $N \subset P$). This proves

(1) *for each $z$ in $\partial P$, there is some $g$ in $G$ with $g \neq I$, $g(z) \in \partial P$.*

In fact, with convexity, (1) is equivalent to local finiteness: see Theorem 9.2.6.

Now consider any $g$ ($\neq I$) in $G$. Clearly, $\tilde{P} \cap g(\tilde{P})$ is convex. Moreover, $\tilde{P} \cap g(\tilde{P})$ cannot contain three non-collinear points else it contains a non-degenerate triangle and then (because $\partial P$ has zero area) we find that $P \cap g(P) \neq \emptyset$. We deduce that $\tilde{P} \cap g(\tilde{P})$ is a geodesic segment, possibly empty. We can now define the sides and vertices of $P$.

**Definition 9.3.2.** A *side* of $P$ is a geodesic segment of the form $\tilde{P} \cap g(\tilde{P})$ of positive length. A *vertex* of $P$ is a single point of the form $\tilde{P} \cap g(\tilde{P}) \cap h(\tilde{P})$ for distinct $I$, $g$ and $h$.

**Warning.** A side of $P$ is not necessarily a side in the usual conventional sense. If we call a *maximal* geodesic segment in $\partial P$ an *edge* of $P$, then an edge may contain infinitely many sides of $P$. From a different point of view, we allow the interior angles of $P$ at the vertices to assume the value $\pi$.

Now $G$ is countable and only finitely many images of $\tilde{P}$ can meet any compact subset of $\Delta$. Thus

(2) *$P$ has only countably many sides and vertices:*
(3) *only finitely many sides and vertices can meet any given compact subset of $\Delta$.*

Clearly the sides and vertices of $P$ lie in $\partial P$. In fact,

(4) *$\partial P$ is the union of the sides of $P$.*

Observe that with this definition of sides, this apparently obvious statement is false for domains which are not locally finite: see Example 9.2.5.

To prove (4), consider any $w$ on $\partial P$. Each sufficiently small circle centre $w$ must contain points in $P$ and points (other than $w$) not in $P$ so there are points $w_n$ in $\partial P$ tending to $w$. A compact neighbourhood of $w$ meets only finitely many images of $\tilde{P}$ so there is some $g$ and infinitely many $n$ with $w_n \in \tilde{P} \cap g(\tilde{P})$. This implies that $\tilde{P} \cap g(\tilde{P})$ is a side containing $w$.

It follows from (4) that every vertex of $P$ actually lies on a side of $P$. Much more is true, namely

(5) *each vertex lies on exactly two sides and it is the common end-point of each.*

To verify (5), let $w$ be an interior point of the side $\tilde{P} \cap g(\tilde{P}) = s$. Choose a point $z$ in $P$ and form the triangle with vertex $z$ and opposite side $s$: the open triangle lies in $P$. A similar construction yields an open triangle in $g(P)$ with side $s$: this shows that a vertex cannot be the interior point of a side and two sides meet, if at all, in a vertex.

Now by (3), (4) and the preceding remarks, every vertex $v$ lies on a finite, positive number of sides and it is the common end-point of these sides. A trivial convexity argument of the type outlined above shows that this number cannot be one, nor can it exceed two. This proves (5): it also proves

(6) *any two sides meet, if at all, in a vertex and this is then a common end-point of each.*

Note that (5) and (6) imply that the intersection of three sides is empty.

Another useful property of fundamental polygons is that if $G = \{I, g_1, g_2, \ldots\}$ acts on $\Delta$ then

(7) *Euclidean diameter* $(g_n D) \to 0$ *as* $n \to \infty$.

If this were not so, we could find $z_n$ and $w_n$ in $g_n(D)$ with

$$z_n \to z, \, w_n \to w, \, z \neq w, \, |z| = |w| = 1.$$

This would imply that the $g_n(D)$ accumulate on the geodesic $[z, w]$ contrary to the local finiteness of $D$.

We turn our attention now to the pairing of sides of $P$ by certain elements of $G$. Let $G^*$ be the set of elements $g$ in $G$ for which $\tilde{P} \cap g(\tilde{P})$ is a side of $P$ and let $S$ be the set of sides of $P$. Clearly, each $g$ in $G^*$ produces a unique side $s$ in $S$ (namely, $s = \tilde{P} \cap g(\tilde{P})$) and every side arises in this way, thus formally there is a surjective map

$$\Phi : G^* \to S$$

given by

$$\Phi(g) = \tilde{P} \cap g(\tilde{P}).$$

In fact, $\Phi$ is a bijection for if $\Phi(g) = \Phi(h)$ then

$$\tilde{P} \cap g(\tilde{P}) = \tilde{P} \cap h(\tilde{P})$$

and this cannot occur for sides unless $g = h$ as (6) shows.

The existence of $\Phi^{-1} : S \to G^*$ shows that to each side $s$ there is associated a unique $g_s$ in $G^*$ with

$$s = \tilde{P} \cap g_s(\tilde{P}).$$

Then

$$g_s^{-1}(s) = \tilde{P} \cap g_s^{-1}(\tilde{P}) = s',$$

say, and, as this has positive length, this too is a side. Note that if $s' = (g_s)^{-1}(s)$ then

$$g_{s'} = (g_s)^{-1}.$$

We have now constructed a map $s \mapsto s'$ of $S$ onto itself and this is called a *side-pairing* of $P$ because

$$(s')' = (g_{s'})^{-1}(s')$$
$$= g_s(s')$$
$$= s.$$

In this way, the set $S$ of sides of $P$ partitions naturally into a collection of pairs $\{s, s'\}$: we do not exclude the possibility that $s = s'$.

The next result is a strengthened version of Theorem 9.2.7 and it is made possible by the polygonal nature of $P$.

**Theorem 9.3.3.** *The side-pairing elements $G^*$ of $P$ generate $G$.*

PROOF. Because of Theorem 9.2.7, it is only necessary to show that if $\tilde{P} \cap h(\tilde{P}) = \emptyset$, then $h$ lies in the group generated by the $g_s$. Consider, then, any $w$ in $\tilde{P} \cap h(\tilde{P})$. First, there is an open disc $N$ with centre $w$ and elements $h_0 (=I), h_1, h_2, \ldots, h_t$ in $G$ such that $h = h_j$ for some $j \neq 0$, and

$$w \in h_0(\tilde{P}) \cap \cdots \cap h_t(\tilde{P});$$
$$N \subset h_0(\tilde{P}) \cup \cdots \cup h_t(\tilde{P}).$$

One can show (alternatively one can decrease the radius of $N$ and assume) that $N$ contains no vertices of any $h_j(\tilde{P})$ except possibly $w$ and no sides of the $h_j(\tilde{P})$ except those that contain $w$ (see (3)). By (4), the boundary of $P$ in $N$ therefore consists of one side only or two distinct sides emanating from $w$. The same is true of each of the other $h_j(\tilde{P})$ thus we have one of the situations illustrated in Figure 9.3.1 (after relabelling $h_1, \ldots, h_t$).

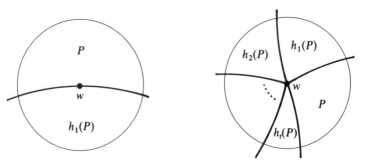

Figure 9.3.1

Formally, we require the fact that (after relabelling) two consecutive polygons in the list

$$h_0(P) = P, h_1(P), h_2(P), \ldots, h_t(P), P = h_0(P)$$

have a side in common. From this it follows that

$$\tilde{P} \cap h_j^{-1} h_{j+1}(\tilde{P})$$

is a geodesic segment of positive length and hence is a side. Thus

$$h_{j+1} = h_j g_s$$

for some side-pairing $g_s$ and we find that $h$ is in the group generated by the $g_s$. $\quad\square$

We come now to a detailed examination of the way in which the images of $\tilde{P}$ tesselate a neighbourhood of any point in $\Delta$. Clearly we may confine our attention to points in $\partial P$ and the situation for these points is completely described in the preceding proof. We now summarize the results.

Take any $w$ in $\partial P$: so $w \in \tilde{P} \cap h(\tilde{P})$ for some $h$. There exist $h_1, \ldots, h_t$ as in the proof of Theorem 9.3.3 and if $g(w) \in \partial P$, then $w \in g^{-1}(\tilde{P})$ and so $h = h_j$ for some $j$. We can now introduce some terminology.

**Definition 9.3.4.** (i) a *cycle* $C$ in $\tilde{P}$ is the intersection of a $G$-orbit with $\tilde{P}$: this is necessarily a finite set $\{z_1, \ldots, z_n\}$ and the *length* $|C|$ of $C$ is $n$.
(ii) If $C$ is a cycle, say $\{z_1, \ldots, z_n\}$ of points in $\Delta$, then the stabilizers $G_j$ of $z_j$ are conjugate to each other and are finite cyclic subgroups of $G$. The *order* of the cycle $C$, which we denote by $\text{Ord}(C)$, is the common order of the $G_j$.
(iii) Let $C$ be the cycle in (ii) and let $P$ subtend an angle $\theta_j$ at $z_j$. The *angle sum* $\theta(C)$ of the cycle $C$ is defined to be $\theta_1 + \cdots + \theta_n$.

The following result is of fundamental importance.

**Theorem 9.3.5.** *For every Fuchsian group $G$, every convex fundamental polygon $P$ and every cycle $C$,*

$$\theta(C) = 2\pi/\text{ord}(C).$$

PROOF. Without doubt, the most efficient description of the proof is by means of cosets. Let $C = \{z_1, \ldots, z_n\}$ so that for some $g_1 \,(=I), g_2, \ldots, g_k$ we have $g_j(z_j) = z_1$. It follows that $g_j(P)$ has $z_1$ as a vertex and the angle of $g_j(P)$ at $z_1$ is $\theta_j$.

Next, $z_1 \in h(\tilde{P})$ if and only if $h^{-1}(z_1)$ is some $z_j$ and this is so if and only if for some $j$, $h(g_j)^{-1}$ fixes $z_1$. Now let $G_1$ be the stabilizer of $z_1$: thus $z_1 \in h(\tilde{P})$ if and only if for some $j$, $h \in G_1 g_j$. Referring now to Figure 9.3.1, we have

$$\{h_0, h_1, \ldots, h_t\} = G_1 g_1 \cup \cdots \cup G_1 g_k, \qquad (h_0 = I),$$

and these are precisely the images of $\tilde{P}$ which contain $z_1$. As the elements of $G_1$ are rotations about $z_1$, each $f$ in $G_1 g_j$ is such that $f(\tilde{P})$ subtends an angle $\theta_j$ at $z_1$: thus

$$
\begin{aligned}
2\pi &= [\text{order}(G_1)](\theta_1 \cdots + \theta_k)\\
&= [\text{ord}(C)]\theta(C).
\end{aligned}
\qquad\qquad\square
$$

Let us examine, in detail, the consequences of Theorem 9.3.5. Suppose first that $z$ is not fixed by any elliptic element of $G$: the cycle $C$ containing $z$ (in $\tilde{P}$) is then said to be an *accidental cycle*. These cycles are characterized by $\text{ord}(C) = 1$ so

$$
\theta(C) = \theta_1 + \cdots + \theta_n = 2\pi \qquad (n = |C|).
$$

If $n = 1$, then $\theta_1 = 2\pi$ and $z \in P$. If $n = 2$ then $\theta_1 = \theta_2 = \pi$ (by Theorem 7.16.1, each $\theta_1$ satisfies $0 \le \theta_j \le \pi$) and $z$ is then an interior point of a side. The converse statements are also true so if $z$ is an *accidental vertex* (a vertex in an accidental cycle $C$) then $|C| \ge 3$.

Next, suppose that $z$ is fixed by an elliptic element in $G$ and that the stabilizer of $z$ has order $q$; thus $\text{ord}(C) = q$. Then

$$
\theta(C) = \theta_1 + \cdots + \theta_n = 2\pi/q.
$$

A special case of great interest is when $|C| = 1$ (so $z$ is not equivalent to any other point on $\partial P$): then $\theta_1 = 2\pi/q$. If $|C| = 1$ and $q = 2$, then

$$
\theta(C) = \theta_1 = \pi.
$$

It is easy to see that in this case, the stabilizer of $z$ is $\{I, g\}$, $g^2 = I$ and $z$ is an interior point of the side

$$
S = \tilde{P} \cap g(\tilde{P}).
$$

Note that in this case,

$$
s' = g^{-1}(s) = s
$$

because $g = g^{-1}$. Conversely, if, in general, $s = s'$ then it is easy to see that $g_s$ is of order two with fixed point on $s$ (consider the effect of $g_s$ on the geodesic containing $s$ and note that $P \cap g_s(P) = \varnothing$).

Because the elliptic fixed points of $G$ demand special attention, it is often convenient to regard all elliptic fix-points as vertices. This is only at variance with the earlier definition in as far as it concerns elliptic fix-points of order two. It is a matter of convention which definition we adopt and the matter is completely settled by stating whether or not elliptic fix-points of order two are vertices of $P$: equivalently, it is settled by stating whether or not we insist that $s \ne s'$. A trivial example should clarify this point.

**Example 9.3.6.** Let $g(z) = -z$ and $G = \{I, g\}$. We may take $P$ as the upper half of $\Delta$ and, according to the two conventions, either:

(1) $P$ has one side, namely $(-1, 1)$, and no vertices; or
(2) $P$ has two sides, namely $(-1, 0)$, $(0, 1)$, and one vertex, namely $\{0\}$. $\square$

We now discuss (as far as we can) the Euclidean boundary of $P$ on $\{|z| = 1\}$: we denote this by $E$. Now $E$ may have uncountably many components but there can only be countably many components of positive (Euclidean) length: we call these the *free sides* of $P$ and these are closed non-degenerate intervals on $\{|z| = 1\}$.

Note that if $w \in E$, then there exist $z_n$ in $P$ converging to $w$. For any $z$ in $P$, the segment $[z, z_n]$ lies in $P$ and obviously, $[z, w) \subset \bar{P}$. The same is true for all points sufficiently close to $z$ and as $P$ is convex, we deduce that $[z, w) \subset P$.

A point $w$ of $E$ need not lie on any side or any free side of $P$; for example, there may be infinitely many sides of $P$ accumulating at, but not containing, the point $w$. We can say very little in this case and we confine the discussion to end-points of two sides.

**Definition 9.3.7.** A point $v$ in $E$ is a *proper vertex* of $P$ (at infinity) if $v$ is the end-point of two sides of $P$: $v$ is an *improper vertex* of $P$ if it is the end-point of a side and free side of $P$. In both cases, we say that $v$ is an *infinite vertex* of $P$.

These vertices are illustrated in Figure 9.3.2.

For each $z$ in $E$, the cycle of $z$ is (as before) $G(z) \cap E$. If $z$ is an ordinary point, the cycle of $z$ must be finite (otherwise infinitely many images of $P$ meet any neighbourhood of $z$ and by (7), $z$ would then be a limit point). By the same token, $z$ must also be a proper or improper vertex at infinity.

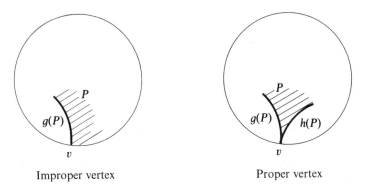

Improper vertex          Proper vertex

Figure 9.3.2

Definition 9.3.4 clearly generalizes to this situation and if $C$ is the cycle of $z$ we have

$$\text{ord}(C) = 1, \qquad \theta_1 + \cdots + \theta_n = \pi:$$

this is the counterpart of Theorem 9.3.5. This holds equally well if $z$ is the interior point of a free side where $|C| = 1$, $\theta_1 = \pi$. Note that otherwise, $\theta_j$ can only take the values 0 or $\pi/2$ so if $|C| = 2$, then necessarily $\theta_1 = \theta_2 = \pi/2$ and so $z$ is the common end-point of a free side of $P$ and a free side of some $g(P)$. We shall return to this case in the next section.

There is only one other case that we can profitably discuss.

**Theorem 9.3.8.** *Let $v$ be any point of $E$ that is fixed by some non-trivial element of $G$. Then $v$ is fixed by a parabolic element of $G$ (and not by any hyperbolic element). Further, the cycle $C$ of $v$ on $E$ is a finite cycle each point of which is a proper vertex of $P$.*

PROOF. First, $v$ cannot be fixed by an elliptic element in $G$ as $|v| = 1$. If $v$ is fixed by a hyperbolic element $h$ of $G$, let $A$ be the axis of $h$ and construct any $[z, v)$ in $P$. Take $z_n$ on $[z, v)$ with $z_n \to v$. Then there exist points $a_n$ on $A$ with

$$\rho(z_n, a_n) \to 0.$$

For each $n$, there is some power of $h$, say $h_n$, such that $h_n(a_n)$ lies on a compact sub-arc of $A$. Thus the points $h_n(z_n)$ lie in a compact subset $K$ of $\Delta$: as the $h_n$ are distinct, this contradicts the fact that $P$ is locally finite. We deduce that $v$ cannot be fixed by hyperbolic elements: thus the elements fixing $v$ must be parabolic.

Obviously the cycle of points on $E$ determined by $v$ contains only parabolic fix-points. If the cycle is infinite, say $v, v_1, v_2, \ldots$ then there are distinct $g_n$ with $g_n(v) = v_n$. If $K$ is any horocyclic region based at $v$, then $g_n(K)$ is a horocyclic region based at $v_n$ and this must meet the convex $P$ as $v \in E$. We deduce that $P$ meets infinitely many images of $K$ and this violates Corollary 9.2.9: it follows that the cycle determined by $v$ (or by any parabolic fix-point) is finite.

Finally, we must show that $v$ is a proper vertex of $P$: the same must then be true (by the same argument) for all points in the cycle of $v$.

Choose any horocyclic region $K$ at $v$. By Corollary 9.2.9, $\tilde{P}$ meets only a finite number of images of $K$, say

$$K, g_1(K), \ldots, g_t(K)$$

based at $v, v_1, \ldots, v_t$ $(v_j = g_j(v))$ respectively. If $v_j \notin E$, then $\tilde{P}$ is disjoint from some Euclidean neighbourhood of $v_j$ so $\tilde{P} \cap g_j(\tilde{K})$ is a compact subset of $\Delta$. By decreasing $K$ as necessary, we may suppose that for each $j$, the point $v_j$ lies in $E$. This shows that cycle of $v$ is now $\{v, v_1, \ldots, v_t\}$.

Without loss of generality, we assume that $G$ acts on $H^2$, that $v = \infty$ and that the stabilizer of $\infty$ is generated by $p: z \mapsto z + 1$. Of course, $K$ is now of the form $\{x + iy: y > k\}$: we may assume that $k > 1$. If

$$a = \inf\{\text{Re}[z]: z \in P\}, \quad b = \sup\{\text{Re}[z]: z \in P\},$$

then $a \leq b \leq a + 1$ (else $b - a > 1$ and $P$ contains, by convexity, a triangle of width exceeding one so $P \cap p(P) \neq \emptyset$). Now $K$ meets $h(\tilde{P})$ if and only if $h^{-1}(K)$ is $K$ or some $g_j(K)$ and this is so if any only if for some $j$ and $n$,

$$h^{-1} = g_j p^n \quad (g_0 = I).$$

It follows that $\infty$ lies on the boundary of $h(P)$ (because $h(v_j) = \infty$). Exactly as above, $h(P)$ lies in a vertical strip of width one and hence there are at most three (consecutive) values of $n$ for which $p^{-n}g_j^{-1}(\tilde{P})$ $(=h(\tilde{P}))$ meets $\tilde{P}$. We deduce that only finitely many images of $\tilde{P}$ can intersect $\tilde{P} \cap K$. This means that $\tilde{P} \cap K$ meets only finitely many sides of $P$ and so, in a sufficiently small horocycle at $\infty$, the boundary of $P$ consists only of two vertical geodesics. $\square$

*Remark.* We end with a remark concerning the elliptic and parabolic conjugacy classes in $G$. Let $g$ be any parabolic element of $G$ with fixpoint, say $v$. Then Corollary 9.2.9 implies that for some $h$ in $G$, the point $h(v)$ lies in $E$ and, of course, is fixed by the parabolic element $hgh^{-1}$ which is conjugate to $g$. By Theorem 9.3.8, there are two sides of $P$ ending at $h(v)$. We conclude that every parabolic element of $G$ is conjugate to some parabolic element which fixes a proper vertex of $P$: in this sense the fundamental polygon $P$ contains representatives of all conjugacy classes of parabolic elements. The same is true of elliptic elements: the proof is trivial and is omitted.

EXERCISE 9.3

1. Let

$$D = \{z \in \Delta: \rho(z, 0) < r\}$$

and suppose that $A_1, \ldots, A_n$ are pairwise disjoint, convex, open subsets of $D$ which satisfy

(i) $\tilde{D} = \tilde{A}_1 \cup \ldots \cup \tilde{A}_n$;
(ii) $0 \in \tilde{A}_1 \cap \ldots \cap \tilde{A}_n$.

Prove that for a suitable choice of $\theta_j$ (with $\theta_1 = \theta_{n+1}$),

$$A_j = \{z \in D: \theta_j < \arg(z) < \theta_{j+1}\}.$$

2. Let $s_1, s_{-1}, s_2, s_{-2}, \ldots$ be pairwise disjoint closed sub-arcs of $\{|z| = 1\}$ and (for convenience) assume that each $s_j$ subtends an angle less than $\pi$ at the origin. Let $L_j$ be the geodesic with the same end-points as $s_j$: thus $\Delta - L_j$ is the union of the two half-planes $H_j$ (containing the origin) and $H'_j$ (with boundary $L_j \cup s_j$).

Let $g_1, g_2, \ldots$ be conformal isometries of $\Delta$ with

$$g_j(L_j) = L_{-j}, \qquad g_j(H_j) = H'_{-j}$$

and define

$$G = \langle g_1, g_2, \ldots \rangle, \qquad D = \bigcap_j H_j$$

Show

(i) each $g_j$ is hyperbolic;
(ii) if $g \in G$ and $g \neq I$, then $g(D) \cap D = \varnothing$.

Now suppose that there exists a positive $\delta$ such that for all $j$,

$$\{z \in H_j : \rho(z, L_j) < \delta\} \subset D.$$

Show that

$$\{z \in \Delta : \rho(z, \tilde{D}) < \delta\} \subset \bigcup_{g \in G} g(D)$$

and deduce that

$$\bigcup_{g \in G} g(\tilde{D}) = \Delta$$

(so $D$ is a convex fundamental domain for $G$). Show also that $D$ is locally finite.

3. Use Question 2 to show that if $D$ is a convex fundamental polygon for a Fuchsian group, then the Euclidean closure of $D$ on $\{|z| = 1\}$ may have uncountably many components (arrange the $s_j$ in a manner analogous to the construction of a Cantor set).

4. In the notation of Question 2, let $s_1$ and $s_{-1}$ be given by $|\arg(z) - \pi| \leq \pi/4$ and $|\arg(z)| \leq \pi/4$ respectively. By constructing $s_j, s_{-j}$ accumulating at the end-points of $s_1$ but not at $s_{-1}$, show that an improper vertex of a convex fundamental polygon for $G$ may be a limit point of $G$.

## §9.4. The Dirichlet Polygon

In this section we describe a particular construction of a convex fundamental polygon and this establishes the existence of such polygons for any Fuchsian group. Let $G$ be a Fuchsian group acting in $\Delta$ and let $w$ be any point of $\Delta$ that is not fixed by any elliptic element of $G$. For each $g$ in $G$ ($g \neq I$) define

$$L_g(w) = \{z \in \Delta : \rho(z, w) = \rho(z, gw)\}$$

and

$$H_g(w) = \{z \in \Delta : \rho(z, w) < \rho(z, gw)\}$$
$$= \{z \in \Delta : \rho(z, w) < \rho(g^{-1}z, w)\}.$$

Note that $L_g(w)$ is a geodesic (not containing $w$) and that $H_g(w)$ is the half-plane which contains $w$ and which is bounded by $L_g(w)$. In fact, $L_g(w)$ is the common boundary of $H_g(w)$ and $H_{g^{-1}}(gw)$: see Figure 9.4.1.

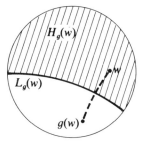

Figure 9.4.1

**Definition 9.4.1.** The *Dirichlet polygon* $D(w)$ for $G$ with *centre* $w$ is defined by

$$D(w) = \bigcap_{g \in G, g \neq I} H_g(w).$$

Sometimes $D(w)$ is called the Poincaré (or normal) polygon for $G$. Dirichlet used the construction in 1850 for Euclidean spaces and it was subsequently exploited by Poincaré for hyperbolic spaces.

In view of the two descriptions for $H_g$, we can either describe $D(w)$ as the set of points $z$ which are closer to $w$ than to any other image of $w$ or as the set of points $z$ which are, among all their images, closest to $w$. Observe that

$$z \in H_g(w) \quad \text{if and only if } w \in H_{g^{-1}}(z)$$

so we have a symmetry expressed by

$$z \in D(w) \quad \textit{if and only if } w \in D(z).$$

If $h$ is any isometry of the hyperbolic plane, then

$$h(H_g(w)) = H_{hgh^{-1}}(hw)$$

and, consequently (using $D_G(w)$ for $D(w)$),

$$h(D_G(w)) = D_{hGh^{-1}}(hw).$$

In particular, if $h \in G$, then

$$h(D(w)) = D(hw).$$

**Theorem 9.4.2.** *The Dirichlet polygon $D(w)$ is a convex fundamental polygon for $G$.*

PROOF. As each $H_g(w)$ is convex and contains $w$ we see that $D(w)$ is convex and non-empty.

The rest of the proof depends on the crucial fact that only finitely many of the $L_g(w)$ can meet any given compact subset of $\Delta$: this is a direct consequence of the fact that if $G = \{g_0, g_1, \ldots\}$ then

$$\rho(w, L_{g_n}(w)) = \tfrac{1}{2}\rho(w, g_n w)$$
$$\to +\infty$$

as $n \to +\infty$.

Now select any $z$ in the closure of $D(w)$. It follows that there is a compact disc $K$ with centre $z$ such that for all $g$, either $K \subset H_g(w)$ or $z \in L_g(w)$ and, moreover, the latter can only occur for a finite set of $g$. Of course, if $z \in D(w)$ then the second possibility cannot occur at all so $K \subset D(w)$ and this proves that $D(w)$ is open. More generally, we see that the boundary of $D(w)$ is contained in the union of the $L_g(w)$, hence

$$\text{h-area}(\partial D(w)) = 0.$$

Next, we prove that there is a fundamental set $F$ with

$$D(w) \subset F \subset \tilde{D}(w).$$

From each orbit $G(z)$, we select *exactly* one point $z^*$ which satisfies

$$\rho(w, z^*) \leq \rho(w, gz)$$

for all $g$ in $G$: such a choice is possible as $G(z)$ does not accumulate at $w$. The set of selected points is $F$: clearly $F$ contains $D(w)$ for if $z \in D(w)$ then we have no choice but to choose $z^* = z$.

To prove that $F \subset \tilde{D}(w)$, select any $z$ in $F$ and consider the segment $[w, z)$. As $w \in D(w)$, no $L_g(w)$ passes through $w$. If $L_g(w)$ meets the segment $(w, z)$ then

$$\rho(z, w) > \rho(z, gw) = \rho(g^{-1}z, w)$$

contrary to the fact that $z \in F$. Thus no $L_g(w)$ meets $(w, z)$ and so $(w, z) \subset D(w)$. It follows that $z \in \tilde{D}(w)$ so $F \subset \tilde{D}(w)$.

We have now shown that $D(w)$ is a convex fundamental domain for $G$: it remains to show that $D(w)$ is locally finite. Let $K$ be any compact disc with centre $w$ and radius $r$ and suppose that $g(\tilde{D}(w))$ meets $K$: thus there is some $z$ in $\tilde{D}(w)$ with $\rho(gz, w) \leq r$. As $z \in \tilde{D}(w)$, we have

$$\rho(w, gw) \leq \rho(w, gz) + \rho(gz, gw)$$
$$\leq r + \rho(z, w)$$
$$\leq r + \rho(gz, w)$$
$$\leq 2r$$

and this can only be true for a finite set of $g$.                                            $\square$

By virtue of Theorem 9.4.2, *all of the results established in Sections 9.1, 9.2 and 9.3 are valid for Dirichlet polygons.* For example, *the quotient space*

$$\tilde{D}(w)/G$$

*is independent (topologically) of the choice of w, provided, of course, that w is not an elliptic fixed-point of G*: this exceptional case is discussed in Section 9.6.

In the particular case of the fundamental polygon $D(w)$, we can say a little more about the structure of the boundary. For example, we have the following elementary but important result.

**Theorem 9.4.3.** *Let $\{z_1, \ldots, z_n\}$ be any cycle on the boundary of the Dirichlet polygon $D(w)$. Then*

$$\rho(z_1, w) = \rho(z_2, w) = \cdots = \rho(z_n, w).$$

PROOF. Consider, for example $z_1$ and $z_2$ on the boundary of $D(w)$ with $z_2 = h(z_1)$. As $[w, z_1) \subset D(w)$ we see that

$$[hw, z_2) = h[w, z_1)$$
$$\subset h(D(w))$$
$$= D(hw).$$

It follows that $z_2$ is equidistant from $w$ and $hw$ and so

$$\rho(w, z_2) = \rho(hw, z_2)$$
$$= \rho(w, h^{-1}z_2)$$
$$= \rho(w, z_1). \qquad \square$$

Each side of $D(w)$ is of the form

$$s = \tilde{D}(w) \cap g(\tilde{D}(w))$$
$$= \tilde{D}(w) \cap \tilde{D}(gw)$$

and in view of our earlier description, this must be contained in $L_g(w)$. Thus *the sides of $D(w)$ are segments of the bisectors $L_g(w)$.* For similar reasons, *the vertices are the boundary points of $D(w)$ where two or more bisectors $L_g(w)$ meet.*

Let us now illustrate some of these ideas by discussing a specific example.

**Example 9.4.4.** Let $G$ be the Modular group acting in $H^2$: we shall show that the open polygon $P$ illustrated in Figure 9.4.2 is the Dirichlet polygon with centre $iv$ for any $v > 1$. Accordingly, let $w = iv$ with $v > 1$ and, for brevity, write $D$ for $D(iv)$ and similarly for $L_g$ and $H_g$.

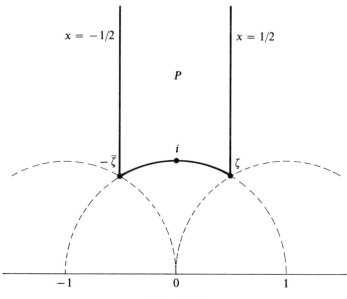

Figure 9.4.2

First, the isometries

$$f(z) = z + 1, \qquad g(z) = -1/z$$

are in $G$ and (as the reader can easily verify) the three geodesic sides of $P$ are $L_f$, $L_{f^{-1}}$, $L_g$. This shows that $D \subset P$.

If $D \neq P$, then some side of $D$ crosses $P$ and so there is some $z$ in $D$ with $h(z) \in h(D) \cap P$. It follows that $z$, $hz \in P$ and we shall now show that this cannot happen. Suppose that

$$h(z) = \frac{az + b}{cz + d}, \qquad \begin{pmatrix} a & b \\ c & d \end{pmatrix} \in \mathrm{SL}(2, \mathbb{Z}).$$

Then

$$\begin{aligned}
|cz + d|^2 &= c^2|z|^2 + 2\,\mathrm{Re}[z]cd + d^2 \\
&> c^2 + d^2 - |cd| \\
&= (|c| - |d|)^2 + |cd|
\end{aligned}$$

as $|z| > 1$ and $|\mathrm{Re}[z]| < \frac{1}{2}$. This lower bound is an integer: it is non-negative and is zero if and only if $c = d = 0$ and this is excluded because $ad - bc = 1$. We deduce that $|cz + d| > 1$ (note that the strict inequality holds) and so

$$\mathrm{Im}[hz] = \frac{\mathrm{Im}[z]}{|cz + d|^2} < \mathrm{Im}[z].$$

Exactly the same argument holds with $z$, $h$ replaced by $hz$, $h^{-1}$ and a contradiction is reached: thus $D = P$.                                     $\square$

As $D$ is a convex fundamental polygon, the material in Section 9.3 is available. We can either view $D$ as having three sides, namely,

$$s_1 = [\zeta, \infty), \qquad s_2 = [-\bar{\zeta}, \infty), \qquad s_3 = [-\bar{\zeta}, \zeta],$$

with the side-pairing $f(s_2) = s_1$, $g(s_3) = s_3$ or we may adopt the alternative convention regarding the fix-points of elliptic elements of order two. If this convention is adopted, we replace $s_3$ by the two sides

$$s_4 = [-\bar{\zeta}, i], \qquad s_5 = [i, \zeta],$$

with $g(s_4) = s_5$, $g(s_5) = s_4$ and we consider $i$ to be a vertex of $D$.

As $P$ is a fundamental polygon for $G$ so is the polygon $P_1$ illustrated in Figure 9.4.3 (we have merely replaced a vertical strip of $P$ by the $f$-image of this strip). Note that in this case, $P_1$ has (according to convention) five or six (but never four) sides: these are (in the case of six sides)

$$s_1 = [-\bar{w}, \infty), \qquad s_2 = f(s_1) = [1 - \bar{w}, \infty);$$
$$s_3 = [-\bar{w}, i], \qquad s_4 = g(s_3) = [i, w];$$
$$s_5 = [w, \zeta], \qquad s_6 = fg(s_5) = [\zeta, 1 - \bar{w}].$$

The cycles of vertices of $P_1$ are the sets

$$\{\infty\}, \{i\}, \{\zeta\}, \{-\bar{w}, w, 1 - \bar{w}\}.$$

Note that the last cycle is an accidental cycle and the angle subtended by $P$ at the vertex $w$ is $\pi$ (regardless of the convention being used).

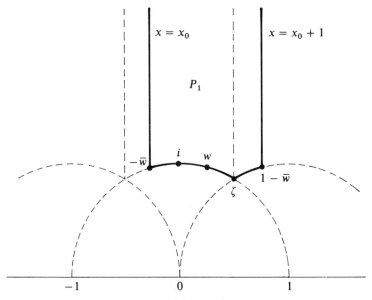

Figure 9.4.3

Returning now to the general situation, it must be expected that certain properties of $D(w)$ depend on the choice of $w$ and, having said this, that there must be certain optimal choices of $w$. The last result in this section describes some such optimal choices.

**Theorem 9.4.5.** *Let $G$ be a Fuchsian group and let $D(w)$ be the Dirichlet polygon with centre $w$. Then for almost all choices of $w$:*

(1) *every elliptic cycle on $\partial D(w)$ has length 1;*
(2) *every accidental cycle on $\partial D(w)$ has length 3;*
(3) *every improper vertex that is an ordinary point is in a cycle of length 2;*
(4) *every proper vertex has cycle length 1 and is a parabolic fix-point;*
(5) *every parabolic cycle has length 1 and is a proper vertex.*

PROOF. The proof of each part follows the same pattern: if the condition $(k)$ fails, then $w$ must lie in some exceptional set $E_k$ with area zero. If $w$ lies outside the set $\bigcup E_j$ of zero area, then all five conditions are satisfied. We write $D$ for $D(w)$.

The verification of (1) is easy. Let $E_1$ be the union of geodesics which are equidistant from two (distinct) elliptic fixed points. Clearly $E_1$ has zero area. If $u, v$ are distinct elliptic fixed points in the same cycle, then $\rho(u, w) = \rho(v, w)$ (Theorem 9.4.3) and so $w \in E_1$.

For the remainder of the proof we need the following simple lemma.

**Lemma 9.4.6.** *Let $R(z)$ be any non-constant rational function of $z$. Then*

$$E = \{z: R(z) \text{ is real}\}$$

*has zero area.*

PROOF OF LEMMA 9.4.6. At every point of the extended plane apart from a finite set $z_1, \ldots, z_n$, the function $R$ is locally a homeomorphism satisfying some Lipschitz condition. Thus each $z \ (\neq z_j)$ has a neighbourhood $N$ with $E \cap N$ having area zero and a countable number of these $N$ cover the plane with $z_1, \ldots, z_n$ deleted.                                                                $\square$

We return to the proof of Theorem 9.4.3. For all $f, g, h$ in $G$, distinct from each other and from $I$, we define

$$R(z) = \frac{(z - gz)(fz - hz)}{(z - fz)(gz - hz)}$$

(for brevity, we prefer not to mention explicitly the dependence of $R$ on $f, g, h$). Note that $R$ may be constant (for example, if $f$, $g$ and $h$ fix 0 and $\infty$). Let

$$E_2 = \bigcup \{z: R(z) \text{ is real}\}$$

the union being over all triples $(f, g, h)$ for which $R$ is not constant. By Lemma 9.4.6, $E_2$ has area zero and we shall show now that if (2) fails, then $w \in E_2$.

Suppose, then that (2) fails so that there are four distinct points $u, f^{-1}u,$ $g^{-1}u, h^{-1}u$ lying in some accidental cycle on $\partial D$. Theorem 9.4.3 implies that

$$\rho(w, u) = \rho(fw, u) = \rho(gw, u) = \rho(hw, u)$$

so the distinct points $w, fw, gw, hw$ lie on a hyperbolic circle with centre $u$. It follows that the cross-ratio $R(w)$ is real so $w \in E_2$ unless, of course, $R$ is constant.

We now show that $R$ cannot be constant. If $R$ is constant, say $\lambda$, then by selecting some $z$ not fixed by $g, f, f^{-1}h, g^{-1}h$ we see that $\lambda \neq 0, \infty$. Now let $z$ tend to a fixed point $v$ of $g$: the numerator of $R$ tends to zero, hence so does the denominator and so $f$ or $h$ also fixes $v$. Suppose, then, that $g$ and $f$ have a common fixed point (the same argument will be valid for $h$). As $g$ and $f$ lie in the Fuchsian group $G$, we see that $\langle g, f \rangle$ is a cyclic group generated, say, by $p$. Clearly $p$ is hyperbolic, parabolic or elliptic depending on whether the orbit of any point under $\langle g, f \rangle$ lies on a hypercycle, or horocycle or a hyperbolic circle respectively (these possibilities are mutually exclusive). By assumption, then, $p$ is elliptic and fixes the centre of the unique hyperbolic circle through $w, gw, fw$. We deduce that $fu = gu = u$ which is a contradiction as $u$ lies in an accidental cycle. This proves (2).

A similar argument establishes (3), (4) and (5). Suppose first that $v$ is a proper vertex of $D$ so there are two sides

$$s_1 = \tilde{D} \cap g(\tilde{D}), \qquad s_2 = \tilde{D} \cap h(\tilde{D})$$

ending at $v$. As $s_1$ is in the geodesic bisecting the segment $[w, gw]$ (and similarly for $s_2$) it follows from Section 7.28 that $v, w, gw, hw$ lie on a horocycle based at $v$.

Now consider the function

$$R_1(z) = [v, z, gz, hz]$$
$$= \frac{(v - gz)(z - hz)}{(v - z)(gz - hz)}.$$

As a horocycle is a Euclidean circle, $R_1(w)$ is real. It follows that either $R_1$ is not constant and $w$ lies in the corresponding exceptional set of zero area or $R_1$ is constant. We must show, therefore, that in the latter case each of (3), (4) and (5) hold.

Suppose, then, that $R_1$ is constant, say $\lambda$ where (as before) $\lambda \neq 0, \infty$. Letting $z$ tend to $v$ we see that $g$ or $h$ fixes $v$: by symmetry, we may assume that $gv = v$. Then, by Theorem 9.3.8, $g$ is parabolic. This implies that the side $g^{-1}(s_1)$ of $D$ also ends at $v$ and so is precisely the side $s_2$. This means that $h = g^{-1}$ and $v$ is a parabolic fixed point in a cycle of length one and this establishes (4) and (5) as every parabolic fixed point on $\partial D$ is a proper vertex (Theorem 9.3.8).

Finally, any improper vertex $v$ that is an ordinary point belongs to a finite cycle $v_1 (=v), v_2, \ldots, v_n$ and $D$ has angles $\theta_1, \ldots, \theta_n$ at these points where each $\theta_j$ is zero or $\pi/2$ and $\sum \theta_j = \pi$. Using (4), we see that $\theta_j$ cannot be zero: thus $n = 2$ and this is (3).                                                                                $\square$

### EXERCISE 9.4

1. Develop the theory of Dirichlet polygons (or, strictly speaking, polyhedra) for discrete subgroups of SL(2, $\mathbb{C}$) acting in $H^3$.

## §9.5. Generalized Dirichlet Polygons

Let $G$ be any group of Möbius transformation which acts discontinuously in some $G$-invariant open subset $\Sigma$ of the extended complex plane. We suppose that $\infty \in \Sigma$ and that $\infty$ is not fixed by any non-trivial element in $G$. These assumptions ensure that every non-trivial $g$ in $G$ has an isometric circle $I_g$. Let $H_g$ denote the exterior of $I_g$: then it can be shown that

$$F_G = \bigcap_{g \in G, g \neq I} H_g$$

is essentially a fundamental domain for $G$ (it need not be connected: it may be necessary to remove some boundary points of $F_G$). This is called the *Ford fundamental region* and it is apparently Euclidean (rather than hyperbolic) in character.

Consider now a Fuchsian group $G$ acting on $\Delta$ and suppose that $\infty$, and therefore the origin as well, is not fixed by any non-trivial $g$ in $G$. We can construct both $F_G$ and also the Dirichlet polygon $D_G(0)$ with centre 0 and we shall see shortly that

$$\Delta \cap F_G = D_G(0) \tag{9.5.1}$$

Note that this identifies two sets one of which is Euclidean in character and in no sense conjugation invariant, while the second is of essentially hyperbolic character and is conjugation invariant. The explanation of this lies in the inversive geometry of the extended complex plane and in this geometry, the two apparently different constructions appear as different cases of one single construction which we shall now describe.

Let $P$ be any model (e.g. $\Delta$ or $H^2$) of the hyperbolic plane constructed as an open disc in the extended complex plane. Let $\partial P$ be the circle at infinity. Select any $\zeta$ in the extended complex plane. Each conformal isometry $g$ of $P$ can be written as

$$g = \sigma_2 \sigma_1,$$

where $\sigma_j: P \to P$ denotes reflection in a geodesic $L_j$. We extend each $L_j$ a circle and we insist that the circle $L_2$ contains $\zeta$. Provided that $g$ does not

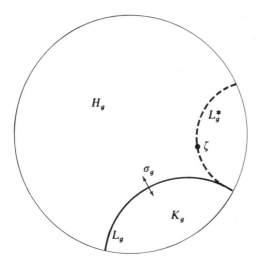

Figure 9.5.1

fix $\zeta$ (which we assume), this determines $\sigma_1, \sigma_2, L_1, L_2$ uniquely and henceforth these special choices will be denoted by

$$\sigma_g, \sigma_g^*, L_g, L_g^*,$$

respectively.

By definition, $\zeta \in L_g^*$. It follows that $\zeta \notin L_g$ else $\sigma_g, \sigma_g^*$ (and therefore $g$) fixes $\zeta$: thus there is a unique hyperbolic half-plane $H_g$ bounded by $L_g$ and containing $\zeta$. Let $K_g$ be the other half-plane bounded by $L_g$: see Figure 9.5.1 where these are illustrated for a parabolic $g$.

**Definition 9.5.1.** Let $G$ be a Fuchsian group acting on $P$ and suppose that $\zeta$ (in the extended complex plane) is not fixed by any non-trivial element of $G$. Then

$$\Pi_G(\zeta) = \bigcap_{g \in G, g \neq I} H_g$$

is called the *generalized Dirichlet polygon* with *centre* $\zeta$.

If $\tau$ denotes reflection in $\partial P$, then $L_g^*$ (extended) contains $\zeta$ if and only it contains $\tau(\zeta)$ (because $L_g^*$ is orthogonal to $\partial P$): thus we have the invariance condition

$$\Pi_G(\zeta) = \Pi_G(\tau\zeta). \tag{9.5.2}$$

In particular, if $P = \Delta$, then

$$\Pi_G(\infty) = \Pi_G(0). \tag{9.5.3}$$

**Theorem 9.5.2.** *In addition to the assumptions made in Definition 9.5.1, suppose that $\zeta$ is an ordinary point of G. Then $\Pi_G(\zeta)$ is a fundamental domain for G in P.*

*If $\zeta \in P$, then $\Pi_G(\zeta)$ is the Dirichlet polygon $D_G(\zeta)$. If $\zeta = \infty$, then $\Pi_G(\zeta)$ is the region exterior to the isometric circles of all elements of G. Finally, for all h, we have*

$$h(\Pi_G(\zeta)) = \Pi_{hGh^{-1}}(h\zeta). \tag{9.5.4}$$

*Remark.* We have deliberately used $P$ for the hyperbolic plane rather than $\Delta$ or $H^2$ in order to cope adequately with the point $\infty$. For example, note that $\infty \in \partial P$ if $P = H^2$ but not if $P = \Delta$.

PROOF. If $\sigma$ denotes reflection in a circle $L$ in $\hat{C}$, then $h\sigma h^{-1}$ is the reflection in $h(L)$. This fact leads directly to (9.5.4) and this is now available to simplify the rest of the proof.

In the case when $\zeta \in P$, we may use (9.5.4) and thereby assume that $P = \Delta$ and $\zeta = 0$. Thus $0 \in L_g^*$ and so for $z$ on $L_g$ we have

$$\begin{aligned} \rho(z, g^{-1}0) &= \rho(z, \sigma_g \sigma_g^* 0) \\ &= \rho(\sigma_g z, \sigma_g^* 0) \\ &= \rho(z, 0). \end{aligned}$$

This identifies $L_g$ with the hyperbolic bisector of the segment $[0, g^{-1}0]$ and so $\Pi_G(\zeta)$ is the corresponding Dirichlet polygon.

If $\zeta = \infty$, then $L_g^*$ is a Euclidean straight line. Thus $g$ acts as a Euclidean isometry on $L_g$ and so $L_g$ is necessarily the isometric circle of $g$. As $\infty \in H_g$, we see that $\Pi_G(\zeta)$ is region exterior to all isometric circles.

It remains to prove that $\Pi_G(\zeta)$ is a fundamental domain for G in P. This is true if $\zeta \in P$ as we have already identified $\Pi_G(\zeta)$ as a Dirichlet polygon. It is also true if $\zeta \notin P \cup \partial P$ because of (9.5.2).

The remaining case is when $\zeta \in \partial P$ and here we may use (9.5.4) and assume that $P = H^2$ and $\zeta = \infty$. First,

$$\begin{aligned} g^{-1} &= \sigma_g \sigma_g^* \\ &= \sigma_g^*(\sigma_g^* \sigma_g \sigma_g^*), \end{aligned}$$

and the bracketed term denotes reflection in $\sigma_g^*(L_g)$. Thus (by the uniqueness of the decomposition of $g^{-1}$)

$$\begin{aligned} L_{g^{-1}} &= \sigma_g^*(L_g) \\ &= g(L_g). \end{aligned}$$

This means that $g(H_g)$ and $g(K_g)$ are separated by $L_{g^{-1}}$. As

$$\begin{aligned} g^{-1}(\zeta) &= \sigma_g \sigma_g^*(\zeta) \\ &= \sigma_g(\zeta) \end{aligned}$$

lies in $K_g$, we find that $\zeta \in g(K_g)$: thus

$$g(H_g) = K_{g^{-1}}$$

and hence

$$g(\Pi_G(\zeta)) \cap \Pi_G(\zeta) = \varnothing.$$

Clearly, this implies that the images of $\Pi_G(\zeta)$ by distinct $g$ and $h$ are disjoint.

Now consider any $z$ in $H^2$. As $\infty$ is an ordinary point of $G$, the orbit of $z$ lies in some (compact) disc in the Euclidean plane and so there is necessarily some point $z'$ in the orbit of $z$ with

$$\text{Im}[z'] \geq \text{Im}[gz]$$

for all $g$ in $G$. As the action of $g$ on points in $K_g$ increases the imaginary part, we see that $z'$ lies on or outside each $L_g$. This in turn implies that the ray $(z', \infty)$ lies outside every $L_g$ and so lies in $\Pi_G(\zeta)$. We deduce that every $z$ is equivalent to some point in the closure of $\Pi_G(\zeta)$.                                           □

*Remark.* The proof that $\Pi_G(\infty)$ is a fundamental domain can be written in Euclidean terms involving computations of derivatives: for example,

$$H_g = \{z : |g^{(1)}(z)| < 1\}.$$

It seems preferable, though, to use the intrinsic method given above.

In conclusion, observe that if $P = \Delta$ and $\zeta = 0$, then from (9.5.3) and Theorem 9.5.2 we have

$$D_G(0) = \Pi_G(\infty)$$
$$= \Delta \cap F_G,$$

which is (9.5.1).

### EXERCISE 9.5

1. Using the notation in the text, let $\sigma'_g$ denote the reflection in $L'_g (= g(L_g))$. Prove that

$$L^*_{(g^{-1})} = L^*_g, \qquad g^{-1} = \sigma^*_g \sigma'_g.$$

2. Prove that $g$ is elliptic, parabolic or hyperbolic according as $L_g, L^*_g$ are intersecting, parallel or disjoint respectively. Show that

   (i) if $g$ is elliptic then it fixes the common point of $L_g$ and $L^*_g$;
   (ii) if $g$ is parabolic then it fixes the common point of tangency of $L_g, L^*_g$;
   (iii) if $g$ is hyperbolic then its fixed points are inverse points with respect to $L_g, L^*_g$ and $L_{g^{-1}}$.

3. Let $g$ be parabolic and not fixing $\infty$ and let $r_n$ be the radius of the isometric circle of $g_n$. Prove that $r_n = r_1/|n|$.

4. Let $I$ and $I'$ be the isometric circles of some hyperbolic $g$ (not fixing $\infty$) and $g^{-1}$. Show that $g(I) = I'$. Compare the images of $I$ and $I'$ under $g^n$ with the isometric circles of $g^n$ and $g^{-n}$.

## §9.6. Fundamental Domains for Coset Decompositions

Let $G$ be a Fuchsian group acting in $\Delta$ and let $H$ be a subgroup of $G$. It is often convenient to construct a fundamental domain for $G$ with special reference to $H$ and with this in mind, we suppose that $G$ has a coset decomposition

$$G = \bigcup_n g_n H. \tag{9.6.1}$$

The essence of the construction is to find an $H$-invariant set $\Sigma$ such that the sets $g_n(\Sigma)$ tessellate $\Delta$: if the $H$-images of some $D$ tessellate $\Sigma$, then $D$ is a fundamental domain for $G$ in $\Delta$.

Suppose now that the set $\Sigma$ is stable under the action of $H$: that is, $g(\Sigma) = \Sigma$ when $g \in H$ and $g(\Sigma) \cap \Sigma = \varnothing$ otherwise. Each coset $g_n H$ determines the corresponding set $g_n(\Sigma)$ uniquely (and independently of the choice of representative $g_n$) and if $m \neq n$, then

$$g_n(\Sigma) \cap g_m(\Sigma) = \varnothing \tag{9.6.2}$$

because $(g_m)^{-1}g_n \notin H$. We make one other assumption, namely

$$\bigcup_n g_n(\tilde{\Sigma}) = \Delta.$$

These last two statements are reminiscent of the definition of a fundamental domain, however here they are with reference to the action of coset representatives rather than all elements of $G$.

**Theorem 9.6.1.** *Let $G$ be a Fuchsian group acting in $\Delta$ and let $H$ be a subgroup of $G$ with coset decomposition (9.6.1). Suppose that $\Pi$ is a convex fundamental polygon for $H$ and that a convex open polygon $\Sigma$ of $\Delta$ satisfies*

(1) *$\Sigma$ is stable under the action of $H$ and*

(2) $$\bigcup_n g_n(\tilde{\Sigma}) = \Delta.$$

*Then $\Pi \cap \Sigma$ is a fundamental domain for $G$ in $\Delta$.*

PROOF. First, $\Pi \cap \Sigma$ is open and convex and its boundary has zero area. It is necessary to show that

$$\bigcup_{g \in G} g(\tilde{\Pi} \cap \tilde{\Sigma}) = \Delta \tag{9.6.3}$$

and, if $f$ and $g$ are distinct elements of $G$, then

$$g(\Pi \cap \Sigma) \cap f(\Pi \cap \Sigma) = \varnothing. \tag{9.6.4}$$

If $z \in \Delta$, then by (2), there is some $n$ with $g_n^{-1}(z)$ in $\tilde{\Sigma}$. Only finitely many $H$-images of $\Pi$ meet some neighbourhood of $g_n^{-1}(z)$ and so for some $h_n$ in $H$,

$$h_n g_n^{-1}(z) \in \tilde{\Pi}.$$

Also, as $g_n^{-1}(z) \in \tilde{\Sigma}$, we have

$$h_n g_n^{-1}(z) \in h_n(\tilde{\Sigma}) = \tilde{\Sigma}$$

and so $h_n g_n^{-1}(z) \in \tilde{\Pi} \cap \tilde{\Sigma}$: this verifies (9.6.3).

Finally, suppose that (9.6.4) fails so that

$$f(\Pi \cap \Sigma) \cap g(\Pi \cap \Sigma) \neq \varnothing.$$

From (9.6.1) we can write

$$f = g_n h_n, \qquad g = g_m h_m \qquad (h_j \in H),$$

and then

$$g_n(\Sigma) \cap g_m(\Sigma) = f(\Sigma) \cap g(\Sigma) \neq \varnothing.$$

We deduce from (9.6.2) that $g_n = g_m$ so

$$h_n(\Pi) \cap h_m(\Pi) \supset g_n^{-1}(f(\Pi \cap \Sigma) \cap g(\Pi \cap \Sigma))$$
$$\neq \varnothing.$$

As $\Pi$ is a fundamental domain for $H$ we deduce now that $h_n = h_m$ so $f = g$. $\square$

We consider three examples: in these, $H$ is a parabolic, elliptic and hyperbolic cyclic subgroup of $G$.

**Example 9.6.2.** Suppose that $H = \langle h \rangle$ where $h$ is parabolic. By considering a conjugate group we may suppose that $G$ acts on $H^2$ and that $h(z) = z + 1$. Every element in $G - H$ has an isometric circle and we let $\Sigma$ denote the set of points having some neighbourhood not meeting any isometric circle. It is easy to see that the hypotheses of Theorem 9.6.1 are satisfied (as a guide, see Section 9.5 or, for full details, see [52], p. 58) and so a fundamental domain for $G$ is (for example) the set of $z$ outside all isometric circles and lying in some strip $\{x + iy: y > 0, x_0 < x < x_0 + 1\}$.

**Example 9.6.3.** Suppose that $H = \langle h \rangle$ where $h$ is elliptic. We may suppose that $G$ acts on $\Delta$ and that $h(z) = e^{2\pi i/n} z$. Again, we take as $\Sigma$ the points in $\Delta$ which are exterior to all isometric circles: equivalently, we follow the construction of the Dirichlet polygon with centre 0 and define $\Sigma$ as the intersection of the half-planes

$$\{z \in \Delta: \rho(z, 0) < \rho(z, g0)\}$$

taken over all $g$ with $g0 \neq 0$. A fundamental domain for $H$ is a sector of $\Delta$, say

$$\Pi = \{z : \theta < \arg z < \theta + 2\pi/n\}$$

(for any $\theta$) and $\Pi \cap \Sigma$ is a fundamental domain for $G$.

**Example 9.6.4.** Suppose that $H = \langle h \rangle$ where $h$ is hyperbolic. We shall also suppose that $h$ is a simple hyperbolic element so the axis $A$ of $h$ is stable under $H$. If $g \notin H$ then $A$ and $g(A)$ are disjoint and the set

$$K_g = \{z : \rho(z, A) < \rho(z, gA)\}$$

is a half-plane. It is easy to see that $\Sigma$ defined by

$$\Sigma = \bigcap_{g \notin H} K_g$$

satisfies the conditions of Theorem 9.6.1 and taking any suitable $\Pi$ (for example, the region bounded by geodesics $L$ and $gL$ orthogonal to $A$) we obtain a fundamental domain for $G$.

EXERCISE 9.6

1. Verify the details of Examples 9.6.2, 9.6.3 and 9.6.4.

2. Show that any cycle on the boundary of the fundamental domain $D$ constructed in Example 9.6.2 necessarily lies on some horocycle based at $\infty$.

## §9.7. Side-Pairing Transformations

Let $G$ be a Fuchsian group and $P$ a convex fundamental polygon for $G$. We have seen that the side-pairing elements of $P$ generate $G$ (Theorem 9.3.3); this short section is devoted to characterizing those primitive elements of $G$ which can arise as side-pairing elements of some choice of $P$.

Each primitive elliptic element and each primitive parabolic element in $G$ pair sides of some fundamental domain (indeed, of some Dirichlet polygon): this follows from Examples 9.6.2 and 9.6.3 or from Theorem 9.4.5 and Corollary 9.2.9. The problem, then, is to characterize the primitive, side-pairing hyperbolic elements in $G$.

**Theorem 9.7.1.** *Let $g$ be a primitive hyperbolic element of a Fuchsian group $G$ and let $A$ be the axis of $g$. Then $g$ pairs sides of some convex fundamental polygon $P$ if and only if for all $h$ in $G$, either $h(A) = A$ or $h(A) \cap A = \varnothing$.*

PROOF. Suppose first, that $h(A) = A$ or $h(A) \cap A = \varnothing$ and define

$$H = \{h \in G : h(A) = A\}.$$

Then $H$ contains all powers of $g$: the only other elements that $H$ can contain are elliptic elements of order two with fixed point on $A$. Exactly as in Example 9.6.4, we can construct a set $\Sigma$ satisfying the conditions of Theorem 9.6.1. We may assume that $G$ acts on $H^2$ and that $g(z) = kz$. If $H$ is cyclic it is generated by $g$ and we can take

$$\Pi = \{z \in H^2 : 1 < |z| < k\}:$$

if $H$ is not cyclic it is generated by $g$ and some elliptic element of order two which we may assume fixes $i\sqrt{k}$ and we then take

$$\Pi = \{z \in H^2 : 1 < |z| < k, \operatorname{Re}[z] > 0\}.$$

In both cases, $g$ pair sides of $\Pi \cap \Sigma$ that contain arcs of $|z| = 1$ and $|z| = k$, respectively.

To prove the necessity of the condition on $h$ and $A$ we suppose that $g$ pairs two sides $s$ and $s'$ of some $P$. Choose a point $w$ in the relative interior of $s$ and not fixed by any non-trivial element of $G$; let $\gamma = [w, gw]$. Then $\gamma$ lies in $P$ apart from its end-points $w$ and $gw$. The curve

$$\Gamma = \bigcup_n g^n(\gamma)$$

is a simple $g$-invariant curve in $\Delta$ which (because $\gamma$ is compact) has as end-points the fixed points $u$ and $v$ of $g$. Note that the axis $A$ of $g$ also has these properties.

Now suppose that $h(A) \cap A \neq \varnothing$, thus there is some $h$ in $G$ such that the geodesics $A$ and $h(A)$ cross or are equal (they cannot be parallel by Theorem 5.1.2). Suppose that $A$ crosses $h(A)$. This means that the curves $\Gamma$ and $h(\Gamma)$ also cross each other, say at the point $\zeta$ in $\Delta$. It follows that for some $z_1$ and $z_2$ in $\gamma$ and some $m$ and $n$, we have

$$\zeta = g^n z_1 = h g^m z_2$$

so

$$z_1 = g^{-n} h g^m z_2.$$

Now the only two distinct points of $\gamma$ which are $G$-equivalent are $w$ and $gw$ so either $z_1 = z_2$ or $z_1 = gz_2$ or $z_2 = gz_1$. In all cases some $g^i h g^j$ fixes some point of $\gamma$. By construction, no point of $\gamma$ is fixed by any non-trivial element of $G$ so $h$ is some power of $g$. This implies that $h(A) = A$. $\square$

In view of Definition 8.1.5, we have shown that the only side-pairing elements of $G$ are elliptic, parabolic and simple hyperbolic elements.

## §9.8. Poincaré's Theorem

Any Fuchsian group $G$ acting on the unit disc $\Delta$ has a convex fundamental polygon $P$. The action of $G$ on $P$ tesselates $\Delta$ and there is a collection of side-pairing maps $g_s$ which generate $G$. Moreover, the sum of the interior angles of $P$ at points of a cycle is a certain submultiple of $2\pi$ (Theorem 9.3.5). Poincaré's Theorem is concerned with reversing this process and so provides a method of *constructing* Fuchsian groups. Suppose that one starts with a polygon $P$ and a collection of side-pairing maps. We use these maps to generate a group $G$. Next, we formulate the notion of a cycle (at this stage we do *not* know whether or not this cycle is the intersection of $P$ with a $G$-orbit) and we impose a suitable angle condition on each cycle. The aim is to prove that $G$ is discrete and that $P$ is a fundamental domain for $G$.

As these ideas arise in other geometries and in other dimensions it seems worthwhile to proceed in a fairly general manner. We shall include hypotheses as they are needed and only at the end shall we give a definitive statement of the result. The argument that we shall use may be summarized as follows. First, we construct a space $X^*$ which *is* tesselated by the group action: then we attempt to identify this tesselation of $X^*$ with the $G$-images of the polygon $P$ in the original space.

We begin by constructing a tesselated space. Let $X$ be any non-empty set. We assume

(A1) *$P$ is an abstract polygon in $X$.*

By this, we mean that $P$ is a non-empty subset of $X$ which has associated with it a non-empty collection of non-empty subsets $s_i$ of $X$ called the *sides* of $P$. The union of the sides is denoted by $\partial P$: we insist that $P$ and $\partial P$ are disjoint and we write

$$\tilde{P} = P \cup \partial P.$$

We also assume

(A2) *there is a side-pairing $\Phi$ of $P$.*

Explicitly, this means that there is an involution (or self-inverse) map $s \mapsto s'$ of the set of sides of $P$ onto itself and associated with each pair $(s, s')$, there is a bijection $g_s$ of $X$ onto itself with

$$g_s(s) = s'$$

and

$$g_{s'} = (g_s)^{-1}.$$

Now let $G$ be the group generated by the $g_s$ and form the Cartesian product $G \times \tilde{P}$. It is helpful to think of $G \times \tilde{P}$ as a collection of disjoint copies

$$(g, \tilde{P}) = \{(g, x): x \in \tilde{P}\}$$

of $\tilde{P}$ indexed by $G$ (like pieces of a jigsaw separated from each other) and to think of $(g, x)$ as the point $g(x)$ viewed from within $g(P)$. We now join these copies together along common edges as dictated to us by the group $G$. Observe first that the map $hg_s h^{-1}$ may be viewed as a map of the side $h(s)$ of $h(P)$ onto both:

(i) the side $h(g_s s)$ of $h(P)$; and
(ii) the side $(hg_s)(s)$ of $(hg_s)(P)$.

Writing $g = hg_s$, we therefore wish to identify $(g, s)$ with $(h, g_s s)$. This identification is achieved by defining the relation $\sim$ on $G \times \tilde{P}$ by

$$(g, x) \sim (h, y)$$

if and only if either:

(i) $g = h, x = y$; or
(ii) $x \in s, y = g_s(x), g = hg_s$.

This relation is symmetric and reflexive (but not necessarily transitive) and it extends to an equivalence relation $*$ on $G \times \tilde{P}$ by defining

$$(g, x) * (h, y)$$

if and only if for some $(g_j, x_j)$ we have

$$(g, x) = (g_1, x_1) \sim (g_2, x_2) \sim \cdots \sim (g_n, x_n) = (h, y).$$

The equivalence class containing $(g, x)$ is denoted by $\langle g, x \rangle$ and the quotient space (of equivalence classes) is denoted by $X^*$. Note that if

$$\langle g, x \rangle = \langle h, y \rangle$$

then

$$g(x) = h(y) \tag{9.8.1}$$

and

$$\langle fg, x \rangle = \langle fh, y \rangle. \tag{9.8.2}$$

In addition, if $x \in P$, then

$$g = h, \qquad x = y. \tag{9.8.3}$$

These facts holds for $\sim$ and hence for $*$.

Each $f$ in $G$ induces a map $f^*: X^* \to X^*$ by the rule

$$f^*: \langle g, x \rangle \mapsto \langle fg, x \rangle$$

and this is well defined by (9.8.2). It is clear that

$$(f^{-1})^* = (f^*)^{-1}$$

and

$$(fh)^* = f^* h^*$$

so the group $G^*$ of all such $f^*$ is a group of bijections of $X^*$ onto itself and $f \mapsto f^*$ is a homomorphism of $G$ onto $G^*$. In fact, this is an isomorphism for if $f^* = h^*$, select $x$ in $P$ and observe that

$$\begin{aligned}
\langle f, x \rangle &= f^*\langle I, x \rangle \\
&= h^*\langle I, x \rangle \\
&= \langle h, x \rangle.
\end{aligned}$$

As $x \in P$, (9.8.3) implies that $f = h$.

If we now define

$$\langle P \rangle = \{ \langle I, x \rangle : x \in P \}$$

and similarly for $\langle \tilde{P} \rangle$ we find that the action of $G^*$ on $\langle P \rangle$ tesselates $X^*$ in the sense that

$$\bigcup_{g^*} g^*\langle \tilde{P} \rangle = X^* \tag{9.8.4}$$

and, if $g^* \neq h^*$, then

$$g^*\langle P \rangle \cap h^*\langle P \rangle = \varnothing : \tag{9.8.5}$$

the proof is trivial.

The relevance of this tesselation to the original problem is easily explained. By (9.8.1) there is a natural map $\alpha: X^* \to X$ given by

$$\alpha\langle g, x \rangle = g(x)$$

and we have the following result.

**Proposition 9.8.1.** (i) *If $\alpha$ is surjective, then*

$$\bigcup_{g \in G} g(\tilde{P}) = X.$$

(ii) *If $\alpha$ is injective, then for distinct $g$ and $h$ in $G$,*

$$g(P) \cap h(P) = \varnothing.$$

Again, the proof (which uses only (9.8.4) and (9.8.5)) is trivial and is omitted. Note that so far, there has been no mention of topology.

We now introduce topologies: explicitly we make the following assumptions.

(A3) $X$ *is a metric space with metric $d$;*
(A4) *the $g_s$ are isometries of $X$ onto itself;*
(A5) $P$ *is open and connected.*

In order to analyse the map $\alpha$ and so use Proposition 9.8.1, we introduce the natural maps

$$\beta: G \times \tilde{P} \to X^*,$$

$$\gamma: G \times \tilde{P} \to X,$$

given by

$$\beta(g, x) = \langle g, x \rangle,$$

$$\gamma(g, x) = g(x).$$

Note that

$$\gamma = \alpha\beta \tag{9.8.6}$$

so the following figure is commutative.

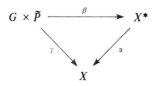

Figure 9.8.1

We give $G$ the discrete topology, $G \times \tilde{P}$ the product topology and $X^*$ the quotient topology. The quotient map $\beta$ is automatically continuous. Next, $\gamma$ is continuous for if $A$ is open in $X$, then

$$\gamma^{-1}(A) = \bigcup_g \{g\} \times (g^{-1}(A) \cap \tilde{P})$$

and this is open in $G \times \tilde{P}$. Finally, as $\gamma$ is continuous, then so is $\alpha$ because $\alpha^{-1}(A)$ is open in $X^*$ if and only if $\beta^{-1}\alpha^{-1}(A)$ is open in $G \times \tilde{P}$.

Each $f$ in $G$ induces a map $\tilde{f}: G \times \tilde{P} \to G \times \tilde{P}$ by the rule

$$\tilde{f}: (g, x) \mapsto (fg, x).$$

Trivially, the $\tilde{f}$ are homeomorphisms of $G \times \tilde{P}$ onto itself, the group of such $\tilde{f}$ is isomorphic to $G$ and

$$\beta\tilde{f} = f^*\beta,$$

$$\gamma\tilde{f} = f\gamma,$$

so

$$\alpha f^*\beta = \gamma\tilde{f} = f\gamma. \tag{9.8.7}$$

In addition, if $A$ is an open subset of $X^*$, then

$$\beta^{-1}(f^*)^{-1}(A) = (\tilde{f})^{-1}\beta^{-1}(A),$$

which is open in $G \times \tilde{P}$. We deduce that $(f^*)^{-1}(A)$ is open in $X^*$ so $f^*$ is continuous. As $(f^*)^{-1} = (f^{-1})^*$, we see that the $f^*$ are homeomorphisms of $X^*$ onto itself.

The final assumption replaces the intuitive angle condition by a formal, dimension free condition which enables us to express the formal details of the proof easily. We require a condition which guarantees that at each point $x$

of $\partial P$ there is a (local) tesselation of some neighbourhood of $x$. This condition must express the fact that the geometry of this local tesselation is consistent with the equivalence relation $*$ and nothing less than this can be adequate.

In order to express this condition concisely, suppose for the moment that

$$\langle I, x \rangle = \{(g_1, x_1), \ldots, (g_n, x_n)\}.$$

Then some $(g_j, x_j)$ is $(I, x)$ and

$$g_1(x_1) = \cdots = g_n(x_n) = I(x) = x.$$

If

$$N_j = \{y \in \tilde{P}: d(y, x_j) < \varepsilon\},$$

(this is the ball in $\tilde{P}$ with centre $x_j$ and radius $\varepsilon$), then the sets $g_j(N_j)$ are subsets of $g_j(\tilde{P})$ and have the point $x$ $(=g_j(x_j))$ in common. As the $g_j$ are isometries,

$$g_j(N_j) \subset \{y \in X: d(y, x) < \varepsilon\}$$
$$= B(x, \varepsilon),$$

say, and we wish to impose the condition that for all sufficiently small $\varepsilon$ the sets $g_j(N_j)$ tesselate $B(x, \varepsilon)$. Formally, we assume

(A6) *Each $x$ in $\tilde{P}$ has a finite equivalence class*

$$\langle I, x \rangle = \{(g_1, x_1), \ldots, (g_n, x_n)\}.$$

*In addition, for all suffiiently small $\varepsilon$,*

$$\bigcup_{j=1}^{n} g_j(N_j) = B(x, \varepsilon)$$

*and, moreover, for each $w$ in $B(x, \varepsilon)$, the set of points in $\bigcup (g_j, N_j)$ which map by $\gamma$ to $w$ is an equivalence class.*

Observe that the result that we are seeking can be expressed by saying that the set of points in $G \times \tilde{P}$ which map by $\gamma$ to any $w$ in $X$ is an equivalence class (so $\alpha$ is a bijection). Thus (A6) appears as a natural local version of the desired global result. Also, observe that as $\langle f, x \rangle$ is the image of $\langle I, x \rangle$ under $f^*$, each equivalence class is finite.

Let us write

$$W = \bigcup_{j} (g_j, N_j), \qquad V = \beta(W).$$

The condition (A6) implies that $\gamma(W) = B(x, \varepsilon)$ and also that $W$ is a union of equivalence classes. In other words,

$$\beta^{-1}(V) = \beta^{-1}(\beta W) = W$$

and we deduce that *V is open in $X^*$.*

To complete the details of the proof we need the following result.

**Proposition 9.8.2.** *The sets $f^*(V)$ are a base for the topology of $X^*$.*

PROOF. We know that the sets $f^*(V)$ are open. Suppose that $A$ is an open subset of $X^*$ and that $\langle f, x \rangle \in A$. Writing $\langle I, x \rangle$ as in (A6) we find that

$$\langle f, x \rangle = \{(fg_1, x_1), \ldots, (fg_n, x_n)\}.$$

As $\beta$ is continuous,

$$\beta^{-1}(A) = \bigcup_{h \in G} (h, A_h),$$

where each $A_h$ is open in $\tilde{P}$. As $(fg_j, x_j)$ is in $\beta^{-1}(A)$, we see that $x_j \in A_h$ when $h = fg_j$ so for these $h$, we have $A_h \neq \varnothing$. Now choose $\varepsilon$ sufficiently small so that (A6) is applicable and that $N_j \subset A_h$ when $h = fg_j$ (this is possible as $j$ takes only the values $1, \ldots, n$ and these $A_h$ are open and non-empty in $\tilde{P}$). Clearly, this means that

$$\tilde{f}(W) = \bigcup_j (fg_j, N_j) \subset \beta^{-1}(A)$$

and so

$$f^*(V) = f^*\beta(W)$$
$$= \beta\tilde{f}(W)$$
$$\subset A.$$

As $(I, x) \in W$, so $\langle f, x \rangle$ (which is $\beta\tilde{f}(I, x)$) lies in $f^*\beta(W)$ and this is $f^*(V)$. $\qquad\square$

We proceed now with the general discussion. First, by (9.8.7),

$$\alpha f^*(V) = \alpha f^*\beta(W)$$
$$= f\gamma(W)$$
$$= B(fx, \varepsilon).$$

Thus $\alpha$ maps each $f^*(V)$ to an open set and so $\alpha: X^* \to X$ *is an open map.*

Next, if $u$ and $v$ are in $f^*(V)$ and if $\alpha(u) = \alpha(v)$, then choose points $u'$ and $v'$ in $\tilde{f}(W)$ with $\beta(u') = u$, $\beta(v') = v$. Thus

$$\gamma(u') = \alpha\beta(u') = \alpha\beta(v') = \gamma(v')$$

and so by (A6) (after referring the problem back to $W$), $u'$ and $v'$ are in the same equivalence class: hence

$$u = \beta(u') = \beta(v') = v.$$

We deduce that $\alpha$ is a bijection, and hence *a homeomorphism, of each* $f^*(V)$ *onto* $f\gamma(W)$.

Next, $X^*$ is Hausdorff. To see this, take distinct points

$$\langle f, x \rangle = \{(f_1, x_1), \ldots, (f_n, x_n)\},$$

$$\langle g, y \rangle = \{(g_1, y_1), \ldots, (g_m, y_m)\}$$

in $X^*$: as these are distinct points in $X^*$, they are disjoint subsets of $G \times \tilde{P}$.

Now choose the $N_i$ corresponding to $\langle I, x \rangle$ as in (A6) and let $M_j$ be the corresponding sets for $\langle I, y \rangle$. We may choose the $N_i$ and $M_j$ so that the sets

$$\bigcup_i (f_i, N_i), \qquad \bigcup_j (g_j, M_j) \tag{9.8.8}$$

are disjoint in $G \times \tilde{P}$ (if $f_i \neq g_j$ then $(f_i, N_i)$ and $(g_j, M_j)$ are disjoint: if $f_i = g_j$ then $x_i \neq y_j$ and we can insist that $N_i$ and $M_j$ are disjoint). Because the disjoint sets (9.8.8) are each a union of equivalence classes, it follows that their $\beta$-images are disjoint (and open): thus $X^*$ is *Hausdorff*.

Finally, $X^*$ is connected. Indeed, as $\tilde{P}$ is connected, so are $(g, \tilde{P})$ and its $\beta$-image $\langle g, P \rangle$. Observe that if $x \in s'$ then

$$\langle g, x \rangle = \langle gg_s, (g_s)^{-1}x \rangle$$

so

$$\langle g, \tilde{P} \rangle \cap \langle\langle gg_s, \tilde{P} \rangle \neq \varnothing.$$

We deduce that

$$\langle g, \tilde{P} \rangle \cup \langle gg_s, \tilde{P} \rangle$$

is connected: hence so is $X^*$ as each $g$ is a product of the $g_s$. The next result is a summary of the results obtained so far.

**Proposition 9.8.3.** $X^*$ *is Hausdorff and connected. Also, every $x^*$ in $X^*$ has an open neighbourhood $N^*$ such that the restriction of $\alpha$ to $N^*$ is a homeomorphism of $N^*$ onto an open subset of $X$.*

Let us specialize now to the case of major interest to us. We suppose that $(X, d)$ is the hyperbolic plane with the hyperbolic metric (the argument will work equally well in the Euclidean plane or in the sphere $S^2$), that $P$ is a hyperbolic polygon (possibly with vertices and free sides on the circle at infinity: these are *not* in $X$) and finally, that $\Phi$ is some given set of side-pairing isometries. Our aim is to deduce that $G$ is discrete and that $P$ is a fundamental polygon for $G$. Note that (A1)–(A5) hold and that there is no need to check (A6) at points on the circle at infinity.

The condition (A6) is easily restated in a simpler form. If $x$ is in $P$, choose $\varepsilon$ so that the open disc $N$ with centre $x$ and radius $\varepsilon$ lies in $P$. For each $y$ in $N$, the equivalence class $\langle I, y \rangle$ contains only $(I, y)$ and (A6) holds trivially for this choice of $x$. Next, suppose that $x$ is in the *interior* of a side $s$. Then $x$ is on a *unique* side of $P$ and this leads immediately to the fact that $\langle I, x \rangle$ contains precisely $(I, x)$ and $(g_s^{-1}, g_s x)$. It is clear that (A6) holds in this case too (with $W$ being the union of two semi-discs) *provided that $g_1(N_1) \cup g_2(N_2)$ is a neighbourhood of $x$* (which we shall assume implicitly).

We now see that (A6) may be rewritten in terms of the vertices of $P$ alone: indeed, (A6) is now equivalent to

(A6)' *for each vertex $x$ of $P$, there are vertices $x_0$ $(=x)$, $x_1, \ldots, x_n$ of $P$ and elements $f_0$ $(=I), \ldots, f_n$ of $G$ such that the sets $f_j(N_j)$ are non-overlapping sets whose union is $B(x, \varepsilon)$ and such that each $f_{j+1}$ is of the form $f_j g_s$ for some $s$ ($j = 1, \ldots, n; f_{n+1} = I$).*

We also assume

(A7) *the $\varepsilon$ in (A6) can be chosen independently of $x$ in $\tilde{P}$.*

This last assumption ensures that each curve in $X$ can be lifted to a curve in $X^*$ (for each point can be referred back to $\tilde{P}$ and then lifted for at least a distance $\varepsilon$) and so $(X^*, \alpha)$ is a smooth unlimited covering surface of $X$, $\alpha$ mapping $X^*$ onto $X$. As $X$ is simply connected, the Monodromy Theorem implies that $\alpha$ is now a homeomorphism and the desired result follows from Proposition 9.8.1. We have proved

**Theorem 9.8.4** (Poincaré's Theorem). *For a polygon $P$ with a side-pairing $\Phi$ satisfying (A6)' and (A7), $G$ is discrete and $P$ is a fundamental polygon for $G$.*

*Remark.* If $P$ has no vertices in $X$, then (A6)' is automatically satisfied. However, (A7) need not hold.

**Example 9.8.5.** Let $P$ be a polygon with $r$ sides and angles $\pi/n_j$ at the vertices $v_j$ in $X$ ($j = 1, \ldots, r$). For each side $s$ let $g_s$ be the reflection across $s$: denote these maps by $g_1, \ldots, g_r$. Then (A6)' holds (see Figure 9.8.2) and (A7) holds (essentially because $\tilde{P}$ is compact). Thus $P$ is a fundamental polygon for $G$.

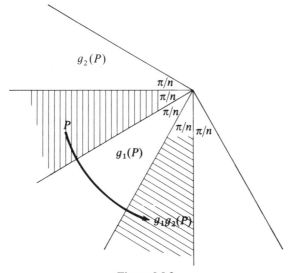

Figure 9.8.2

Later, we shall need the following result and this is closely related to Poincaré's Theorem. Let $P$ be an open hyperbolic polygon in $\Delta$ and let $\Phi$ be a side-pairing of $P$. We shall always assume that if $x$ is an interior point of a side $s$ with corresponding point $y = g_s(x)$, then for any choice of neighbourhoods $N_x$ and $N_y$ *relative to $P$* of $x$ and $y$ respectively, the set

$$N_x \cup (g_s)^{-1}(N_y)$$

is a neighbourhood of $x$ (thus $N_x$ and $(g_s)^{-1}N_y$ lie on different sides of $s$ near $x$).

Finally, for each $z$ and each $g$, define

$$\theta(z) = \sum_{g \in G} \theta_g(z),$$

where $\theta_g(z)$ is the angle subtended at $z$ by $g(P)$. If $z \in g(P)$ then $\theta_g(z) = 2\pi$; if $z \notin g(\bar{P})$, then $\theta_g(z) = 0$.

**Theorem 9.8.6.** *Let $P$ be a hyperbolic polygon with compact closure in $\Delta$ and let $\Phi$ be a side-pairing satisfying the assumption given above. If the group $G$ generated by the side-pairing elements is discrete, then $\theta(z)$ is constant, say $2\pi k$, on $\Delta$, $k$ is an integer and*

$$\text{h-area}(P) = k \ \text{h-area}(\Delta/G).$$

PROOF. Let $V$ be the set of all images of all vertices of $P$: by the discreteness of $G$, $V$ contains only isolated points in $\Delta$. Let $B$ be the union of all images of $\partial P$: by discreteness, $B$ is a closed subset of $\Delta$ and obviously, $V \subset B$.

The set $\Delta - B$ is open so is a disjoint union of domains, say $\Delta_j$. By definition, each $\Delta_j$ either lies inside $g(P)$ or is disjoint from $g(P)$ so $\theta_g$ is $2\pi$ throughout $\Delta_j$ or it is zero throughout $\Delta_j$. We deduce that $\theta(z)$ is constant on each $\Delta_j$, say equal to $2\pi k_j$ there where $k_j$ is an integer.

Next, consider $w$ in $B - V$. The side-pairing assumptions ensure that there are pairs of distinct elements $(g_j, g_j^*)$, say, $(j = 1, \ldots, n)$ such that $w$ lies interior to a common side of $g_j(P)$ and $g_j^*(P)$ and that for all other $g$, $\theta_g(z)$ is constant (0 or $2\pi$) near (that is, in a neighbourhood of) $w$. By discreteness, we can choose one such neighbourhood, say $N$, valid for all other $g$. Each term

$$\theta_{g_j}(z) + \theta_{g_j^*}(z)$$

is constant (namely $2\pi$) near $w$: thus $\theta(z)$ is constant near $w$. We conclude that $\theta/2\pi$ is continuous and integer valued on the domain $\Delta - V$. As $V$ contains only isolated points, $\theta$ is constant on $\Delta - V$. A similar argument holds for $w$ in $V$; however, this is of no consequence.

Finally, let $Q$ be any open fundamental polygon for the discrete $G$ and for any set $A$, let $\chi_A$ be the characteristic function of $A$. For almost all $z$ in $\Delta$, we have

$$\sum_g \chi_{g(Q)}(z) = 1.$$

Also, for almost all $z$ in $\Delta$,

$$k = \sum_g \theta_g(z)/2\pi$$

$$= \sum_g \chi_{g(P)}(z).$$

Thus (writing $\mu$ for hyperbolic area and taking all integrals over $\Delta$) we have

$$\mu(P) = \int \chi_P(z) \left[ \sum_g \chi_{g(Q)}(z) \right] d\mu(z)$$

$$= \sum_g \int \chi_P(z)\chi_{g(Q)}(z)\, d\mu(z)$$

$$= \sum_g \int \chi_{g^{-1}(P)}(w)\chi_Q(w)\, d\mu(w)$$

$$= \sum_{g^{-1}} \int \chi_{g^{-1}(P)}(w)\chi_Q(w)\, d\mu(w)$$

$$= \int \chi_Q(w) \left[ \sum_h \chi_{h(P)}(w) \right] d\mu(w)$$

$$= k\mu(Q). \qquad \qquad \square$$

In fact, Theorem 9.8.6 says that if we identify the sides of $P$ we obtain a branched covering of the compact space $\Delta/G$: thus the covering is a $k - 1$ map for some $k$.

### EXERCISE 9.8

1. The proof of Poincaré's Theorem remains valid if (A7) is replaced by: (A7)' there exists a positive $\varepsilon$ such that for all $x$ in $\tilde{P}$ there is a single valued branch of $\alpha^{-1}$ in $B(x, \varepsilon)$.

   Show that the application in Example 9.8.5 remains valid if some of the $v_j$ now lie on the circle at infinity (the $v_j$ are now fixed by a parabolic elements in $G$ and a horocyclic disc at $v_j$ is suitably tesselated).

2. Generalize the ideas in Question 1 to include arbitrary polygons with some vertices on the circle at infinity provided that these are fixed by parabolic elements in $G$ (but see Question 3).

3. The condition concerning parabolic elements in Question 2 is essential. Show (for example) that $g: z \mapsto 2z$ is a side-pairing of

$$P = \{z \in H^2 : 1 < \mathrm{Re}[z] < 2\}$$

   but that $P$ is not a fundamental domain for $G$ ($= \langle g \rangle$) in $H^2$.

   Show, however, that Poincaré's Theorem is applicable to this $P$ and $G$ if $G$ is considered to act on the first quadrant with the metric $ds = (|z|/xy)|dz|$.

4. Let $X = \mathbb{C} - \{0\}$ with metric $ds = |dz|/|z|$. For $\theta$ in $(0, 2\pi)$, let

$$P = \{z \in X : 1 < |z| < 3, 0 < \arg(z) < \theta\}.$$

Divide $\partial P$ into four sides in the obvious way and generate $G$ from the side-pairing isometries

$$g(z) = 3z, \qquad h(z) = e^{i\theta}z.$$

Examine the case $\theta = 2\pi p/q$ where $(p, q) = 1$ by reference to the covering surface $X^*$ (which exists even if $X$ is not simply connected). One can view this as a multiple tesselation of $X$.


## §9.9. Notes

There are other constructions of fundamental polygons and, in particular, of polygons which relate to a particular defining relation (a product of commutators) of a group with compact quotient space. For further information see, for example, [46], [47], [52], [70], [85], [86] and [114]. For other information on convex fundamental polygons, see [71], [72], [73] [83]: for recent treatments of Poincaré's Theorem (Section 9.8) see [24] [48] and [62]. Theorem 9.8.6 occurs in [48].

# CHAPTER 10
# Finitely Generated Groups

## §10.1. Finite Sided Fundamental Polygons

We recall that a side $s$ of a convex fundamental polygon $P$ is a segment of the form $\tilde{P} \cap g(\tilde{P})$ (except that this set may be considered as two sides when $g$ is elliptic and of order two). By an *edge* of $P$ we mean a maximal geodesic segment in $\partial P$. We must distinguish carefully between sides and edges and to convince the reader of the necessity of this, we begin with an example in which one edge contains infinitely many sides.

**Example 10.1.1.** We work in $H^2$. For $n = 0, 1, 2, \ldots$, let $C_n$ be the geodesic with end-points $1 + 4n$ and $3 + 4n$ and let $C'_n$ be its reflection in the imaginary axis. For each $n$, let $g_n$ be the hyperbolic element that preserves $H^2$ and that maps the exterior of $C_n$ onto the interior of $C'_n$ and let $G$ be the group generated by the $g_n$. By Poincaré's Theorem (Section 9.8), the region exterior to all of the $C_n$ and $C'_n$ is a fundamental domain for $G$.

Now let $D$ be the region in the second quadrant exterior to all of the $C'_n$ and let

$$D_n = \{x + iy : x > 0, y > 0, 4n < |z| < 4(n + 1), |z - (4n + 2)| \geq 1\}:$$

see Figure 10.1.1. It is clear that

$$P = D \cup \left( \bigcup_{n=0}^{\infty} g_n[D_n \cup C_n] \right)$$

is a convex fundamental polygon for $G$ and that the positive imaginary axis is a single edge $e$ of $P$. However, for each $n$,

$$g_n^{-1}(\tilde{P}) \cap \tilde{P} = [4in, 4i(n + 1)]$$

and so the edge $e$ contains infinitely many sides of $P$.

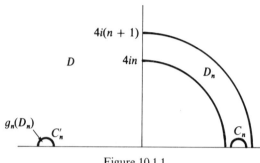

Figure 10.1.1

In view of the preceding example, and as we are about to make claims about the number of sides of a polygon, the distinction between edges and sides must be kept clear.

**Theorem 10.1.2.** *Let G be a non-elementary Fuchsian group with Nielsen region N. Then the following statements are equivalent:*

(1) *G is finitely generated;*
(2) *for any convex fundamental polygon P of G, h-area $(P \cap N) < +\infty$;*
(3) *there exists a convex fundamental polygon of G with finitely many sides;*
(4) *every convex fundamental polygon of G has finitely many sides.*

PROOF. Obviously, (4) implies (3). Now assume that (3) holds and let $P$ be a finite sided convex fundamental polygon. Each closed free side $\lambda_i$ $(i = 1, \ldots, m)$ of $P$ lies in the interior of an interval of discontinuity $\sigma_i$ which determines a half-plane $H_i$ containing $N$ (see Section 8.5). Then

$$P_1 = P \cap H_1 \cap \cdots \cap H_m$$

is a finite sided polygon with no free sides and so has finite h-area. As $P_1$ contains $P \cap N$, we see that (2) holds for this choice of $P$. However, as $N$ is $G$-invariant, it is easy to see that h-area $(P \cap N)$ is independent of the choice of $P$ and (2) follows.

Next, we prove that (2) implies (1). First, write $Q = P \cap N$. It is clear that $N$ meets $\partial P$ and as $N$ is $G$-invariant, $\tilde{Q}$ contains all or none of the points in a cycle on $\partial P$. Suppose that $\tilde{Q}$ contains (perhaps as a proper subset) cycles $C_1, \ldots, C_t$ of vertices in $\Delta \cap \partial P$ and also points $w_1, \ldots, w_n$ on $\partial \Delta$ and let $Q_0$ be the polygon whose set of vertices is

$$C_1 \cup \cdots \cup C_t \cup \{w_1, \ldots, w_n\}$$

($Q_0$ is the convex hull of a finite set of points on the boundary of a convex set). By convexity, $Q_0 \subset Q$ so the sum of the interior angles of $Q_0$ at its vertices is not greater than the sum of the interior angles of $Q$ at the same points. If the cycle $C_j$ has length $\ell_j$ and order $q_j$, we deduce from Section 7.15 that

h-area$(Q) \geq$ h-area$(Q_0)$

$$\geq \pi[n + \ell_1 + \cdots + \ell_t - 2] - 2\pi\left(\frac{1}{q_1} + \cdots + \frac{1}{q_t}\right)$$

$$= \pi(n - 2) + \pi \sum_j \left(\ell_j - \frac{2}{q_j}\right).$$

It is convenient to adopt here the convention that an elliptic fixed point of order two and cycle length one is *not* a vertex. With this convention, each cycle $C_j$ is either accidental (and $\ell_j \geq 3$, $q_j = 1$) or elliptic (and $\ell_j q_j \geq 3$). In all cases, then,

$$\ell_j - \frac{2}{q_j} \geq \frac{\ell_j}{3} \geq \tfrac{1}{3}$$

so

$$3n + t < 6 + (3/\pi)\, \text{h-area}(Q).$$

It follows that only finitely many sides of $P$ (in either convention) meet $\tilde{Q}$.

Now let $g_s$ be those side-pairing maps of $P$ for which the corresponding sides meet $\tilde{Q}$: there are only finitely many such $g_s$ and it is only necessary to show that these generate $G$. We select any $g$ in $G$ and, by the convexity and invariance of $N$, we can join two points, one in $Q$ and the other in $g(Q)$, by a segment $\sigma$ in the convex set $N$. We may assume that $\sigma$ does not meet any image of any vertex of $P$; then $\sigma$ crosses, in turn, the images

$$P, g_1(P), \ldots, g_n(P),$$

where $g = g_n$. As

$$g_{j+1}(\tilde{P}) \cap g_j(\tilde{P}) \cap N$$

is a geodesic segment, so is

$$\tilde{P} \cap (g_{j+1})^{-1} g_j(\tilde{P}) \cap N,$$

(because $N$ is $G$-invariant) and so each $(g_{j+1})^{-1} g_j$ is some $g_s$. This proves that (2) implies (1).

Next, we prove that (1) implies (3): a proof that (3) implies (4) will be given in the next section and this will then complete the proof of Theorem 10.1.2. We assume that (1) holds and let $D$ be the Dirichlet polygon with centre at the origin (which we may assume is not an elliptic fixed point of $G$). The idea of this part of the proof is first to show that the Euclidean boundary of $D$ on $\partial\Delta$ has only finitely many components (so the surface $\Delta/G$ has only finitely many "ends"). This allows us to express $D$ in the form

$$D = K \cup D_1 \cup \cdots \cup D_n,$$

where $K$ is a compact subset of $\Delta$ and where each $D_j$ is a subdomain of $\Delta$ whose boundary on $\partial\Delta$ is connected. It is then only necessary to show that only finitely many sides of $D$ meet each $D_j$ for this is certainly true of the compact set $K$. Actually, it is not essential for the proof to show that the boundary of $D_j$ on $\partial\Delta$ is connected but the proof of this is very easy and it can only increase the general understanding of the ideas involved.

The side-pairing elements of $D$ generate $G$ and so each element of some finite generating set (which exists by (1)) is a finite word in the side-pairing elements of $D$. It follows that a finite number of side-pairing elements generate $G$: let these be $g_1, \ldots, g_t$.

Choose some $r$ in $(0, 1)$ such that the disc $\{|z| \le r\}$ contains arcs (of positive length) of each of the sides of $P$ paired by $g_1, \ldots, g_t$. Let

$$K = \tilde{D} \cap \{z : |z| < r\}$$

(it is more convenient to take this $K$ rather than its compact closure) and let

$$G(K) = \bigcup_{g \in G} g(K).$$

Observe that for each $j$, the set $K \cup g_j(K)$ is connected ($K$ is convex) and hence so is each

$$K \cup g_{j_1}(K) \cup g_{j_1}g_{j_2}(K) \cup \cdots \cup (g_{j_1} \cdots g_{j_q})(K).$$

Because the $g_i$ generate $G$, this implies that $G(K)$ is connected.

We may choose $r$ in $(0, 1)$ so that the circle $\{|z| = r\}$ does not meet any vertex of $D$ and so that it is not tangent to any side of $D$. Then

$$\tilde{D} \cap \{z : |z| = r\} = \sigma_1 \cup \cdots \cup \sigma_s,$$

where the $\sigma_j$ are pairwise disjoint closed arcs of $\{|z| = r\}$ lying, apart from their end-points, entirely in $D$. Note that by Theorem 9.4.3, the collection of end-points of all the $\sigma_j$ are also paired by the side-pairing maps. This implies that each end-point of each $\sigma_j$ is the end-point of some $h(\sigma_i)$ for a unique $h$ and unique $\sigma_i$. The same is true of each $h(\sigma_i)$ and of each subsequent image of the $\sigma_i$, and we deduce that each $\sigma_j$ lies in a simple arc $\Gamma_j$ comprising of images of the $\sigma_i$. Because there are only finitely many $\sigma_i$, the arcs $\Gamma_j$ contain images of the same $\sigma_i$ and the uniqueness of the construction of the $\Gamma_j$ implies that $\Gamma_j$ is invariant under some non-trivial element $h_j$ of $G$. Note that $\Gamma_j$ consists of the images of a compact arc under iterates of $h_j$. If $h_j$ is elliptic (and hence of finite order) then $\Gamma_j$ is a Jordan curve in $\Delta$. If $h_j$ is hyperbolic, then $\Gamma_j$ is a cross-cut of $\Delta$ with the fixed points of $h_j$ as its end-points. If $h_j$ is parabolic, then $\Gamma_j$ is a closed Jordan curve in $\Delta$ apart from its initial (and equal final) point which is the fixed point of $h_j$. Note that a point of $K$ cannot be equivalent to any point of any $\sigma_j$ so $G(K)$ does not meet any $\Gamma_j$.

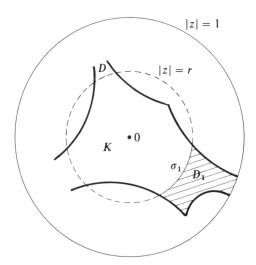

Figure 10.1.2

Now let $D_j$ be the union of $\sigma_j$ (a cross-cut of $D$) and the component of $D-\sigma_j$ that does not contain the origin: see Figure 10.1.2. Observe that $\Gamma_j$ separates $D_j$ and $G(K)$ in $\Delta$.

It is easy to see that $\bar{D}_j \cap \partial\Delta$ is connected. Indeed, if $u$ and $v$ are distinct points in this set, construct a curve by joining $u$ to $ru$ (on $\sigma_j$) radially, then $ru$ to $rv$ in $\sigma_j$ and finally, $rv$ to $v$ radially. This curve, which we denote by $\tau_j$, lies in $D$ and does not meet $G(K)$. If $h$ ($\neq I$) is in $G$, then $h(D)$ does not meet $\tau_j$ and so lies on the same side of $\tau_j$ as does $G(K)$. We deduce that the region $\Sigma_j$ illustrated in Figure 10.1.3. does not meet any $h(D)$, $h \neq I$, and so lies in $D$. This shows that $\bar{D}_j \cap \partial\Delta$ is connected.

We now return to the classification of the $h_j$ stabilizing $\Gamma_j$ and complete the proof. If $h_j$ is elliptic, then $\Gamma_j$ is a Jordan curve in $\Delta$ so one component of $\Delta-\Gamma_j$ has a compact closure in $\Delta$. If this component is $D_j$, then only finitely many sides of $D$ meet $\tilde{D}_j$. If this component is not $D_j$, then it contains $G(K)$ and so $G$ is finite: then the Dirichlet polygon for $G$ obviously only has a finite number of sides.

Suppose next, that $h_j$ is hyperbolic so $\Gamma_j$ is a cross-cut of $\Delta$. One component of $\Delta-\Gamma_j$ contains $G(K)$ (and hence the orbit of the origin) and so every limit point of $G$ lies in the closure of this component. The other component of $\Delta-\Gamma_j$ contains $D_j$ and there are no limit points on the open arc of $\partial\Delta$ that bounds this component. However, $D_j$ lies in $D$ and so lies between the geodesic bisecting the segment $[0, h_j0]$ and the geodesic bisecting $[0, (h_j)^{-1}0]$ and these separate $D_j$ from the fixed points of $h_j$. We deduce that the Euclidean closure of $D_j$ lies in the set of ordinary points of $G$. As the Euclidean diameters of images of $D$ tend to zero (Section 9.3) we see that $\tilde{D}_j$ can meet only a finite number of images of $\tilde{D}$ and hence only a finite number of sides of $D$.

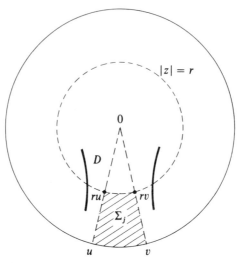

Figure 10.1.3

Finally, suppose that $h_j$ is parabolic. In this case, $\bar{D}_j \cap \partial\Delta$ consists of a single point, namely the fixed point of $h_j$. However, we know that two sides of $D$ end at a parabolic fixed point (Theorem 9.3.8) so in this case too, only finitely many sides of $D$ meet $\tilde{D}_j$.

Subject to proving that (3) implies (4) (in Section 10.2) the proof of Theorem 10.1.2 is complete.                                                   □

## §10.2. Points of Approximation

Consider a Fuchsian group $G$ acting in $\Delta$. Let $\zeta$ be a limit point of $G$ so there are distinct $g_n$ in $G$ with $g_n(0)$ converging to $\zeta$. How fast (in Euclidean terms) can $g_n(0)$ converge to $\zeta$? Clearly,

$$|\zeta - g_n(0)| \geq 1 - |g_n(0)|$$

with equality if, for example, $g_n$ is the $n$th iterate of some hyperbolic $g$ with axis equal to the Euclidean diameter $[-\zeta, \zeta]$ of $\Delta$. We conclude that the fastest rate of convergence (to within a constant factor) occurs when

$$|\zeta - g_n(0)| = O(1 - |g_n(0)|)$$

as $n \to +\infty$. As the terms

$$\|g_n\|^2, \ 2\cosh \rho(0, g_n 0), \ 2/(1 - |g_n(0)|)$$

are asymptotic to each other as $n \to \infty$, this fastest rate can be described in hyperbolic terms, namely,

$$|\zeta - g_n(0)| = O(1/\cosh \rho(0, g_n 0)),$$

or in terms of matrices, namely,

$$|\zeta - g_n(0)| = O(\|g_n\|^{-2}).$$

Moreover, it is easy to see that we can replace the origin in the first two expressions by any $z$ in $\Delta$. This is implicit in the next result which provides yet another interpretation of this fastest rate of convergence.

**Theorem 10.2.1.** *Let $G$ be a Fuchsian group acting in $\Delta$, let $\zeta$ be a limit point of $G$ and let $g_1, g_2, \ldots$ be distinct elements of $G$. Then the following statements are equivalent:*

(1) *for each $w$ in $\Delta$,*

$$|\zeta - g_n(w)| = O(\|g_n\|^{-2});$$

(2) *for each $w$ in $\Delta$ and each geodesic half-ray $L$ ending at $\zeta$,*

$$o(g_n(w), L) = O(1);$$

(3) *for each geodesic half-ray $L$ ending at $\zeta$ there is a compact subset $K$ of $\Delta$ such that for all $n$,*

$$(g_n)^{-1}(L) \cap K \neq \varnothing.$$

PROOF. In general, $\rho(gw, L) \leq m$ if and only if $g^{-1}(L)$ meets the compact disc $\{z : \rho(z, w) \leq m\}$: thus (2) and (3) are equivalent and for a given $L$, (2) is true or false independently of the choice of $w$. Further if $L_1$ and $L_2$ are geodesic half-rays ending at $\zeta$, then for some $m_1$,

$$L_2 \subset \{z : \rho(z, L_1) \leq m_1\}$$

so (2) is also true or false independently of the choice of $L$. For the remainder of the proof, $L$ will denote the Euclidean radius $[0, \zeta)$ and $L'$ will denote the Euclidean diameter $(-\zeta, \zeta)$. Observe that if $z$ is close to $\zeta$, then $\rho(z, L) = \rho(z, L')$.

Suppose first that (1) holds. Then putting $w = 0$ we obtain

$$|\zeta - g_n(0)| = O(1 - |g_n(0)|).$$

This implies that $g_n(0) \to \zeta$ so for all sufficiently large $n$,

$$\rho(g_n 0, L) = \rho(g_n 0, L').$$

If $z \in \Delta$ then (from Section 7.20) we have

$$\sinh \rho(z, L') = \frac{2|\operatorname{Im}[\bar{\zeta}z]|}{1 - |\bar{\zeta}z|^2}$$

$$\leq \frac{2|\bar{\zeta}z - 1|}{1 - |z|^2}$$

$$\leq \frac{2|z - \zeta|}{1 - |z|}.$$

Putting $z = g_n(0)$ with $n$ large, we obtain (2) in the case $w = 0$. As (2) is independent of the choice of $w$, we see that (1) implies (2).

Next, let $z$ be in $\Delta$ and closer to $\zeta$ than to $-\zeta$ and let $v$ be the foot of the *Euclidean* perpendicular from $z$ to $L'$. Then $\zeta$ is the point on $\partial\Delta$ that is nearest to $v$ and so

$$|z - \zeta| \leq |z - v| + |v - \zeta|$$

$$\leq |z - v| + |v - (z/|z|)|$$

$$\leq 2|z - v| + |z - (z/|z|)|$$

$$\leq 2|z - v| + (1 - |z|).$$

As

$$|z - v| = |\operatorname{Im}[\bar{\zeta}z]|,$$

we deduce that

$$\frac{|z - \zeta|}{1 - |z|} \leq 2 \sinh \rho(z, L') + 1$$

$$\leq 2 \sinh \rho(z, L) + 1.$$

Putting $z = g_n(0)$ and using (2), we find that (2) implies that (1) holds with $w = 0$.

Finally, if $w \in \Delta$ and

$$g(z) = \frac{az + \bar{c}}{cz + \bar{a}}, \qquad |a|^2 - |c|^2 = 1,$$

we obtain (by direct computation)

$$|g(w) - g(0)| \leq \frac{2}{(1 - |w|) \cosh \rho(0, g0)}$$

$$= \frac{4}{(1 - |w|)\|g\|^2}, \tag{10.2.1}$$

because

$$|a| = \cosh \tfrac{1}{2}\rho(0, g0), \qquad |c| = \sinh \tfrac{1}{2}\rho(0, g0).$$

We have seen that (2) implies that (1) holds when $w = 0$. Clearly this with (10.2.1) yields (1) for a general $w$. $\qquad\qquad\square$

In view of the different characterizations of the fastest rate of convergence it is convenient to adopt some suitable terminology.

**Definition 10.2.2.** A limit point $\zeta$ of a Fuchsian group $G$ is a *point of approximation of $G$* if for each $w$ in $\Delta$ there is a sequence of distinct $g_n$ in $G$ with

$$|\zeta - g_n(w)| = O(\|g_n\|^{-2}).$$

**Theorem 10.2.3.** *A point of approximation of a Fuchsian group $G$ cannot lie on the boundary of any convex fundamental polygon for $G$.*

PROOF. Suppose that a point of approximation $\zeta$ lies on the boundary of a convex fundamental polygon $P$. By convexity, we can construct a geodesic half-ray $L$ lying in $P$ and ending at $\zeta$. By Theorem 10.2.1(3), the images $(g_n)^{-1}(P)$ meet a compact set and this violates the fact that $P$ is a locally finite (see Definition 9.3.1). $\qquad\qquad\square$

**Example 10.2.4.** Every parabolic fixed point of a Fuchsian group $G$ lies on the boundary of some Dirichlet region: thus a parabolic fixed point of $G$ cannot be a point of approximation of $G$.

For a finitely generated groups, Theorem 10.2.3 and Example 10.2.4 give a complete description of the limit points of $G$.

**Theorem 10.2.5.** *A Fuchsian group $G$ is finitely generated if and only if each limit point is either a parabolic fixed point of $G$ or a point of approximation of $G$.*

*Remark.* Let us say that the limit set $\Lambda$ *splits* if it contains only parabolic fixed points or points of approximation of $G$. If $G$ is finitely generated, then there exists a finite sided convex fundamental polygon for $G$ (because (1) implies (3) in Theorem 10.1.2). We shall show that the existence of such a polygon implies that $\Lambda$ splits. We will also prove that if $\Lambda$ splits then every convex fundamental polygon for $G$ has finitely many sides and this implies that $G$ is finitely generated (because (4) implies (1) in Theorem 10.1.2). Observe that this reasoning shows that Theorem 10.1.2(3) implies that $\Lambda$ splits and hence that Theorem 10.1.2(4) holds. Thus in proving Theorem 10.2.5 in this way, we also complete the proof of Theorem 10.1.2.

PROOF OF THEOREM 10.2.5. First, suppose that $\Lambda$ splits and let $P$ be any convex fundamental polygon for $G$. If $P$ has infinitely many sides, then these sides must accumulate at some point $\zeta$ on $\partial\Delta$. As the Euclidean diameters of the images of $P$ tend to zero, $\zeta$ must be a limit point on $\partial P$. By Theorem 10.2.3, $\zeta$ cannot be a point of approximation of $G$ and by Theorem 9.3.8, $\zeta$ cannot be a parabolic fixed point of $G$ (else two sides of $P$ end at $\zeta$). This contradicts the fact that $\Lambda$ splits so $P$ can only have finitely many sides.

Now suppose that $P$ is a finite sided convex fundamental polygon for $G$: we may assume that $P$ is a Dirichlet polygon (as the proof of Theorem 10.1.2 shows that in this case, $G$ is finitely generated and then any Dirichlet polygon is finite sided) and we may assume (for simplicity) that the conditions stated in Theorem 9.4.5 hold. By conjugation, we may also suppose that the centre of $P$ is at the origin.

If two sides of $P$, say $s$ and $s'$, have a common end-point $v$ on $\partial\Delta$, then $v$ is a parabolic fixed point of $G$ (Theorem 9.4.5) and the stabilizer of $v$ is generated by a parabolic element $p$ of $G$ which maps $s$ onto $s'$. Now construct an open horocyclic region at $v$ bounded by a horocycle $Q$. Note that there is a compact arc $q$ of $Q$ such that $Q$ is the union of the images $p^n(q)$, $n \in \mathbb{Z}$.

A similar construction holds for the free sides of $P$. Each end-point of a free side is the end-point of some image of some free side. The interval of discontinuity $\sigma$ in which a given free side lies is the countable union of images of the finite number of free sides of $P$: these images are non-overlapping and accumulate only at the end-points of $\sigma$. It follows that some $h$ in $G$ maps one image of a free side in $\sigma$ to another such image, also in $\sigma$, and so $h(\sigma) = \sigma$ (because the intervals of discontinuity are permuted by the elements of $G$). We deduce that $h$ fixes both end-points of $\sigma$ and so is hyperbolic. The geodesic $L$ with the same end-points as $\sigma$ is the axis of $h$ and we may assume that $h$ generates the stabilizer of $L$. Note that there is a compact sub-arc $\ell$ of $L$ such that $L$ is the union of the images $h^n(\ell)$, $n \in \mathbb{Z}$.

The geodesics $L$ and the horocycles $Q$ are finite in number and they separate the boundary points of $P$ on $\partial\Delta$ from a compact subset $P_0$ of $\tilde{P}$. Let $K$ denote the compact set consisting of the union of $P_0$ and the finite number of arcs $q$ and $\ell$.

Now let $\zeta$ be any limit point of $G$ which is not a parabolic fixed point and let $L_0$ be a geodesic half-ray ending at $\zeta$. The initial point of $L_0$ can be mapped to a point in $\tilde{P}$ and the corresponding image of $L_0$ cannot lie entirely in one of the horocyclic or hypercyclic regions constructed above else it ends at a parabolic fixed point or an ordinary point of $G$ respectively. It follows that either $L_0$ meets $P_0$ or, alternatively, $L_0$ meets one of these regions in which case some image of $L_0$ meets one of the arcs $q$ or $\ell$. In both cases an image of $L_0$ meets $K$ and so there is some $z_0$ in $L_0$ with, say, $g_0(z_0)$ in K.

Now let $L_n$ be obtained from $L_0$ by deleting the initial segment of $L_0$ of length $n$. Exactly as for $L_0$, the ray $L_n$, contains some $z_n$ with $g_n(z_n)$ in $K$. Clearly, $z_n \to \zeta$ and the set $\{g_1, g_2, \ldots\}$ is infinite: thus by Theorem 10.2.1, $\zeta$ is a point of approximation and $\Lambda$ splits.                                                    $\square$

EXERCISE 10.2

1. Verify Example 10.2.4 by working in $H^2$ with $\infty$ the parabolic fixed point (use Theorem 10.2.1(2)).

## §10.3. Conjugacy Classes

Any group is partitioned into the disjoint union of its conjugacy classes. The classification of conformal Möbius transformations is invariant under conjugation and so we may speak unambiguously of elliptic, parabolic and hyperbolic conjugacy classes. Within the group of all Möbius transformations, the conjugacy classes are parametrized by the common value of trace$^2$ of their elements but, as we shall now see, this is not true of the conformal group of isometries of the hyperbolic plane.

**Theorem 10.3.1.** *Within the group of all isometries of the hyperbolic plane, two non-trivial conformal isometries are conjugate if and only if they have the same value of* trace$^2$. *Within the group of conformal isometries, the value* trace$^2$ *determines two parabolic or elliptic conjugacy classes or one hyperbolic conjugacy class.*

PROOF. We shall prove the result in detail for the parabolic case only. Using the model $H^2$, any two parabolic isometries are conjugate (in the group of conformal isometries) to, say, $z \mapsto z + p$ and $z \mapsto z + q$ where $p$ and $q$ are real and non-zero. These are conjugate in the group of conformal isometries if and only if for some real $a$, $b$, $c$ and $d$ with $ad - bc = 1$, we have

$$\frac{a(z + p) + b}{c(z + p) + d} = \frac{az + b}{cz + d} + q.$$

Putting $z = -d/c$, we find that $cp = 0$: thus $c = 0$ and $ap = dq$. As $ad = 1$, we have $a^2 p = q$ so $p$ and $q$ must have the same sign. This shows that within the conformal group, trace$^2$ determines two conjugacy classes of parabolic elements. In the full group of isometries, however, the translations $z \mapsto z + 1$ and $z \mapsto z - 1$ are conjugate: indeed if $\alpha$ and $\beta$ denote reflections in $x = 0$ and $x = \frac{1}{2}$ respectively, then

$$\beta\alpha = \alpha(\alpha\beta)\alpha$$

so $\alpha\beta$ and $\beta\alpha$ are conjugate.

The elliptic case is handled similarly using the model $\Delta$ and two rotations fixing the origin. The hyperbolic case is best handled in $H^2$ with two hyperbolic elements fixing 0 and $\infty$. In this case, each element is conjugate to its inverse because there is a conformal isometry, namely $z \mapsto -1/z$, interchanging 0 and $\infty$.                                    $\square$

We are now going to examine in detail the conjugacy classes in a Fuchsian group.

**Theorem 10.3.2.** *Let $G$ be a Fuchsian group and let $v_1, v_2, \ldots$ be the parabolic and elliptic fixed points on the boundary of some convex fundamental polygon for $G$. Suppose that $g_j$ generates the stabilizer of $v_j$: then any elliptic or parabolic element of $G$ is conjugate to some power of some $g_j$.*

PROOF. If $g$ is elliptic or parabolic with fixed point $v$, then some $h$ in $G$ maps $v$ to some point on $\partial P$. Thus for some $j$, we have $h(v) = v_j$ and then $hgh^{-1} \in \langle g_j \rangle$. □

**Corollary 10.3.3.** *If $G$ is finitely generated, then $G$ has a finite number of maximal cyclic subgroups $\langle g_1 \rangle, \dots, \langle g_n \rangle$ such that any elliptic or parabolic element in $G$ is conjugate to exactly one element in exactly one of these subgroups.*

We need only observe that if $g$ is elliptic or parabolic and if two powers of $g$ are conjugate, say if

$$hg^n h^{-1} = g^m,$$

then $h$ has the same fixed points as $g$ and so is itself a power of $g$: thus $n = m$. Note that if $g$ is parabolic and fixes $v$, then $h$ also fixes $v$ and so cannot be hyperbolic.

Later, we shall need information on the number of such conjugacy classes of these maximal cyclic subgroups in a subgroup $G_1$ of $G$ and the following simple result is sufficient for our needs.

**Theorem 10.3.4.** *Let $G$ be a Fuchsian group and $G_1$ a subgroup of index $k$ in $G$. Suppose that $G$ and $G_1$ have $t$ and $t_1$ respectively, conjugacy classes of maximal parabolic cyclic subgroups. Then $t_1 \leq kt$. The same result holds for elliptic elements.*

PROOF. Let $D$ be a Dirichlet polygon for $G$ in which parabolic and elliptic fixed points on $\partial D$ have cycle length one. Thus exactly $t$ parabolic fixed points lie in $\partial D$. Now express $G$ as a coset decomposition, say

$$G = g_1 G_1 \cup \cdots \cup g_k G_1$$

so

$$D^* = (g_1)^{-1}(\tilde{D}) \cup \cdots \cup (g_k)^{-1}(\tilde{D})$$

contains at least one point from each $G_1$-orbit. As $D^*$ has at most $kt$ parabolic fixed points on its boundary, we have $t_1 \leq kt$. The same proof holds for elliptic elements. □

We turn now to the conjugacy classes of hyperbolic elements in a Fuchsian group.

**Theorem 10.3.5.** *Any non-elementary Fuchsian group contains infinitely many conjugacy classes of maximal hyperbolic cyclic subgroups.*

PROOF. Suppose not, then there are hyperbolic elements $h_1, \dots, h_t$ in $G$ such that each hyperbolic element in $G$ is conjugate to some power of some $h_j$. Let $u$ and $v$ be distinct limit points of $G$. By Theorem 5.3.8, there are hyperbolic elements $f_1, f_2, \ldots$ with distinct axes $A_1, A_2, \ldots$ such that $A_n$ has end-points $u_n$ and $v_n$ where $u_n \to u$ and $v_n \to v$.

As each $f_n$ is conjugate to some power of one of a finite number of the $h_j$, we may relabel and assume that $h_j = h_1$ for every $n$. Then

$$f_n = g_n(h_1)^{n_1}(g_n)^{-1},$$

say, and so the elements

$$q_n = g_n h_1 (g_n)^{-1}$$

have distinct axes $A_n$ and the same translation length $T$ as $h_1$. As $A_n$ converges to the geodesic $(u, v)$, this violates discreteness: explicitly, if $z \in (u, v)$, then

$$\sinh \tfrac{1}{2}\rho(z, q_n z) = \sinh(\tfrac{1}{2}T)\cosh \rho(z, A_n)$$
$$\to \sinh(\tfrac{1}{2}T)$$

as $n \to +\infty$ yet the $q_n$ are distinct.                                                  □

Now let the conjugacy classes of hyperbolic elements in a Fuchsian group $G$ be $C_1, C_2, \ldots$. The elements in $C_n$ have a common translation length, say $T_n$.

**Theorem 10.3.6.** *If $G$ is finitely generated then $T_n \to +\infty$ as $n \to +\infty$.*

PROOF. Theorem 10.2.5 and its proof shows that every hyperbolic fixed point of $G$ is a point of approximation and moreover, that there exists a compact subset $K$ of $\Delta$ such that *every* hyperbolic axis has an image which meets $K$. This means that every hyperbolic conjugacy class $C_n$ contains an element $g_n$ with its axis $A_n$ meeting $K$. For some $d$,

$$K \subset \{z \in \Delta : \rho(0, z) \le d\}.$$

From Sections 7.4 and 7.35 we obtain

$$\|g_n\|^2 = 2\cosh \rho(0, g_n 0)$$
$$= 2 + 4\sinh^2 \tfrac{1}{2}\rho(0, g_n 0)$$
$$= 2 + 4\sinh^2(\tfrac{1}{2}T_n)\cosh^2 \rho(0, A_n)$$
$$\le 2 + 4\cosh^2(d)\sinh^2(\tfrac{1}{2}T_n)$$

so $T_n \to +\infty$ as $n \to +\infty$.                                                        □

*Remark.* Using known information about the convergence of series, for example, Theorem 5.3.13, we can obtain more precise information about the rate at which $T_n$ tends to $+\infty$.

There are two types of hyperbolic elements in a Fuchsian group which warrant special attention. First, there are the simple hyperbolic elements (Definition 8.1.5). There are also the *boundary hyperbolic elements* $h$ which are characterized by the fact that they leave some interval of discontinuity on the circle at infinity invariant: of course, these only exist for Fuchsian groups of the second kind.

**Theorem 10.3.7.** *A finitely generated Fuchsian group has only a finite number of conjugacy classes of maximal boundary hyperbolic cyclic subgroups. A finitely generated Fuchsian group can have infinitely many conjugacy classes of primitive simple hyperbolic elements.*

PROOF. A finitely generated group $G$ has a convex fundamental polygon $P$ with only a finite number of free sides, say $s_1, \ldots, s_n$. Each free side $s_j$ lies in an interval of discontinuity $\sigma_j$ whose stabilizer is generated by a boundary hyperbolic element, say $h_j$.

If $h$ is any boundary hyperbolic element, it leaves some interval of discontinuity $\sigma$ invariant and we can construct a half-ray $L$ ending at some interior point of $\sigma$ and lying entirely in some image $f(P)$ (because the images of $P$ do not accumulate at the interior points of $\sigma$). As $f^{-1}(L)$ lies in $P$ and ends at an ordinary point of $G$, it must end in some $s_j$. Thus $f(\sigma) = \sigma_j$ and so $fhf^{-1}$ leaves $\sigma_j$ invariant: this proves that $h$ is conjugate to some power of $h_j$.

Finally, we must exhibit an example of a finitely generated Fuchsian group which contains infinitely many non-conjugate primitive simple hyperbolic elements.

Construct a quadilateral $P$ in $\Delta$ with vertices $v_1, v_2, v_3, v_4$ lying on the circle at infinity. Let $f$ and $g$ be hyperbolic elements pairing the sides of $P$ as illustrated in Figure 10.3.1. By Poincaré's Theorem (see Exercise 9.8.2), the group $G$ generated by $f$ and $g$ is discrete and $P$ is a fundamental polygon for $G$. As $f$ and $g$ pair sides of a convex fundamental polygon, they are

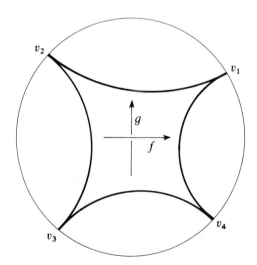

Figure 10.3.1

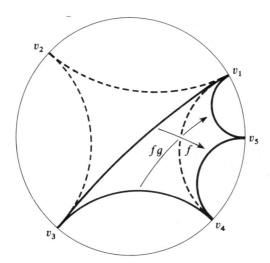

Figure 10.3.2

simple hyperbolic elements of $G$ (Theorem 9.7.1). It is clear from the geometry of the actions that the axes of $f$ and $g$ cross $P$ and this implies that $f$ and $g$ are primitive.

Now let $v_5 = f(v_1)$: then the quadilateral with vertices $v_1, v_3, v_4, v_5$ is also a convex fundamental polygon for $G$, this time with its sides paired by $f$ and $fg$: see Figure 10.3.2. Exactly as above, $f$ and $fg$ are simple, primitive hyperbolic elements.

This process can be repeated to obtain a sequence $g, fg, f^2g, \ldots$ of primitive simple hyperbolic elements of $G$. By conjugation, we may assume that $G$ now acts on $H^2$ and that

$$f = \begin{pmatrix} u & 0 \\ 0 & 1/u \end{pmatrix}, \qquad g = \begin{pmatrix} a & b \\ c & d \end{pmatrix},$$

where $u > 1$. A trivial computation shows that $\text{trace}^2(f^n g) \to +\infty$ as $n \to +\infty$ so the sequence $(f^n g)$ contains infinitely many non-conjugate elements ( note that $a$ is not zero else $g$ and $f$ have a common fixed point). $\square$

EXERCISE 10.3

1. Construct an infinitely generated Fuchsian group $G$ containing infinitely many conjugacy classes of simple primitive hyperbolic elements with the same translation length (see Theorems 10.3.6 and 10.3.7).

2. Verify the details in the text relating to Figures 10.3.1 and 10.3.2 in the proof of Theorem 10.3.7 (use Exercise 9.8.2). Give an alternative construction in which the vertices $v_j$ are replaced by free sides and apply Poincaré's Theorem directly.

## §10.4. The Signature of a Fuchsian Group

Let $G$ be a finitely generated non-elementary Fuchsian group. Any Dirichlet polygon $D$ for $G$ is finite sided and topologically, $\tilde{D}/G$ is a compact surface $S$ of some genus, say $g$, with a certain number of holes removed. As $\Delta/G$ and $\tilde{D}/G$ are homeomorphic (Theorem 9.2.4), the genus $g$ does not depend on the choice of $D$.

Now consider the Nielsen region $N$ for $G$ and corresponding quotient space $\tilde{N}/G$. The argument given in Section 10.3 shows that the boundary of $N$ in $\Delta$ consists of all axes of all boundary hyperbolic elements in $G$. Let $A$ be one such axis with stabilizer generated by $h$ and let $H$ be the component of $\Delta - A$ not containing $N$. Obviously, $H$ is stable with respect to $\langle h \rangle$ so the projection of $H$ into $\Delta/G$ is topologically a cylinder, namely $H/\langle h \rangle$ (Theorem 6.3.3). One end of this cylinder is the simple loop $A/\langle h \rangle$: indeed no image of $A$ can cross $A$ (as the open arc of $\partial \Delta$ which bounds $H$ contains only ordinary points of $G$) and there are no elliptic elements of order two stabilizing $A$ (else $G$ would then have only two limit points).

If we denote the natural projection of $\Delta$ onto $\Delta/G$ by $\pi$, we see that $\pi(\Delta)$ is the disjoint union of $\pi(N)$ together with simple loops of the form $\pi(A)$ and with cylinders of the form $\pi(H)$. The cylinders $\pi(H)$ are joined to $\pi(N)$ across the common boundary loops $\pi(A)$ and there are the same number, say $t$, of these as there are conjugacy classes of maximal boundary hyperbolic cyclic subgroups. It is clear now that the three spaces $\Delta/G$, $\tilde{D}/G$, $\tilde{N}/G$ are homeomorphic to each other.

In addition, $G$ contains only a finite number, say $s$, of conjugacy classes of maximal parabolic cyclic subgroups and each of these corresponds to a puncture on the surface $S$ (consider the quotient of a horocyclic region that is stable under a cyclic parabolic subgroup). Finally, $G$ contains only a finite number, say $r$, of conjugacy classes of maximal elliptic cyclic subgroups: let these have orders $m_1, \ldots, m_r$ respectively. We introduce terminology to summarize these facts.

**Definition 10.4.1.** The symbol

$$(g: m_1, \ldots, m_r ; s; t) \qquad (10.4.1)$$

is called the *signature* of $G$: each parameter is a non-negative integer and $m_j \geq 2$.

If there are no elliptic elements in $G$, we simply write $(g: 0; s; t)$. It is possible to state precisely which signatures occur.

**Theorem 10.4.2.** *There is a non-elementary finitely generated Fuchsian group with signature (10.4.1) and $m_j \geq 2$ if and only if*

$$2g - 2 + s + t + \sum_{j=1}^{r} \left(1 - \frac{1}{m_j}\right) > 0. \qquad (10.4.2)$$

The proof that (10.4.2) is a necessary condition for the existence of a group with signature (10.4.1) is a consequence of the following result.

**Theorem 10.4.3.** *Let G be a non-elementary finitely generated Fuchsian group with signature* (10.4.1) *and Nielsen region N. Then*

$$\text{h-area}(N/G) = 2\pi\left\{2g - 2 + s + t + \sum_{j=1}^{r}\left(1 - \frac{1}{m_j}\right)\right\}.$$

If $G$ is also of the first kind, then $N = \Delta$ and $t = 0$: thus we obtain a formula for the area of any fundamental polygon of $G$.

**Corollary 10.4.4.** *Let G be a finitely generated Fuchsian group of the first kind with signature* $(g: m_1, \ldots, m_r; s; 0)$. *Then for any convex fundamental polygon P of G,*

$$\text{h-area}(P) = 2\pi\left[2g - 2 + s + \sum_{j=1}^{r}\left(1 - \frac{1}{m_j}\right)\right].$$

PROOF OF THEOREM 10.4.3. We take $D$ to be the Dirichlet polygon ıor $G$ with centre $w$ so

$$\text{h-area}(D \cap N) = \text{h-area}(N/G).$$

By choosing $w$ appropriately, we may assume that each elliptic and parabolic cycle on $\partial D$ has length one and (by taking $w$ to avoid a countable set of geodesics) we may assume that no cycle of vertices of $D$ lies on the axes of hyperbolic boundary elements.

Clearly, only finitely many distinct images of a hyperbolic axis can meet the closure of any locally finite fundamental domain. As $N$ is bounded by hyperbolic axes (because $G$ is finitely generated), this implies that only finitely many sides of $N$ meet $D$ and so $D \cap N$ is a finite sided polygon. The boundary of $D \cap N$ consists of, say, $2n$ paired sides (which are arcs of paired sides of $P$) and $k$ sides which are not paired (and consist of arcs in $D$ of the axes bounding $N$). The vertices of $D \cap N$ are the $r$ elliptic cycles of length one, the $s$ parabolic cycles of length one, some accidental cycles of $P$ (say $a$ of these) and finally $k$ cycles of length two corresponding to the end-points of the $k$ unpaired sides of $D \cap N$.

Applying Euler's formula (after "filling in" the holes), we obtain

$$2 - 2g = (1 + t) - (n + k) + (r + a + k + s)$$

so

$$n - a = 2g - 1 + r + s + t.$$

Now join $w$ to each vertex of $D \cap N$, thus dividing $D \cap N$ into $2n + k$ triangles. Adding the areas of these triangles, we obtain

$$\text{h-area}(D \cap N) = (2n + k)\pi - 2\pi - 2\pi a - \pi k - \sum_{j=1}^{r} \frac{2\pi}{m_j}$$

$$= 2\pi \left[ n - a - 1 - \sum_{j=1}^{r} \frac{1}{m_j} \right]$$

$$= 2\pi \left[ 2g - 2 + s + t + \sum_{j=1}^{r} \left( 1 - \frac{1}{m_j} \right) \right]. \qquad \square$$

It is evident from the nature of the formula in Theorem 10.4.3 that h-area$(N/G)$ has a positive universal lower bound, valid for all groups $G$. For brevity, write

$$A = (1/2\pi) \, \text{h-area}(N/G)$$

and, in order to compute this lower bound, we may assume that $A < \frac{1}{6}$: this is a convenient number for the following analysis and we shall soon see that there are groups for which $A < \frac{1}{6}$.

If $r = 0$ or if $m_j = 2$ for each $j$, then $A = n/2$ for some integer $n$. As $A > 0$, we find that $A \geq \frac{1}{2}$ so we may assume that $r > 0$ and that some $m_j$ is at least three. Then

$$1 > 6A$$

$$\geq 6 \left[ 2g - 2 + s + t + \left( \frac{r-1}{2} \right) + \frac{2}{3} \right]$$

which yields

$$4g + 2s + 2t + r < 4.$$

Because

$$2 < A + 2$$
$$\leq 2g + s + t + r$$
$$\leq 4g + 2s + 2t + r$$
$$< 4,$$

we obtain

$$2g + s + t + r = 3$$
$$= 4g + 2s + 2t + r$$

so

$$g = s = t = 0, \qquad r = 3.$$

We may now assert that

$$A = 1 - \left( \frac{1}{m_1} + \frac{1}{m_2} + \frac{1}{m_3} \right) > 0.$$

If each $m_j$ is at least three, then one $m_j$ is at least four and then $A \geq \frac{1}{12}$. If not, then $m_3 = 2$, say, and so

$$A = \tfrac{1}{2} - \left( \frac{1}{m_1} + \frac{1}{m_2} \right) > 0.$$

If each of $m_1$ and $m_2$ is at least four, then one is at least five and then $A \geq \frac{1}{20}$. If not, then $m_2 = 3$, say, and

$$A \geq \tfrac{1}{42}$$

with equality when and only when $G$ has signature $(0: 2, 3, 7; 0; 0)$. For future reference we state this as our next result.

**Theorem 10.4.5.** *For every non-elementary Fuchsian group $G$ with Nielsen region $N$*

$$\text{h-area}(N/G) \geq \pi/21.$$

*Equality holds precisely when $G$ has signature* $(0: 2, 3, 7; 0; 0)$ *in which case $N = \Delta$.*

We end this section with the remaining part of the proof of Theorem 10.4.2.

PROOF OF THEOREM 10.4.2. *Sufficiency.* Given the symbol (10.4.1) satisfying (10.4.2), we must construct a Fuchsian group $G$ which has (10.4.1) as its signature.

For any positive $d$, construct the circle given by $\rho(z, 0) = d$ and also a set of $4g + r + s + t$ points $z_j$ equally spaced around this circle (and labelled in the natural way). The arcs $z_j z_{j+1}$ subtend an angle $2\theta$ at the origin where

$$\theta = \frac{2\pi}{8g + 2r + 2s + 2t}.$$

For the first four of these arcs, we construct a configuration with mappings $h_j$ as illustrated in Figure 10.4.1. Note that the points $z_1, \ldots, z_5$ are all images of each other.

This construction is repeated $g - 1$ more times, starting the next stage at $z_5$ and so on: this accounts for $4g$ arcs $z_j z_{j+1}$, an angle $8g\theta$ at the origin and mappings $h_1, \ldots, h_{2g}$.

Using the next $r$ arcs $z_j z_{j+1}$, we construct configurations with mappings $e_i$ as illustrated in Figure 10.4.2 (recall that the integers $m_i$ are available from (10.4.1) and $m_i \geq 2$). Necessarily, $e_i$ is an elliptic element of order $m_i$

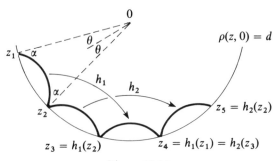

Figure 10.4.1

and fixing $w_i$. This part of the construction accounts for an additional angular measure of $2r\theta$ at the origin. Next, we repeat the construction $s$ times and now on each occasion the corresponding $w_i$ are on $\{|z| = 1\}$: the angle at $w_i$ is zero and the corresponding mappings $p_i$ (for $e_i$) are parabolic.

There are now $t$ remaining arcs, each subtending an angle of $2\theta$ at the origin. On each of these arcs we construct the configurations and hyperbolic mappings $b_i$ as illustrated in Figure 10.4.3 where

$$\theta_1 = \left(\frac{1 + d}{1 + 2d}\right)\theta.$$

We have now constructed a polygon with vertices $z_j, u_i, v_i, w_i$ and with side-pairings given by the $h_i, e_i, p_i$ and $b_i$. The group $G$ generated by these maps may or may not be discrete but in any case, the points $z_1, z_2, \ldots$ lie in the same $G$-orbit. Moreover, the angle sum subtended at these $z_j$ is

$$\phi(d) = 8g\alpha + 2(\beta_1 + \cdots + \beta_{r+s+t}).$$

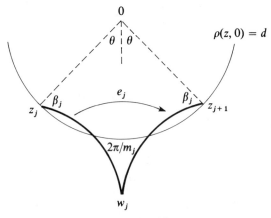

Figure 10.4.2

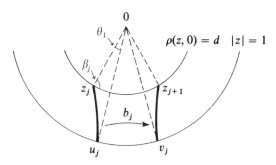

Figure 10.4.3

Each of the angles $\alpha$ and $\beta_j$ depend continuously on the parameter $d$. We shall show that for some choice of $d$ we have $\phi(d) = 2\pi$. Then Poincaré's Theorem (see Exercise 9.8.2) implies that $G$ is discrete and that the constructed polygon is a fundamental domain for $G$. It then remains to verify that $G$ does indeed have the signature (10.4.1).

By elementary trigonometry, we have (using Figures 10.4.1, 10.4.2 and 10.4.3 in turn)

(i) $$\cosh d = \cot \theta \cot \alpha;$$

(ii) $$\cosh d = \frac{\cos \theta \cos \beta_j + \cos(\pi/m_j)}{\sin \theta \sin \beta_j},$$

when $j = 1, \ldots, r$, and a similar expression with $\cos(\pi/m_j)$ replaced by 1 when $j = r + 1, \ldots, r + s$;

(iii) $$\cosh d = \frac{\cos \theta_1 \cos \beta_j + 1}{\sin \theta_1 \sin \beta_j}.$$

Note that as $d \to 0$, so $\alpha \to (\pi/2) - \theta$. In (ii), we have

$$\cos(\theta + \beta_j) = \cos\left(\pi - \frac{\pi}{m_j}\right) + \sin \theta \sin \beta_j(\cosh d - 1)$$

and so as $d \to 0$,

$$\beta_j \to \pi - \frac{\pi}{m_j} - \theta,$$

with the appropriate interpretation of $m_j = +\infty$ when $r < j \leq r + s$. In (iii), we have

$$\theta_1 + \beta_j \to \pi$$

so

$$\beta_j \to \pi - \theta.$$

It follows that as $d \to 0$, so

$$\phi(d) \to 2\pi\left[2g - 2 + s + t + \sum_{j=1}^{r}\left(1 - \frac{1}{m_j}\right)\right] + 2\pi$$

$$> 2\pi.$$

As $d \to +\infty$, the angles $\alpha$ and $\beta_j$ each tend to zero (note that $\theta_1 \to \theta/2$) so in this case, $\phi(d) \to 0$. We deduce that for some choice of $d$, we have $\phi(d) = 2\pi$ and $G$ is then discrete.

It is clear that $G$ has elliptic elements of orders $m_1, \ldots, m_r$ and also $s$ parabolic and $t$ boundary elements and that these do not represent the same conjugacy classes (essentially because they pair adjacent sides of the fundamental polygon). If $\Delta/G$ has genus $g^*$, then by Euler's formula applied to the identified polygon,

$$2 - 2g^* - s - t = 1 - (2g + r + s + t) + (1 + r)$$

so (as expected), $g^* = g$.                                                      □

EXERCISE 10.4

1. Let $G$ be a non-elementary Fuchsian group and suppose that a parabolic element $g$ in $G$ generates the stabilizer of its fixed point $v$. By considering a suitable horocyclic region $H$ based at $v$, show that $\pi(H)$ is conformally equivalent to a punctured disc in $\Delta/G$.

2. Show that there is a positive constant $\delta$ such that if $P$ is any convex fundamental polygon for some non-elementary Fuchsian group $G$, then $P \cap N$ contains a disc of radius at least $\delta$. Obtain an explicit estimate of $\delta$.

3. Let $P$ be the hyperbolic quadilateral in $H^2$ with vertices $-1, 0, 1, \infty$. Show that $P$ is a fundamental domain for the group $G$ generated by

$$g(z) = z + 2, \qquad h(z) = z/(2z + 1).$$

Compute the signature of $G$ and verify the formula for the area of $H^2/G$ explicitly in this case. Find the index of $G$ in the Modular group (this is a particular case of Selberg's Lemma).

# §10.5. The Number of Sides of a Fundamental Polygon

We restrict our discussion in this section to a finitely generated group $G$ of the first kind. In this case, we can omit the last parameter in the signature (10.4.1) and we can consider parabolic elements as elliptic elements with order $m_j = +\infty$. Thus we can shorten the notation for the signature to $(g: m_1, \ldots, m_n)$ or, if $G$ has no elliptic or parabolic elements, to $(g: 0)$.

**Theorem 10.5.1.** *Let $G$ be a finitely generated Fuchsian group of the first kind and let $P$ be any convex fundamental polygon for $G$. Suppose that $P$ has $N$ sides (where no side is paired with itself).*

(i) *If $G$ has signature $(g: m_1, \ldots, m_n)$ where possibly $n = 0$, then*

$$N \le 12g + 4n - 6.$$

*This upper bound is attained by the Dirichlet region with centre $w$ for almost all choices of $w$.*

(ii) *If $G$ has signature $(g: 0)$, then $N \ge 4g$ and this is attained for some $P$.*

(iii) *If $G$ has signature $(g: m_1, \ldots, m_n)$, $n > 0$, then*

$$N \ge 4g + 2n - 2$$

*and this is attained for some $P$.*

PROOF. Suppose that $P$ has elliptic or parabolic cycles $C_1, \ldots, C_n$ and accidental cycles $C_{n+1}, \ldots, C_{n+A}$: either (but not both) of these sets of cycles may be absent. In general, we let $|C|$ denote the number of points in the cycle $C$.

Now

$$|C_j| \ge 1 \quad \text{if } 1 \le j \le n;$$
$$|C_j| \ge 3 \quad \text{if } n < j \le n + A,$$

and

$$N = \sum_{j=1}^{n+A} |C_j|.$$

Thus

$$0 \le A \le (N - n)/3.$$

Euler's formula yields

$$2 - 2g = 1 - (N/2) + n + A \tag{10.5.1}$$

and the inequalities in (i) and (iii) follow by eliminating $A$. The inequality in (ii) follows from (10.5.1) by putting $n = 0$ and observing that as $n = 0$, we have $A \ge 1$.

The polygon $P$ has $N$ sides and hence $N$ vertices. For almost all choices of $w$, the Dirichlet region with centre $w$ has $|C_j| = 1$ for $1 \leq j \leq n$ and $|C_j| = 3$ for $j > n$. Then

$$3A = N - n$$

and so equality holds in (i). The proof of Theorem 10.4.2 (sufficiency) shows that the lower bound of $4g$ in (ii) may be attained. Finally, a similar argument to that used in the same proof shows that the lower bound in (iii) may also be obtained: briefly, one constructs the polygon as though the signature were $(g: m_1, \ldots, m_{n-1})$ and seeks a value of $d$ so that $\phi(d) = 2\pi/m_n$. $\qquad\qquad\qquad\qquad\qquad\qquad\qquad\qquad\square$

In the next section we shall study *Triangle groups*: these are the groups with signatures $(0: p, q, r)$ where (necessarily)

$$\frac{1}{p} + \frac{1}{q} + \frac{1}{r} < 1.$$

Observe that for almost all choices of the centre $w$, the corresponding Dirichlet region has *six* sides: the customary fundamental polygon for such groups is a quadrilateral yet, in some sense, this is the exceptional case.

# §10.6. Triangle Groups

This section is devoted to an important class of Fuchsian groups known as the Triangle groups. Roughly speaking, these are the discrete groups with the more closely packed orbits and the smallest fundamental regions. We begin with a geometric definition that does not mention discreteness.

**Definition 10.6.1.** A group $G$ of isometries of the hyperbolic plane is said to be of *type* $(\alpha, \beta, \gamma)$ if and only if $G$ is generated by the reflections across the sides of some triangle with angles $\alpha$, $\beta$ and $\gamma$.

Of course, such groups exist if and only if $\alpha$, $\beta$ and $\gamma$ are non-negative and satisfy

$$0 \leq \alpha + \beta + \gamma < \pi.$$

Any two such groups of the same type are conjugate in the group of all isometries (because two triangles with the same angles are congruent) and there is no significance to be attached to the order of $\alpha$, $\beta$ and $\gamma$ in the triple $(\alpha, \beta, \gamma)$.

The next example shows that such a group (even if discrete) may be of more than one type.

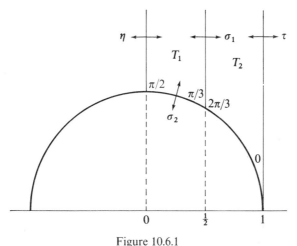

Figure 10.6.1

**Example 10.6.2.** Let $T_1$ and $T_2$ be the two triangles illustrated in Figure 10.6.1: the corresponding groups are

$$G_1 = \langle \sigma_1, \sigma_2, \eta \rangle$$

of type $(0, \pi/2, \pi/3)$ and

$$G_2 = \langle \sigma_1, \sigma_2, \tau \rangle$$

of type $(0, 0, 2\pi/3)$ where $\eta$, $\sigma_1$ and $\tau$ are reflections in the lines $x = 0$, $x = \frac{1}{2}$ and $x = 1$ respectively and $\sigma_2$ is the reflection in $|z| = 1$.

Clearly,

$$\eta \sigma_1 = \sigma_1 \tau$$

so $\eta \in G_2$ and $\tau \in G_1$; thus $G_1 = G_2$. In fact, the subgroup of conformal isometries of this group is the Modular group and so $G_1$ is itself discrete.

Note that

$$\text{h-area}(T_2) = 2\text{h-area}(T_1)$$

so $T_2$ is *not* a fundamental domain for $G_2$. $\qquad\qquad\qquad\qquad\qquad\qquad\square$

Each group $G$ of type $(\alpha, \beta, \gamma)$ has a distinguished subgroup $G_0$ of index two in $G$, namely the subgroup of conformal elements of $G$: we call $G_0$ a *conformal group of type* $(\alpha, \beta, \gamma)$. If $\sigma_1, \sigma_2$ and $\sigma_3$ denote the reflections which generate $G$, then the elements of $G_0$ are precisely the words of even length in the $\sigma_i$ and $G_0$ is generated by, say, $\sigma_1\sigma_2$ and $\sigma_3\sigma_2$ because

$$\sigma_i\sigma_j = (\sigma_j\sigma_i)^{-1}, \qquad \sigma_1\sigma_3 = (\sigma_1\sigma_2)(\sigma_3\sigma_2)^{-1}.$$

Suppose that $\gamma$ is the angle of the triangle at the vertex $v_3$ opposite the side fixed by the reflection $\sigma_3$. Then $\sigma_1\sigma_2$ fixes $v_3$ and it is parabolic if $\gamma = 0$ and elliptic with a rotation of angle $2\gamma$ if $\gamma > 0$. Thus $G_0$ is generated by a pair $f$, $g$ of conformal isometries, each being elliptic or parabolic. It is convenient to consider parabolic elements as elliptic elements of infinite order and we shall frequently adopt this convention in the following discussion.

If $G$ of type $(\alpha, \beta, \gamma)$ (or its corresponding conformal subgroup $G_0$) is discrete, then every elliptic element in $G_0$ is of finite order. Thus if any of $\alpha$, $\beta$ and $\gamma$ are positive, then they are necessarily of the form

$$k\pi/p, \qquad (k, p) = 1 \qquad\qquad (10.6.1)$$

for (coprime) integers $k$ and $p$. This is a *necessary condition* for discreteness but it is *not sufficient*. Indeed, it is easy to see that if $\alpha$, $\beta$ and $\gamma$ are all positive, then the images of the triangle $T$ under $G$ cover the hyperbolic plane. We deduce that if $G$ is discrete then two disjoint copies of $T$ must contain a fundamental region for $G_0$ and so (from Theorem 10.4.5),

$$\text{h-area}(T) \geq \pi/42.$$

It follows that if $\alpha$, $\beta$ and $\gamma$ are of the form (10.6.1) with

$$\pi - (\alpha + \beta + \gamma) < \pi/42$$

(and such angles clearly exist) then $G_0$ is not discrete.

A *sufficient condition* for discreteness is that each of $\alpha$, $\beta$ and $\gamma$ is of the form

$$\pi/p, \qquad 2 \leq p \leq +\infty \qquad\qquad (10.6.2)$$

for some integer $p$: indeed, if this is so then a direct application of Poincaré's Theorem shows that $G$ is discrete. This sufficient condition, however, is *not necessary*: for example, $G_2$ of type $(0, 0, 2\pi/3)$ in Example 10.6.2 is discrete.

The apparent discrepancy between (10.6.1) and (10.6.2) is easily resolved. A group of type $(\alpha, \beta, \gamma)$ is discrete if and only if it is also of some (possibly different) type $(\pi/p, \pi/q, \pi/r)$: for example, $G_2$ in Example 10.6.2 is also of type $(0, \pi/2, \pi/3)$. This result will be proved later in this section.

We shall confine our attention to discrete conformal groups and we adopt the following standard terminology.

**Definition 10.6.3.** A group $G$ is a $(p, q, r)$-*Triangle group* if and only if $G$ is a conformal group of type $(\pi/p, \pi/q, \pi/r)$: we call $G$ a *Triangle group* if it is a $(p, q, r)$-Triangle group for some integers $p$, $q$ and $r$.

Observe that, from the remarks relating to (10.6.2), a Triangle group is necessarily discrete. Now we derive two results concerning Triangle groups.

**Theorem 10.6.4.** *A group G is a $(p, q, r)$-Triangle group if and only if it is a discrete group of the first kind with signature $(0: p, q, r)$.*

**Theorem 10.6.5.** *Let G be a discrete group of conformal isometries of the hyperbolic plane. If G contains a Triangle group $G_0$ as a subgroup, then G itself is a Triangle group.*

PROOF OF THEOREM 10.6.4. Suppose first that $G$ is a $(p, q, r)$ Triangle group. Then $G$ is the conformal subgroup of index two of a discrete group $G^*$ generated by reflections $\sigma_1$, $\sigma_2$ and $\sigma_3$ across the sides of a triangle $T^*$ with angles $\pi/p$, $\pi/q$ and $\pi/r$. Poincaré's Theorem implies that $T^*$ is a fundamental domain for $G^*$ and so

$$T = T^* \cup \sigma_1(T^*)$$

is a fundamental domain for $G$. Clearly, then, $G$ is of the first kind.

The isometries

$$g = \sigma_1\sigma_2, \qquad h = \sigma_1\sigma_3$$

generate $G$ and

$$g^r = h^q = (h^{-1}g)^p = I:$$

see Figure 10.6.2. The images of a neighbourhood of $v_3$ relative to $T$ under iterates of $g$ tesselate a plane neighbourhood of $v_3$ so (as $T$ is a fundamental domain) neither $v_1$ nor $v_2$ are images of $v_3$. This shows that $g$ is not conjugate to any power of $h$ or $h^{-1}g$. By symmetry, then, $G$ has three elliptic or parabolic conjugacy classes of subgroups represented by $\langle g \rangle$, $\langle h \rangle$ and $\langle h^{-1}g \rangle$.

The genus $k$ of $\Delta/G$ is found from Euler's formula, namely

$$2 - 2k = (faces) - (edges) + (vertices)$$
$$= 1 - 2 + 3:$$

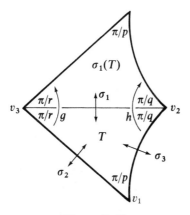

Figure 10.6.2

so $G$ has signature $(0: p, q, r)$. Alternatively, one can show that $k = 0$ by applying the Area formula to $T$.

Now suppose that $G$ is a discrete group of conformal isometries with signature $(0: p, q, r)$. Let $D$ be a convex fundamental polygon for $G$ with, say, cycles of lengths $N_p$, $N_q$ and $N_r$ corresponding to the conjugacy classes associated with $p$, $q$ and $r$. Suppose also that there are $t$ accidental cycles of lengths, say, $M_1, \ldots, M_t$ so $M_j \geq 3$. Observe that as $G$ is of the first kind, $D$ has no free sides.

Select any $w$ in $D$ and join $w$ to each vertex of $D$. Equating areas, we obtain

$$2\pi\left[1 - \left(\frac{1}{p} + \frac{1}{q} + \frac{1}{r}\right)\right]$$

$$= \text{h-area}(D)$$

$$= [N_p + N_q + N_r + M_1 + \cdots + M_t]\pi - 2\pi - 2\pi t - 2\pi\left(\frac{1}{p} + \frac{1}{q} + \frac{1}{r}\right)$$

and so

$$1 = (N_p - 1) + (N_q - 1) + (N_r - 1) + \sum_{j=1}^{t} (M_j - 2).$$

As each of the $t + 3$ terms on the right is a non-negative integer, only two cases arise, namely

*Case 1.* $N_p = N_q = N_r = 1; t = 1, M_1 = 3$; or

*Case 2.* $N_p$, $N_q$, $N_r$ are (in some order) 1, 1, 2 and there are no accidental cycles.

In Case 2, $D$ has four vertices and so is a quadrilateral. Supposing that $N_q = N_r = 1$, we see that $D$ is as illustrated in Figure 10.6.3.

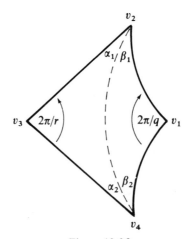

Figure 10.6.3

The cycle corresponding to $N_q$ $(=1)$ is $\{v_1\}$ so the two sides ending at $v_1$ are paired: thus

$$\rho(v_1, v_4) = \rho(v_1, v_2)$$

and so $\beta_1 = \beta_2$. Similarly, $\alpha_1 = \alpha_2$ so

$$\begin{aligned}
\alpha_1 + \beta_1 &= \alpha_2 + \beta_2 \\
&= \tfrac{1}{2}(\alpha_1 + \alpha_2 + \beta_1 + \beta_2) \\
&= \pi/p.
\end{aligned}$$

The properties of isosceles triangles guarantee that the segment $[v_1, v_3]$ is a line of symmetry of the quadrilateral so in this case $G$ is the conformal Triangle group associated with group generated by reflections across the sides of the triangle with vertices $v_1$, $v_2$ and $v_3$.

In Case 1, $D$ is a hexagon with elliptic (or parabolic) vertices $v_1, v_2, v_3$ and a single accidental cycle $\{a_1, a_2, a_3\}$. The side-pairing must occur as in Figure 10.6.4. where we have sub-divided $D$ into the regions $Q$, $T_1$ and $T_2$. As

$$h(a_2) = gf(a_2),$$

we see that $h = gf$ ($a_2$ is not an elliptic or parabolic fixed point). It is now easy to see that $Q \cup h(T_1) \cup g(T_2)$ is a fundamental quadrilateral with vertices $v_1, v_2, v_3, h(v_2)(=g(v_2))$ and this reduces Case 1 to Case 2.   □

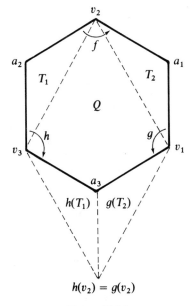

Figure 10.6.4

Observe that this proof shows that a *convex fundamental polygon for a Triangle group is necessarily a quadrilateral or a hexagon*: the reader should now review the remark at the end of Section 10.5.

PROOF OF THEOREM 10.6.5. In view of Theorem 10.6.4, we can work entirely with the signatures of $G$ and $G_0$. As

$$0 < \text{h-area}(\Delta/G) \leq \text{h-area}(\Delta/G_0) \leq 2\pi$$

we see that $G_0$ is of some finite index $k$ in $G$ (Theorem 9.1.3). The case $k = 1$ is trivial so we may assume that $k \geq 2$, hence

$$\text{h-area}(\Delta/G_0) = k \, \text{h-area}(\Delta/G) \tag{10.6.3}$$
$$\geq 2 \, \text{h-area}(\Delta/G).$$

According to Theorem 10.6.4, $G_0$ has some signature $(0: p, q, r)$. Let $G$ have signature $(g: t_1, \ldots, t_n)$: then the Area formula (Corollary 10.4.4) yields

$$1 - \left(\frac{1}{p} + \frac{1}{q} + \frac{1}{r}\right) = k\left[2g - 2 + \sum_{j=1}^{n}\left(1 - \frac{1}{t_j}\right)\right]$$
$$> 0.$$

The left-hand side is at most one: so $g = 0$ or 1. If $g = 1$, then $n \geq 1$ (else the area is zero) and (as $t_j \geq 2$ and $k \geq 2$) we have

$$1 \leq \frac{nk}{2} \leq k \sum_{j=1}^{n}\left(1 - \frac{1}{t_j}\right)$$
$$\leq 1 - \left(\frac{1}{p} + \frac{1}{q} + \frac{1}{r}\right)$$
$$\leq 1.$$

This cannot be so, however, as then equality holds throughout, $G_0$ contains parabolic elements (for then $p = q = r = \infty$) but $G$ does not ($t_1 = \cdots = t_n = 2$).

We deduce that $g = 0$ and (for positive area) $n \geq 3$. This yields

$$k(\tfrac{1}{2}n - 2) \leq k\left[\sum_{j=1}^{n}\left(1 - \frac{1}{t_j}\right) - 2\right]$$
$$= 1 - \left(\frac{1}{p} + \frac{1}{q} + \frac{1}{r}\right)$$
$$\leq 1.$$

As $k \geq 2$ we obtain $n \leq 5$. If $n = 5$, then $k = 2$ and equality again holds throughout: this is excluded exactly as above. Thus $n = 3$ or 4. If $n = 3$, then $G$ has a signature $(0: t_1, t_2, t_3)$ and so is a Triangle group. It only remains to exclude the case $g = 0, n = 4$.

Assume, then, that $g = 0$ and $n = 4$. We may assume that $r \leq q \leq p$ and $p \leq t_4$ (as $G_0$ contains an element of order $p$, so does $G$). Then

$$1 - \frac{3}{p} \geq 1 - \left(\frac{1}{p} + \frac{1}{q} + \frac{1}{r}\right)$$

$$= k\left[\sum_{j=1}^{4}\left(1 - \frac{1}{t_j}\right) - 2\right]$$

$$\geq 2\left[\frac{3}{2} + \left(1 - \frac{1}{p}\right) - 2\right]$$

$$= 1 - \frac{2}{p}.$$

This is false unless $p = \infty$ in which case, equality holds throughout so $k = 2$ and the signatures of $G_0$ and $G$ are

$$(0: \infty, \infty, \infty), \qquad (0: 2, 2, 2, \infty),$$

respectively. This is excluded, however, by Theorem 10.3.4.                □

Finally, we turn our attention to conformal groups of an arbitrary type $(\alpha, \beta, \gamma)$. We observed earlier that these groups are generated by elliptic or parabolic elements $g$ and $h$ which pair the sides of a quadrilateral with a line of symmetry as illustrated in Figure 10.6.5. Conversely, given such a configuration, it is clear that $\langle g, h \rangle$ is a conformal group of type $(\alpha, \beta, \gamma)$. Note that the reflection in $(v_1, v_3)$ interchanges $v_2$ and $v_4$ so $(v_2, v_4)$ is orthogonal to $(v_1, v_3)$.

**Theorem 10.6.6.** *A conformal group of some type $(\alpha, \beta, \gamma)$ is discrete if and only if it is a Triangle group.*

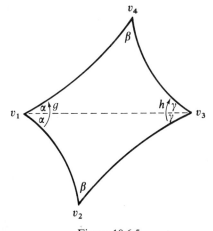

Figure 10.6.5

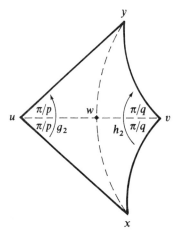

Figure 10.6.6

PROOF OF THEOREM 10.6.6. By definition, a Triangle group is a discrete conformal group of some type $(\alpha, \beta, \gamma)$. Now suppose that $G$ is a discrete conformal group of some type $(\alpha, \beta, \gamma)$: by virtue of Theorem 10.6.5, it is only necessary to construct a Triangle group which arises as a subgroup of $G$. We refer to Diagram 10.6.5 and there are three cases to consider.

*Case* 1: *both g and h are elliptic.*
As $G$ is discrete, $g$ is of finite order $p$ and $h$ is of finite order $q$, say. Thus there is some $g_1$ in $\langle g \rangle$ with angle of rotation $2\pi/p$ and some $h_1$ in $\langle h \rangle$ with angle of rotation $2\pi/q$. Now take conjugates (in $G$), say $g_2$ of $g_1$ and $h_2$ of $h_1$ such that the fixed points $u$ (of $g_2$) and $v$ (of $h_2$) are distinct but otherwise are as close together as possible: this can be achieved because the images of $v_3$ cannot accumulate at $v_1$.

Now construct the quadrilateral illustrated in Figure 10.6.6 by drawing the geodesics at angles $\pi/p$ and $\pi/q$ from $[u, v]$. These geodesics must meet at some points $x$ and $y$ (possibly on the circle at infinity) as otherwise (from Section 7.10)

$$\cosh \rho(u, v) > \frac{1 + \cos(\pi/p) \cos(\pi/q)}{\sin(\pi/p) \sin(\pi/q)}$$

$$\geq \frac{1 + \cos \alpha \cos \gamma}{\sin \alpha \sin \gamma}$$

$$\geq \frac{\cos \beta + \cos \alpha \cos \gamma}{\sin \alpha \sin \gamma}$$

$$= \cosh \rho(v_1, v_3)$$

contrary to our choice of $u$ and $v$. As remarked earlier, $(x, y)$ and $(u, v)$ are othogonal.

Now observe that

$$f = (g_2)^{-1} h_2$$

fixes $x$ and is the reflection $\sigma_{x,v}$ in $[x, v]$ followed by the reflection $\sigma_{x,u}$ in $[x, u]$: indeed,

$$f = (\sigma_{u,v}\sigma_{u,x})^{-1}(\sigma_{u,v}\sigma_{x,v}).$$

If $x$ is on the circle at infinity, then $f$ is parabolic and $\langle g_2, h_2 \rangle$ is a $(p, q, \infty)$ Triangle group. If $x$ is a finite point, then $f$ is elliptic and of finite order so the angle at $x$ is of the form $k\pi/r$ with $(k, r) = 1$ (and $f$ is a rotation of $2\pi k/r$). There is then some anti-clockwise rotation $f_2$ about $x$ of angle $2\pi/r$. If $k \geq 3$ then $f_2(v)$ is nearer to but distinct from $u$, contrary to our choice of $u$ and $v$. Thus $k = 1$ or 2. If $k = 1$, then the angle at $x$ is $\pi/r$ and $\langle g_2, h_2 \rangle$ is a $(p, q, r)$-Triangle group. If $k = 2$, then

$$f_2 = \sigma_{x,w}\sigma_{x,v}$$

and so $\langle h_2, f_2 \rangle$ is a $(2, q, r)$-Triangle group in $G$ associated with the triangle with vertices $x, v, w$. This completes the proof in Case 1.

*Case 2: $g$ is elliptic and $h$ is parabolic.*
We work in $H^2$ and suppose that $h$ fixes $\infty$. The line joining the fixed points of $g$ and h is necessarily a line of symmetry of the quadrilateral so the situation is as illustrated in Figure 10.6.7.

The orbit of $v_1$ contains points of maximal height ($h$ is parabolic fixing $\infty$ and this is essentially Jørgensen's inequality) and this symmetric construction can be carried out using an image of $v_1$ of maximal height instead of $v$ and a rotation of angle $2\pi/p$ about this point ($p$ being the order of $g$) instead of $g$. Because the original angle at $v_1$ is not less than $2\pi/p$, the new diagram provides a quadrilateral exactly as in Figure 10.6.7; however, we may now assume that $v_1$ is of maximal height in its orbit and that the angle at $v_1$ is $2\pi/p$.

If $\theta = 0$, then $\langle g, h \rangle$ is a $(p, \infty, \infty)$-Triangle group. If $\theta > 0$, then $\theta = k\pi/r$ for some coprime $k$ and $r$. If $k \geq 2$ there is an anti-clockwise rotation $f$ of $2\pi/r$ about $v_4$ in $G$ and $f(v_1)$ has greater imaginary part than $v_1$. This cannot be so: thus $k = 1$ and $\langle g, h \rangle$ is a $(p, r, \infty)$-Triangle group in $G$.

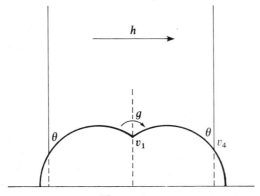

Figure 10.6.7

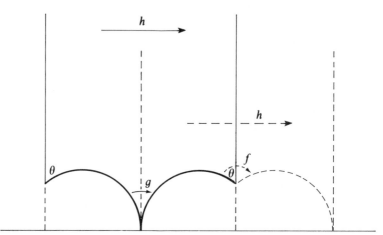

Figure 10.6.8

*Case* 3: *g and h are parabolic.*
We work in $H^2$ and we may assume that $g$ fixes 0 and $h$ fixes $\infty$: the situation is illustrated in Figure 10.6.8 and if $\theta = 0$, then $\langle g, h \rangle$ is a $(\infty, \infty, \infty)$-Triangle group. If $\theta > 0$, we construct the group $\langle f, h \rangle$ where $f = hg^{-1}$ is elliptic and this reduces Case 3 to Case 2.                                                     □

EXERCISE 10.6

1. Show that if $G$ is a Fuchsian group acting on $\Delta$ and if h-area$(\Delta/G) < \pi/3$ then $G$ is a triangle group. Show that the bound of $\pi/3$ cannot be improved.

2. Show that if $G$ is a conformal discrete group of some type $(\alpha, \beta, \gamma)$, then it is of exactly one type $(\pi/p, \pi/q, \pi/r)$.

3. Construct a fundamental quadilateral for a *Hecke group* $H_q$ $(q = 3, 4, \ldots)$ of signature $(0: 2, q, \infty)$ and show that $H_q$ is generated by a parabolic $g$ and an elliptic $h$ of order two.

4. Let $v_1, v_2, v_3$ and $v_4$ be distinct points on and placed in this order around $\{|z| = 1\}$. Let $g$ and $h$ be parabolic elements with

$$g(v_1) = v_1, \qquad g(v_2) = v_4, \qquad h(v_3) = v_3, \qquad h(v_4) = v_2.$$

Show that $g^{-1}h$ is parabolic if and only if the cross-ratio $[v_1, v_2, v_3, v_4]$ takes a specific value. Is $G = \langle g, h \rangle$ discrete? In any event, the quadilateral is not a fundamental domain for $G$ unless $g^{-1}h$ is parabolic.

## §10.7. Notes

For information on finite sided polygons, see [9], [10], [34], [35], [38], [46], [58], [76]. Points of approximation were studied by Hedlund (see [51], p. 181): also, see [8] and [109]. For results on conjugacy classes and subgroups, consult [49] and [97]. For a discussion of Triangle groups see [48] (for angles of the form $\pi a/b$) and [65].

# CHAPTER 11

# Universal Constraints On Fuchsian Groups

## §11.1. Uniformity of Discreteness

This chapter is concerned with the uniformity of discreteness exhibited by Fuchsian groups. As there is no uniformity to be found in the class of elementary groups, these must be regarded as exceptional. The Triangle groups are also, in some respects, exceptional. In general, a sharp quantitative expression for uniform discreteness will take a special form (depending only on the signature) for Triangle groups, and another single form (independent of the signature) for all non-elementary non-Triangle discrete groups. Thus it is the nature rather than the existence of the uniformity which leads one to treat the Triangle groups as a special case.

We shall discuss the following aspects of uniformity.

(1) *The distribution of a cycle of vertices of a fundamental polygon.* What are the geometric constraints relating to a cycle of vertices? What (if anything) can be said about accidental cycles?

(2) *The geometric constraints on the isometries.* For example, how close can two elliptic fixed points be in a discrete group? What are the constraints on the translation lengths of hyperbolic elements?

(3) *The location of canonical regions.* Canonical regions were defined in Section 7.37. The definition does not depend on discreteness: what can be said in the presence of discreteness and what does this imply for the quotient surface?

(4) *The displacement function $\rho(z, gz)$.* This has been discussed earlier (see, for example, Theorem 8.3.1): what can be said when elliptic elements are present?

(5) *The constraints on the corresponding matrix group.* A typical example of this is Jørgensen's inequality.

The results presented here do not form a complete and comprehensive account of uniformity of discreteness. Nevertheless, they indicate from a *geometric point of view* why such results must exist and references are given to further results of this type.

Broadly speaking, our attitude here is to apply simple geometric ideas to obtain universal constraints: these methods may fail for certain (and usually, relatively few) Triangle groups and for these, the reader is invited to supply individual computations.

## §11.2. Universal Inequalities for Cycles of Vertices

We establish here some of the universal constraints which must be satisfied by a cycle of elliptic *or even accidental vertices* on the boundary of a fundamental polygon of a Fuchsian group.

First, consider a Fuchsian group $G$ acting on $H^2$ with $g(z) = z + 1$ in $G$ and generating the stabilizer of $\infty$. In this case, we can construct a fundamental domain as in Section 9.6, this being the region lying exterior to all isometric circles and inside any strip of width one. Note that in this case, each cycle of vertices lies on some horocycle $\text{Im}[z] = \text{constant}$.

By choosing the vertical strip $x_0 < x < x_0 + 1$ suitably, we may assume that the cycle of vertices is $w_j$ $(j = 1, \ldots, n + 1)$ where $w_j = u_j + iv$ and

$$x_0 = u_1 < u_2 < \cdots < u_{n+1} = x_0 + 1.$$

Now construct triangles $T_j$ with angles $\theta_j$ as in Figure 11.2.1.

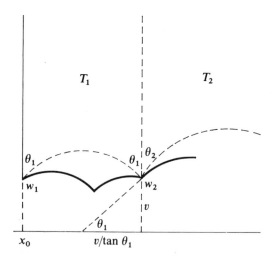

Figure 11.2.1

By noting that the $T_j$ lie in a fundamental region (by convexity) and by considering the angle sum at the cycle $(w_j)$ we have

$$\sum_j \theta_j \leq \pi/q, \qquad (11.2.1)$$

where the cycle $(w_j)$ is of order $q$ (for an accidental cycle, $q = 1$). Clearly, by considering the Euclidean projection of the $T_j$ onto the $x$-axis, we have

$$\sum_j v \cot \theta_j = \tfrac{1}{2}.$$

By Jensen's inequality, (1.2.2), we have (using (11.2.1) first)

$$\cot(\pi/qn) \leq \cot\left(n^{-1} \sum_j \theta_j\right)$$
$$\leq \sum_j n^{-1} \cot \theta_j$$
$$= 1/2vn.$$

This yields the following result.

**Theorem 11.2.1.** *Suppose that* $g : \mapsto z + 1$ *generates the stabilizer of* $\infty$ *in a Fuchsian group G acting on* $H^2$ *and let* $w_1, \ldots, w_n$ *be those vertices in a cycle of order q which lie in some strip* $x_0 \leq x < x_0 + 1$. *Then*

$$\text{Im}[w_j] \leq 1/2n \tan(\pi/qn).$$

For an accidental cycle, we have $q = 1$ and $n \geq 3$: thus we obtain the next result.

**Corollary 11.2.2.** *If* $(w_j)$ *in Theorem 11.2.1 is an accidental cycle, then*

$$\text{Im}[w_j] \leq \tfrac{1}{6} \tan(\pi/3) = 1/2\sqrt{3},$$

*or, in an invariant form,*

$$\sinh \tfrac{1}{2}\rho(w_j, gw_j) \geq \sqrt{3}.$$

**Corollary 11.2.3.** *If* $(w_j)$ *in Theorem 11.2.1 is an elliptic cycle of order q* $(q \geq 3)$ *then*

$$\text{Im}[w_j] \leq \tfrac{1}{2}\tan(\pi/q),$$

*or, equivalently,*

$$\sinh \tfrac{1}{2}\rho(w_j, gw_j) \geq 1/\tan(\pi/q).$$

We shall see in Section 11.3 that the bound in Corollary 11.2.3 is best possible.

We can also obtain inequalities for accidental vertices on the boundary of a Dirichlet polygon.

**Theorem 11.2.4.** *Let G be a non-elementary Fuchsian group and let* $v_1, \ldots, v_n$
*be an accidental cycle on the boundary of the Dirichlet polygon with centre w.*

(i) *If* $n \geq 5$, *then* $\cosh \rho(w, v_j) \geq 1/\tan^2(\pi/n) \geq 1 \cdot 89 \ldots$ ;
(ii) *if* $n = 4$, *then* $\cosh \rho(w, v_j)$ *is not less than some absolute constant* $\mu(>1)$;
(iii) *there is no universal lower bound in the case* $n = 3$.

If $G$ has no elliptic elements, a universal lower bound exists for all values
of $n$.

**Theorem 11.2.5.** *Let G be a non-elementary Fuchsian group without elliptic*
*elements. If* $(v_j)$ *is an accidental cycle of vertices on the boundary of the*
*Dirichlet polygon with centre w, then*

$$\cosh \rho(w, v_j) \geq \sqrt{2}.$$

PROOF OF THEOREM 11.2.5. The cycle $(v_j)$ lies on a circle $C$, say $\{z : \rho(z,w) = r\}$
and contains at least three vertices with, say,

$$v_2 = g(v_1), \qquad v_3 = h(v_1).$$

Let $G_0$ be the group generated by $g$ and $h$. If $G_0$ is elementary, then it is
cyclic with a parabolic or hyperbolic generator $f$. In either case, the points
$v_1$, $v_2$ and $v_3$ cannot lie on a circle so $G_0$ must be non-elementary. By Theorem
8.3.1,

$$\sinh \tfrac{1}{2}\rho(v_1, gv_1) \sinh \tfrac{1}{2}\rho(v_1, hv_1) \geq 1.$$

Now

$$\begin{aligned}
\rho(v_1, gv_1) &= \rho(v_1, v_2) \\
&\leq \rho(v_1, w) + \rho(w, v_2) \\
&= 2r
\end{aligned}$$

and similarly for $h$. We deduce that $\sinh r \geq 1$ as required. $\qquad\square$

PROOF OF THEOREM 11.2.4. We may assume that $G$ acts on $\Delta$ and that $w = 0$
as all terms are invariant under conjugation. Thus the points $v_j$ lie (and can
be assumed to be labelled cyclically) on some circle $\rho(z, 0) = r$. The arcs
$(v_j, v_{j+1})$ (not containing any other $v_i$) subtend an angle $2\alpha_j$ at the origin and

$$\sum_j \alpha_j = \pi.$$

As the cycle length is at least three there is at most one $j$ for which $2\alpha_j \geq \pi$.
If $2\alpha_j < \pi$ then the triangle $T_j$ with vertices $0, v_j, v_{j+1}$ and angles $2\alpha_j, \theta_j, \theta_j$
lies in the Dirichlet polygon and as the angle sum of the cycle is $2\pi$, we have

$$\sum_j \theta_j \leq \pi.$$

Note that from Section 7.12 (by considering one half of $T_j$)

$$\cosh r \tan \theta_j \tan \alpha_j = 1:$$

see Figure 11.2.2.

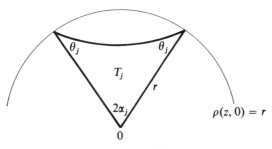

Figure 11.2.2

Now either each $\alpha_j$ is less than $\pi/2$ and then

$$\sum_{j=1}^{n} (\theta_j + \alpha_j) \le 2\pi$$

or exactly one $\alpha_j$ (say $\alpha_n$) is at least $\pi/2$, in which case

$$\sum_{j=1}^{n-1} (\theta_j + \alpha_j) \le 2\pi - \alpha_n \le 3\pi/2.$$

In both cases, some $\theta_k + \alpha_k$ is at most the average value which (as $n \ge 5$) is at most $2\pi/n$. Thus for this $k$ we have (see (1.2.3))

$$\tan \alpha_k \tan \theta_k \le \tan^2\left(\frac{\theta_k + \alpha_k}{2}\right)$$

$$\le \tan^2(\pi/n).$$

This proves (i): note that it provides no information when $n$ is 3 or 4.

The case $n = 4$ is more complicated and the proof of (ii) will be given in Section 11.6.

To prove (iii), construct the polygon $P$ illustrated in Figure 11.2.3. The polygon has four pairs of sides with side-pairing elements $g$, h (each of

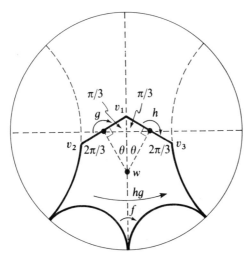

Figure 11.2.3

order two), $hg$ (hyperbolic) and $f$ (parabolic). By Poincaré's Theorem, $P$ is a fundamental domain for the non-elementary Fuchsian group generated by $f, g$ and $h$.

This construction is possible if and only if $\theta < \pi/6$ and then

$$\cosh t \sin(\pi/3) = \cos\theta, \quad 2t = \rho(v_1, v_2):$$

thus $t \to 0+$ as $\theta \to \pi/6$ from below. Observe that each $v_j$ is the same distance from $w$ and that

$$\cosh \rho(w, v_1) \tan(\pi/3) \tan\theta = 1.$$

Thus as $\theta \to \pi/6$, so $\rho(w, v_1) \to 0$.

It remains only to prove that $P$ is actually the Dirichlet polygon $D(w)$ for $\langle f, g, h \rangle$ with centre $w$. Now the sides paired by $f$ are the perpendicular bisectors of the segments $[w, fw]$, and $[w, f^{-1}w]$: a similar statement holds for $hg$. Also the two sides making the edge $[v_1, v_2]$ lie on the perpendicular bisector of the segment $[w, gw]$: a similar statement holds for $[v_1, v_3]$. We deduce that $P$ contains the Dirichlet polygon $D(w)$; as $P$ is a fundamental domain, it must be $D(w)$. $\qquad\square$

**Example 11.2.6.** Given any integer $k$ with $k \geq 2$ we can construct a Fuchsian group $G$ acting on $\Delta$ which has as its fundamental domain a regular polygon with $4k$ sides and all vertices lying in one accidental cycle (see Section 10.4). Referring to the proof of Theorem 11.2.4(i), we find that $\alpha_j = \theta_j = \pi/4k$ so in this case, equality holds in (i). Thus (at least for $n$ of the form $4k$), Theorem 11.2.4(i) is best possible.

Finally, we consider unbounded fundamental polygons (although the idea in the following proof clearly extends to other situations).

**Theorem 11.2.7.** *Let $D$ be a fundamental polygon for a Fuchsian group $G$ and suppose that $D$ contains two points $w_1$ and $w_2$ on the circle at infinity. Let $L$ be the geodesic joining $w_1$ and $w_2$. If $v$ is an elliptic fixed point of $G$ of order $n$, lying on the boundary of $D$, then*

$$\cosh \rho(v, L) \geq 1/\sin(\pi/n) \geq 2/\sqrt{3}.$$

PROOF. The triangle with vertices $w_1$, $w_2$ and $v$ lies in $D$ and so the interior angle of this triangle at $v$ cannot exceed $2\pi/n$. This means that $v$ cannot be too close to $L$: the numerical details are left to the reader. $\qquad\square$

Note that this result implies that no elliptic fixed point on $\partial D$ lies in the lens region between the two hypercycles making an angle $\pi/6$ with $L$.

EXERCISE 11.2

1. Derive an inequality similar to (i) in Theorem 11.2.4 which is applicable to an elliptic cycle of order $q$ on the boundary of the Dirichlet polygon.

2. Is the bound in Corollary 11.2.2 best possible?

3. Let $D$ be a convex fundamental polygon for a Fuchsian group $G$. Show that if there is some $w$ such that the sides of $D$ lie on the bisectors $\{z: \rho(z, w) = \rho(z, gw)\}$, $g \in G$, then $D$ is the Dirichlet polygon with centre $w$.

4. Let $D$ be a convex fundamental polygon for a Fuchsian group $G$ acting in $\Delta$ and suppose that $D$ contains a geodesic $L$. Prove that if $\{v_1, \ldots, v_n\}$ is an accidental cycle on $\partial D$ then
$$\cosh\rho(v_1, L) + \ldots + \cosh\rho(v_n, L) \geq n/\sin\left(\frac{\pi}{n}\right) \geq n^2/\pi.$$

Find a corresponding inequality when the $v_j$ form an elliptic cycle of order $q$.

5. With reference to Figure 11.2.3, show that $f^{-1}hg$ is parabolic (write $f = \alpha\beta$, $hg = \alpha\gamma$ where $\alpha$, $\beta$ and $\gamma$ are reflections).

## §11.3. Hecke Groups

In this section, we study the class of Hecke groups as these play an exceptional role in the following discussions.

**Definition 11.3.1.** A *Hecke group* is a Triangle group with signature $(0: 2, q, \infty)$ for some integer $q$ satisfying $3 \leq q \leq +\infty$.

Let
$$g(z) = -1/z, \qquad h(z) = z + 2\cos(\pi/q):$$
then $\langle g, h \rangle$ has signature $(0: 2, q, \infty)$ and a fundamental domain for $\langle g, h \rangle$ is illustrated in Figure 11.3.1. As any two Triangle groups with the same signature are conjugate, we see that any Hecke group with signature $(0: 2, q, \infty)$ is conjugate to $\langle g, h \rangle$. Note that $hg$ is elliptic of order $q$ and fixes one vertex of the triangle.

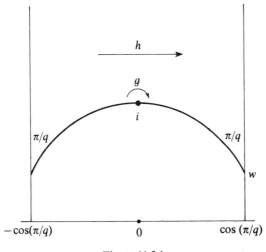

Figure 11.3.1

It is sometimes convenient to normalize the parabolic generator $h$ so that this is the map $z \mapsto z + 1$. Then $g$ becomes

$$g(z) = -1/4z \cos^2(\pi/q)$$

and this is elliptic of order two with fixed point $i/2 \cos(\pi/q)$. Note that with this normalization, the fixed point, say $w$, of order $q$ satisfies

$$\text{Im}[w] = \tfrac{1}{2} \tan(\pi/q):$$

this shows that Corollary 11.2.3 is best possible.

The next two results help in identifying Hecke groups.

**Proposition 11.3.2.** *Let $G$ be a Fuchsian group with parabolic elements. If $G$ has a fundamental domain with h-area less than $\pi$, then $G$ has one of the signatures $(0: 2, q, \infty)$ where $3 \le q \le +\infty$ or $(0: 3, q, \infty)$ where $q = 3, 4$ or $5$.*

PROOF. As the fundamental domain has finite area, $G$ has signature $(k: m_1, \ldots, m_n, \infty)$ say, the $\infty$ being present as $G$ is known to include parabolic elements. From Section 10.4 we deduce that

$$2\pi\left[ 2k - 2 + \sum_{j=1}^{n} \left(1 - \frac{1}{m_j}\right) + 1 \right] < \pi \qquad (11.3.1)$$

and so (as $m_j \ge 2$)

$$4k + n < 3.$$

Thus $k = 0$ and (for positive area) $n = 2$. With this information, (11.3.1) now yields

$$\frac{1}{m_1} + \frac{1}{m_2} > \frac{1}{2}$$

and hence $\min\{m_1, m_2\} \le 3$. The result now follows easily. $\qquad\square$

**Theorem 11.3.3.** *Let $G_0$ be a Hecke group and let $G$ be a Fuchsian group containing $G_0$. Then $G = G_0$.*

PROOF. We may suppose that $G$ acts on $H^2$ so

$$k \text{ h-area}(H^2/G) = \text{h-area}(H^2/G_0), \qquad (11.3.2)$$

where $G_0$ is of index $k$ in $G$. By assumption, $G_0$ has signature $(0: 2, p, \infty)$, say, and so $G$ has one of the signatures described in Proposition 11.3.2.

If $k \ge 2$ then

$$\text{h-area}(H^2/G) \le \pi/2$$

and so $G$ is also a Hecke group (see the proof of Proposition 11.3.2) with signature $(0: 2, p, \infty)$. This contradicts (11.3.2) so $k = 1$ and $G = G_0$.

For an alternative proof, recall that the elliptic fixed points of order $q$ at the vertices of the triangle in Figure 11.3.1 have the largest possible imaginary part for any group containing $G_0$. In particular, their images by elements of $G$ cannot have a larger imaginary part so these fixed points must lie on the boundary of the corresponding fundamental domain $D$ for $G$ (constructed as in Section 9.6). It follows from convexity that $D$ contains the triangle and as $G \supset G_0$, $D$ must be the triangle. Thus $G = G_0$. $\qquad \square$

EXERCISE 11.3

1. With reference to Figure 11.3.1, show that $hg$ is the composition of reflections in the two sides with common vertex $w$ and hence is a rotation about $w$ of angle $2\pi/q$.

2. Show that if $G$ contains parabolic elements and if h-area$(H^2/G) < 2\pi/3$, then $G$ is a Hecke group.

## §11.4. Trace Inequalities

The objective here is to obtain certain algebraic inequalities which must be satisfied by two elements in order that they generate a non-elementary discrete group.

**Theorem 11.4.1.** *Suppose that the two parabolic elements $g$ and $h$ generate a non-elementary Fuchsian group $G$. Then one of the following possibilities must occur:*

(1) trace$[g, h] \geq 18$;
(2) trace$[g, h] = 2 + 16 \cos^4(\pi/r)$ *and $G$ has signature* $(0: 2, r, \infty)$;
(3) trace$[g, h] = 2 + 16 \cos^4(\pi/2r)$ *and $G$ has signature* $(0: r, \infty, \infty)$.

PROOF. By conjugation, we may suppose that $G$ acts on $H^2$ and that

$$h(z) = z + 1, \qquad g(z) = z/(cz + 1).$$

By using $g^{-1}$ if necessary, we may suppose that $c > 0$. As

$$\text{trace}[g, h] = \text{trace}[h, g] = 2 + c^2, \qquad (11.4.1)$$

the three possibilities are equivalent to

(1) $c \geq 4$;
(2) $c = 2 + 2 \cos(2\pi/r)$;
(3) $c = 2 + 2 \cos(\pi/r)$.

Jørgensen's inequality, namely $c \geq 1$, holds so assuming that (1) fails, we have $1 \leq c < 4$. Now construct the quadrilateral with sides formed by the isometric circles of $g$ and $g^{-1}$ and the lines $x = \frac{1}{2}$ and $x = -\frac{1}{2}$: see Figure 11.4.1. Observe that $1 \leq c < 4$ implies that the point $w$ does exist.

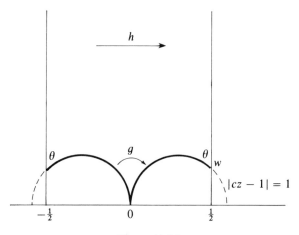

Figure 11.4.1

By considering reflections in $x = 0$, $x = \frac{1}{2}$ and $|cz - 1| = 1$, we find that $hg^{-1}$ is a rotation of angle $2\theta$ about $w$. Thus for some $k$ and $r$ (which we may assume are coprime) we have

$$\theta = k\pi/r, \qquad c = 2 + 2\cos(k\pi/r).$$

If $k = 1$ or $r = \infty$, then Poincaré's Theorem is applicable, the quadrilateral is a fundamental polygon for $G$ and $G$ has signature $(0: r, \infty, \infty)$: this is Case (3).

If $k \geq 2$ and $r$ is finite, then there is some $f$ in $G$ which is a rotation of angle $2\pi/r$ about $w$. In this case, construct the quadrilateral in Figure 11.4.2. Observe that as $k \geq 2$ we have $\pi/r = \theta/k \leq \theta/2$ so (by elementary trigonometry) $\phi \geq \pi/2$.

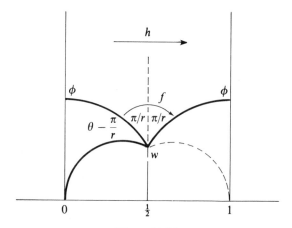

Figure 11.4.2

Now the images under $\langle h, f \rangle$ (and hence under $G$) of the quadrilateral with angles $0, \phi, \phi, 2\pi/r$ cover the hyperbolic plane (because any curve from $w$ can be covered by images of the quadrilateral a small, but fixed, distance at a time) and so $G$ has a fundamental domain of area at most $\pi - (2\pi/r)$. Proposition 11.3.2 implies that $G$ is a Triangle group with signature, say, $(0: r_1, s, \infty)$ where $r$ divides $r_1$. Thus

$$2\pi\left(1 - \frac{1}{r_1} - \frac{1}{s}\right) \le 2\pi - 2\phi - \frac{2\pi}{r}$$

$$\le \pi - \frac{2\pi}{r}$$

$$\le \pi - \frac{2\pi}{r_1}.$$

This shows that $s = 2$: thus equality holds throughout and so $\phi = \pi/2$. Because $\phi = \pi/2$, we have $\theta = 2\pi/r$: then $k = 2, r = r_1$ and this is Case (2).

$\square$

**Theorem 11.4.2.** *Suppose that $h$ is parabolic and that $g$ and $h$ generate a non-elementary Fuchsian group $G$. Then*

(1) $\text{trace}[g, h] \ge 3$;
(2) *if* $3 \le \text{trace}[g, h] < 6$ *then $G$ has signature* $(0: 2, q, \infty)$ *and*

$$\text{trace}[g, h] = 4 + 2\cos(2\pi/q);$$

(3) *if* $\text{trace}[g, h] < 18$ *then $G$ contains elliptic elements.*

PROOF. We may assume that $G$ acts on $H^2$ and that

$$h(z) = z + 1, \qquad g(z) = \frac{az + b}{cz + d},$$

where $ad - bc = 1$ and $c > 0$. As (11.4.1) holds, we see that (1) is simply Jørgensen's inequality. In order to prove (2), we assume that $\text{trace}[g, h] < 6$ or, equivalently, $c < 2$. This means that $G$ has a fundamental domain lying outside the isometric circle of $g$ and inside a vertical strip of width one: see Figure 11.4.3. As the isometric circle of $g$ has a Euclidean diameter greater than one, we see that $H^2/G$ has area less than $\pi$ and so $G$ has one of the signatures given in Proposition 11.3.2.

Now observe that $g = \sigma_2\sigma_1$ where $\sigma_1$ is the reflection in $L_1$ given by $|cz + d| = 1$ and where $\sigma_2$ is the reflection in a vertical line $L_2$. For any choice of the integer $n$, let $L_3$ be the line $L_2$ translated by a (Euclidean) distance $n/2$. Then

$$h^n g = (\sigma_3\sigma_2)(\sigma_2\sigma_1)$$

$$= \sigma_3\sigma_1$$

so $\sigma_3\sigma_1$ is in $G$.

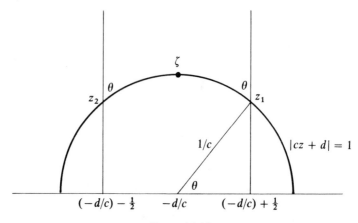

Figure 11.4.3

Now let $L$ be the vertical geodesic orthogonal to $L_1$: see Figure 11.4.4. By choosing $n$ to minimize the Euclidean distance between $L$ and $L_3$, we see that $L_3$ meets $L_1$ at a point $w$ in an angle $\phi$, say. Thus $h^n g$ fixes $w$ and is a rotation about $w$ of angle $2\phi$. Clearly, if $t$ is the distance between $L$ and $L_3$, then $t < \frac{1}{4}$ and

$$\cos \phi = ct$$

$$< \tfrac{1}{2}:$$

thus $\phi > \pi/3$. Also, $\phi \le \pi/2$.

Let $p$ be the order of the fixed point $w$. Then $\phi = k\pi/p$ say, with $(k, p) = 1$ and hence

$$\tfrac{1}{3} < k/p \le \tfrac{1}{2}. \tag{11.4.2}$$

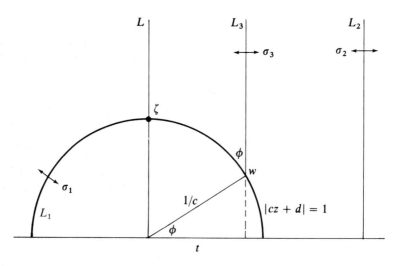

Figure 11.4.4

Also, by Corollary 11.2.3, we have

$$\tan(\pi/p) \geq 2 \, \text{Im}[w]$$
$$= 2(1/c) \sin \phi$$
$$\geq \sin(\pi/3)$$
$$= \sqrt{3}/2 \tag{11.4.3}$$

so $p = 2, 3$ or $4$. With these values of $p$, the only solutions of (11.4.2) lead to $\phi = \pi/2, p = 2, w = \zeta$ so $G$ contains as elliptic element $f$ of order two which fixes $\zeta$.

If $G$ has one of the signatures $(0: 3, q, \infty)$ where $q = 3, 4$ or $5$, then $q = 4$ (as $f \in G$). But in this case, $w$ is a fixed point of order four so

$$2 \, \text{Im}[w] \leq \tan(\pi/4)$$
$$= 1$$

contrary to (11.4.3). Thus $G$ must be a Hecke group with signature $(0: 2, q, \infty)$ say. The elements $h$ and $f$ generate $G$ and pair the sides of the triangle illustrated in Figure 11.4.3 (with $f$ interchanging the sides $[\zeta, z_1]$ and $[\zeta, z_2]$). From consideration of areas, we have

$$\pi - 2\pi/q \leq \pi - 2\theta$$

and so $\theta \leq \pi/q$. On the other hand, the minimum angle of rotation in $G$ is $2\pi/q$ so $2\theta \geq 2\pi/q$. This gives $\theta = \pi/q$ and

$$c = 2 \cos(\pi/q)$$

which is (2).

If $G$ has no elliptic elements then $c \geq 4$ (see the proof of Theorem 8.3.1) so (3) holds. $\qquad\square$

Similar results hold for elliptic elements in place of parabolic elements.

**Theorem 11.4.3.** *Let $g$ be a rotation of angle $2\pi/n$ $(n \geq 3)$ about some point in the hyperbolic plane and suppose that $f$ and $g$ generate a non-elementary Fuchsian group. Then apart from certain Triangle groups (which are listed in the proof),*

(1) $\text{trace}[f, g] \geq 2 + 4 \cos^2(\pi/n) \geq 3$;
(2) $|\text{trace}^2(g) - 4| + |\text{trace}[f, g] - 2| \geq 4$.

*Remark.* If $f$ and $g$ lie in a non-Triangle discrete group, then either $\langle f, g \rangle$ is elementary or (1) and (2) hold: see Theorem 10.6.5.

*Remark.* The inequality (2) is meaningful for all $n \geq 3$: this is not true if the lower bound is replaced by one.

PROOF. We may suppose that $f$ and $g$ act on $\Delta$ and that in terms of matrices,

$$g = \begin{pmatrix} e^{i\pi/n} & 0 \\ 0 & e^{-i\pi/n} \end{pmatrix}, \quad f = \begin{pmatrix} a & \bar{c} \\ c & \bar{a} \end{pmatrix},$$

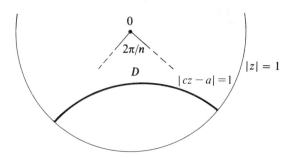

Figure 11.4.5

where $|a|^2 - |c|^2 = 1$. Now $\langle f, g \rangle$ has a fundamental domain lying within the region $D$ illustrated in Figure 11.4.5: here, $D$ is the region outside the isometric circle of $f^{-1}$ and within a sector of angle $2\pi/n$ situated symmetrically with respect to the isometric circle.

The exceptional groups are those for which $D$ is bounded. In this case, the signature $(k: m_1, \ldots, m_s)$ satisfies

$$2\pi\left[2k - 2 + \sum_{j=1}^{s}\left(1 - \frac{1}{m_j}\right)\right] < \pi - 2\pi/n \qquad (11.4.4)$$

$$\leq \pi - 2\pi/m_s,$$

say, where $n$ divides $m_s$. Thus $k = 0$ and $s = 3$. A more detailed investigation of (11.4.4) now yields the exceptional cases

$$m_1 = 2 \quad \text{or} \quad (m_1, m_2) = (3, 3), (3, 4) \text{ or } (3, 5).$$

Assume now that $\langle f, g \rangle$ is not one of these exceptional groups. Then $D$ is unbounded and, noting that the isometric circle of $f^{-1}$ is the bisector of $[0, f0]$, we may use the Angle of Parallelism formula to obtain

$$\cosh \tfrac{1}{2}\rho(0, f0) \sin(\pi/n) \geq 1,$$

or, equivalently,

$$\sin^2(\pi/n) \sinh^2 \tfrac{1}{2}\rho(0, f0) \geq \cos^2(\pi/n).$$

As

$$|c| = \sinh \tfrac{1}{2}\rho(0, f0)$$
$$\geq \cot(\pi/n),$$

a computation yields first (1) and then (2). $\qquad \square$


EXERCISE 11.4

1. Verify that (in the proof of Theorem 11.4.1) the assumption $c > 0$ ensures that $g$ acts in the direction shown in Figure 11.4.1.

2. Suppose that $c > 4$ and let $G$ be generated by

$$h(z) = z + 1, \qquad g(z) = \frac{z}{cz + 1}.$$

Show that $G$ is discrete and find its signature.

Prove analytically and geometrically (which is much shorter) that $gh^{-1}$ is hyperbolic with translation length $T$ where the hyperbolic distance between $x = \frac{1}{2}$ and the isometric circle of $g$ is $\frac{1}{2}T$.

3. Suppose that $G$ is a Fuchsian group acting on $H^2$ which contains

$$h(z) = z + 1, \qquad g(z) = \frac{az + b}{cz + d} \qquad (c \neq 0, \ ad - bc = 1)$$

Prove that h-area$(H^2/G) \geq \pi/3$.

Show that the triangle bounded by the isometric circle of $g$ and the two vertical lines

$$x = (-d/c) - \tfrac{1}{2}, \qquad x = (-d/c) + \tfrac{1}{2}$$

contains a fundamental domain for $G$ and deduce that $|c| \geq 1$ (this is Jørgensen's inequality).

4. As in the proof of Theorem 11.4.2, assume that $c < 2$. Show that $G$ contains an element of order two as follows.

(i) Let

$$f(z) = \frac{az + b}{cz + d} \qquad (ad - bc = 1)$$

be in $G$ with the smallest (positive) value of $|c|$ possible. By considering the matrix for $f^2$, show that either $f$ has order two or that $|\text{trace}(f)| \geq 1$.

(ii) Show that for a suitable $n$, $|\text{trace}(h^n f)| < 1$ so $h^n f$ is of order two.

## §11.5. Three Elliptic Elements of Order Two

Let $f$, $g$ and $h$ be elliptic elements of order two with distinct fixed points $u$, $v$ and $w$ respectively. If $u$, $v$ and $w$ are collinear then the group $G$ generated by $f$, $g$ and $h$ is elementary for it leaves the geodesic containing these points invariant. We shall assume that $u$, $v$ and $w$ are not collinear: let $\alpha$, $\beta$ and $\gamma$ be the angles and $a$, $b$ and $c$ be the lengths of the sides of the triangle with vertices $u$, $v$ and $w$: see Figure 11.5.1.

The three vertices of the triangle determine a positive number $\lambda$ which is defined by

$$\begin{aligned}
\lambda &= \sinh a \ \sinh b \ \sin \gamma \\
&= \sinh b \ \sinh c \ \sin \alpha \qquad\qquad (11.5.1) \\
&= \sinh c \ \sinh a \ \sin \beta,
\end{aligned}$$

the equality of these expressions being a consequence of the Sine Rule. If we

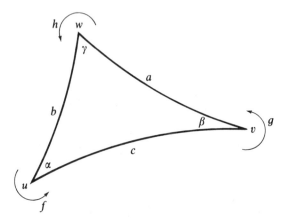

Figure 11.5.1

view the side $[u, v]$ as the base of the triangle lying, say, on the geodesic $L_w$, then the height of the triangle is $\rho(w, L_w)$ where

$$\sinh \rho(w, L_w) = \sinh a \sin \beta.$$

Thus we may also write (in the obvious way)

$$\lambda = \sinh(base) \times \sinh(height),$$

regardless of the choice of which side is the base.

The quantity $\lambda$ is related to the elliptic elements $f$, $g$ and $h$ as follows.

**Theorem 11.5.1.** *The absolute value of the trace of any of the isometries*

$$fgh, \; hfg, \; ghf, \; hgf, \; fhg, \; gfh$$

*is equal to* $2\lambda$.

PROOF. First, $|\text{trace}(fgh)|$ is invariant under cyclic permutations of $f$, $g$ and $h$: for example,

$$|\text{trace}(fgh)| = |\text{trace } h(fgh)h^{-1}|$$
$$= |\text{trace}(hfg)|.$$

Also,

$$|\text{trace}(fgh)| = |\text{trace}(fgh)^{-1}|$$
$$= |\text{trace}(hgf)|$$

so $|\text{trace}(fgh)|$ is invariant under any permutation of $f$, $g$ and $h$.

Now let $L$ be the geodesic through $u$ and $v$. Construct

(i) the geodesic $L_1$ through $w$ and orthogonal to $L$;
(ii) the geodesic $L_2$ through $w$ and orthogonal to $L_1$;
(iii) the geodesics $L_3$ and $L_4$ orthogonal to $L$ with

$$\rho(L_1, L_3) = \rho(u, v) = \rho(L_1, L_4):$$

see Figure 11.5.2.

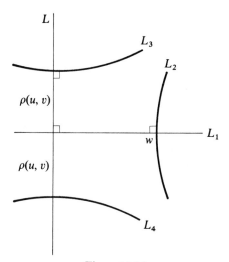

Figure 11.5.2

Denoting reflection in $L_j$ by $\sigma_j$, we have

$$\sigma_2\sigma_1 = h, \qquad \sigma_1\sigma_3 = fg \ (\text{or } gf)$$

as $fg$ is a hyperbolic element with axis $L$ and translation length $2\rho(u, v)$. It follows from Theorem 7.38.1 that

$$\tfrac{1}{2}|\text{trace}(hfg)| = \tfrac{1}{2}|\text{trace}(\sigma_2\sigma_3)|$$
$$= (L_2, L_3).$$

Now the inversive product $(L_2, L_3)$ is $\cosh \rho(L_2, L_3)$ when $L_2$ and $L_3$ are disjoint and it is $\cos \phi$ when $L_2$ and $L_3$ meet at an angle $\phi$ (possibly zero). In all cases (see Theorem 7.17.1, Lemma 7.17.3 and Theorem 7.18.1(iii)) we have

$$(L_2, L_3) = \sinh \rho(L, L_2) \sinh \rho(L_1, L_3)$$
$$= \sinh \rho(w, L) \sinh \rho(u, v)$$
$$= \lambda. \qquad \square$$

We shall now examine how the value of $\lambda$ determines the nature of the group generated by $f$, $g$ and $h$.

**Theorem 11.5.2.** *Let $f$, $g$ and $h$ be elliptic elements of order two which generate a non-elementary group $G$ and let $\lambda$ be given by (11.5.1).*

(1) *If $\lambda > 1$ then $G$ is discrete and has signature $(0: 2, 2, 2; 0; 1)$.*
(2) *If $\lambda = 1$ then $G$ is discrete and has signature $(0: 2, 2, 2; 1; 0)$.*
(3) *If $\lambda < 1$ then $G$ is discrete only if $\lambda$ is one of the values*

$$\cos(\pi/q), q \geq 3; \qquad \cos(2\pi/q), q \geq 5; \qquad \cos(3\pi/q), q \geq 7:$$

*the possible signatures for $G$ are*

$$(0: 2, 2, 2, q; 0; 0), (0: 2, 3, q; 0; 0), (0: 2, 4, q; 0; 0).$$

A construction of a fundamental domain for each discrete $G$ will arise in the proof and it will be apparent that every value of $\lambda$ given in Theorem 11.5.2 does give rise to a discrete group. Thus we can derive the following universal bound.

**Corollary 11.5.3.** *If $f$, $g$ and $h$ are elliptic elements of order two which generate a non-elementary discrete group, then*

$$|\operatorname{trace}(fgh)| \geq 2\cos(3\pi/7)$$

*and this is best possible.*

PROOF OF THEOREM 11.5.2. We suppose first that $\lambda > 1$. Then we can construct the polygon illustrated in Figure 11.5.3 where $u'$ and $v'$ are images of $u$ and $v$ respectively under some power of the hyperbolic element $fg$ with axis $L$. Note that

$$\rho(L_3, L_4) = 2\rho(u, v).$$

The elements fixing $u'$ and $v'$ are, say,

$$(fg)^m f(fg)^{-m}, \quad (fg)^n g(fg)^{-n},$$

respectively. The side-pairing maps of the polygon in Figure 11.5.3 generate $G$ and by Poincaré's Theorem, the polygon is a fundamental domain for $G$. In this case, $G$ has signature $(0:2, 2, 2; 0; 1)$. This proves (1): an obvious modification gives (2) with $\lambda = 1$ precisely when $L_2$ is tangent to $L_3$ and $L_4$ on the circle at infinity.

The case when $\lambda < 1$ is more difficult: here $L_2$ meets $L_3$ and $L_4$ at an

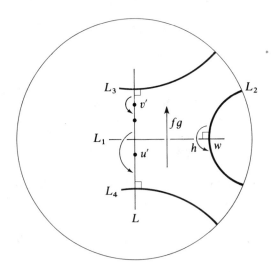

Figure 11.5.3

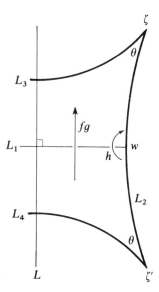

Figure 11.5.4

angle $\theta$ say and we consider the polygon illustrated in Figure 11.5.4. Note that as discussed earlier, we have $\lambda = \cos\theta$.

Suppose now that $G$ is discrete. Then $hgf$ (or $hfg$) satisfies

$$hgf = (\sigma_2\sigma_1)(\sigma_1\sigma_3)$$
$$= \sigma_2\sigma_3$$

and this is rotation of angle $2\theta$ about $\zeta$. Let $\zeta$ be an elliptic fixed point of order $q$ so $\theta = \pi p/q$ for some integer $p$, $(p, q) = 1$.

If $p = 1$, we obtain a fundamental polygon for $G$ and in this case $G$ has signature $(0: 2, 2, 2, q; 0; 0)$ and $\lambda = \cos(\pi/q)$ where $q \geq 3$.

From now on we may assume that $p \geq 2$. The $G$-images of the compact quadrilateral cover the hyperbolic plane (there is a positive $r$ such that each point of the quadrilateral lies in a disc of radius $r$ covered by $G$-images) so by considering areas we have

$$2\pi\left[2k - 2 + \sum_{j=1}^{s}\left(1 - \frac{1}{m_j}\right)\right] \leq \pi - 2\pi p/q,$$

where $G$ has signature $(k: m_1, \ldots, m_s)$. This gives

$$4k - 4 + s < 1.$$

For positive area, we also have

$$0 < 2k - 2 + s$$

and so the only possibilities are $k = 0$ and $s = 3$ or $4$.

In fact, $s = 3$. To see this, assume that $s = 4$. As $G$ contains an element of order $q$, we may suppose that $q$ divides $m_4$. Then as $p \geq 2$, $m_j \geq 2$ and $q \leq m_4$, we have

$$2\left[\frac{1}{2} - \frac{1}{m_4}\right] \leq 2\left[2 - \sum_{j=1}^{4} \frac{1}{m_j}\right]$$
$$\leq 1 - \frac{2p}{q}$$
$$\leq 1 - \frac{4}{m_4}.$$

This implies that $m_4 = \infty$ and hence that $G$ contains parabolic elements: however this cannot be so as the quadrilateral is compact and contains points from every orbit. Thus $s = 3$ and $G$ is a Triangle group.

Let us now write the signature of $G$ as $(0: l, m, n)$ where $q$ divides $n$. By Theorem 9.8.6, there is a positive integer $N$ such that the quadrilateral contains $N$ images of each point in the plane. Thus by considering areas,

$$2\pi N\left[1 - \left(\frac{1}{l} + \frac{1}{m} + \frac{1}{n}\right)\right] = \pi - \frac{2\pi p}{q}.$$

As $\theta = \pi p/q$ and as $\zeta$ and $\zeta'$ are in the same orbit we find that $N \geq p$ (consider points close to $\zeta$). Thus

$$2p\left[1 - \left(\frac{1}{l} + \frac{1}{m} + \frac{1}{n}\right)\right] \leq 2N\left[1 - \left(\frac{1}{l} + \frac{1}{m} + \frac{1}{n}\right)\right]$$
$$\leq 1 - \frac{2p}{q}$$
$$\leq 1 - \frac{2p}{n}. \qquad (11.5.2)$$

The inequality between the first and last terms yields (as $p \geq 2$)

$$\frac{1}{l} + \frac{1}{m} \geq \frac{2p - 1}{2p} \geq \frac{3}{4}$$

and the solutions of this are

$$(l, m, p) = (2, 3, 2), (2, 3, 3), (2, 4, 2).$$

If $(l, m, p) = (2, 4, 2)$, then equality holds throughout (11.5.2) so $q = n$: thus in this case $G$ has signature $(0: 2, 4, q)$ where $q \geq 5$ and $\lambda = \cos(2\pi/q)$.

If $(l, m, p) = (2, 3, 3)$, equality again holds throughout (11.5.2) so $q = n$, $G$ has signature $(0: 2, 3, q)$ where $q \geq 7$ and $\lambda = \cos(3\pi/q)$.

For the remaining case, namely $(l, m, p) = (2, 3, 2)$ we need a slightly different argument. First, the elliptic fixed points $u', v', w, \zeta$ and $\zeta'$ lie in at

most two orbits (none can lie in the orbit of order three). This means that
$N \geq 3$ and using the middle terms of (11.5.2) we have

$$6\left(\frac{1}{6} - \frac{1}{n}\right) \leq 1 - \frac{4}{q}$$

so $q = n$ (because $n/q$ is an integer). This actually completes the proof of
Theorem 11.5.2 as stated as this does not assert which signatures correspond
to which values of $\lambda$. Briefly, there exist integral solutions of the above
equations which do not correspond to discrete groups and a more detailed
analysis yields all possibilities. For example, the middle terms of (11.5.2)
yield

$$N = 3 + \frac{6}{n - 6}$$

so

$$(n, N) = (7, 9), (8, 6), (9, 5) \text{ or } (12, 4).$$

However, $N$ must be a multiple of three (consider the fixed points of order
three in the quadrilateral).                                        □

As an illustration of the possible cases, consider the quadrilateral
illustrated in Figure 11.5.5 where $\rho(u', w) = \rho(u', v')$. Let $\alpha$, $\beta$ and $\gamma$ be
reflections as shown. The three rotations of order two (fixing $w$, $u'$ and $v'$
respectively) are $\alpha\beta$, $(\alpha\gamma)^2$ and $\gamma(\alpha\beta)\gamma$ and these generate the same group as
$\alpha\beta$, $\beta\gamma$ and $\gamma\alpha$, namely a Triangle group with signature $(0: 2, 4, q)$.

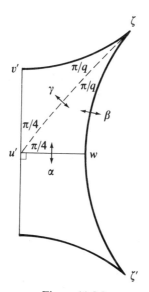

Figure 11.5.5

EXERCISE 11.5

1. Suppose that $f$, $g$ and $h$ are elliptic elements of order two with collinear fixed points $u$, $v$ and $w$ respectively. Find a necessary and sufficient condition for $\langle f, g, h \rangle$ to be discrete in terms of $\rho(u, v)$ and $\rho(v, w)$.

2. Give a proof of Theorem 11.5.1 using matrices (take $f$, $g$ and $h$ fixing $i$, $ti$ and $u + iv$ respectively in $H^2$).

## §11.6. Universal Bounds on the Displacement Function

Our aim is to obtain lower bounds of

$$M(g, h) = \inf_z \max \{\sinh \tfrac{1}{2}\rho(z, gz), \sinh \tfrac{1}{2}\rho(z, hz)\}$$

and

$$P(g, h) = \inf_z \sinh \tfrac{1}{2}\rho(z, gz) \sinh \tfrac{1}{2}\rho(z, hz)$$

for various choices of $g$ and $h$ subject to $\langle g, h \rangle$ being discrete and non-elementary. Observe that

$$M(g, h)^2 \geq P(g, h).$$

Obviously, a lower bound on $P(g, h)$ is preferable for it shows that if one of the sinh terms is small then the other term is correspondingly large: this does not follow from a lower bound on $M(g, h)$. If $g$ or $h$ is elliptic, then $P(g, h) = 0$ so one must use $M(g, h)$.

The inequality

$$M(g, h) \geq m$$

means that for every $z$, either $g$ or $h$ moves $z$ at least a distance $2 \sinh^{-1}(m)$. It is known that in every case,

$$M(g, h) \geq 0 \cdot 131846 \ldots :$$

the existence of a lower bound was established by Marden: this lower bound, which is best possible, was obtained by Yamada and is given in Theorem 11.6.14.

The evaluation of the best lower bounds for $M(g, h)$ and $P(g, h)$ is intimately connected with the geometric constraints on $g$ and $h$ and both the numerical bounds and the geometric constraints appear in this section. At this point, the reader should recall Theorem 8.3.1: *if $\langle g, h \rangle$ is discrete, non-elementary and has no elliptic elements, then $P(g, h) \geq 1$ and this lower bound is best possible.*

We shall obtain different lower bounds depending on the classification of $g$ and $h$. First, we assume that one of these is parabolic.

**Theorem 11.6.1.** *Let $g$ and $h$ be isometries and suppose that $\langle g, h \rangle$ is discrete and non-elementary.*

(1) *If $g$ and $h$ are parabolic, then $P(g, h) \geq \frac{1}{4}$. If, in addition, $\langle g, h \rangle$ is not a Triangle group, then $P(g, h) \geq 1$.*

(2) *If $g$ is parabolic and $h$ is hyperbolic, then $P(g, h) \geq \frac{1}{4}$. If, in addition, $\langle g, h \rangle$ is not a Triangle group, then $P(g, h) \geq \frac{1}{2}$.*
*All four bounds are best possible.*

PROOF. Let $g$ be parabolic and let $h$ be parabolic or hyperbolic. We may suppose that $g$ and $h$ act on $H^2$ and that

$$g(z) = z + 1, \qquad h(z) = \frac{az + b}{cz + d}, \qquad ad - bc = 1.$$

As $\langle g, h \rangle$ is non-elementary, $c \neq 0$. Now $h$ has two real, finite, possibly coincident, fixed points $u$ and $v$ and

$$|z - h(z)| \cdot |cz + d| = |c| \cdot |z - u| \cdot |z - v|$$
$$\geq |c| y^2.$$

From Theorem 7.2.1 we obtain

$$\sinh \tfrac{1}{2}\rho(z, gz) \sinh \tfrac{1}{2}\rho(z, hz) = |z - h(z)| \cdot |cz + d| / 4y^2$$
$$\geq |c| / 4. \qquad (11.6.1)$$

From Jørgensen's inequality, $|c| \geq 1$ so in both cases, $P(g, h) \geq \frac{1}{4}$.

Suppose now that $\langle g, h \rangle$ is not a Triangle group. If $h$ is parabolic, Theorem 11.4.1 yields $|c| \geq 4$ and so $P(g, h) \geq 1$. If $h$ is hyperbolic, then from Theorem 11.4.2 we deduce that $|c| \geq 2$ so $P(g, h) \geq \frac{1}{2}$. This establishes (1) and (2); the following examples show that these lower bounds are best possible. ☐

**Example 11.6.2.** The isometries $g$, $h$ and $f$ given by

$$g(z) = z + 1, \qquad h(z) = \frac{z}{z + 1}, \qquad f(z) = \frac{2z + 3}{z + 2}$$

are parabolic, parabolic and hyperbolic respectively and generate a discrete group (a subgroup of the Modular group). A computation using (11.6.1) with $z = iy$ gives

$$\sinh \tfrac{1}{2}\rho(z, gz) \sinh \tfrac{1}{2}\rho(z, hz) = \tfrac{1}{4}$$

and

$$\sinh \tfrac{1}{2}\rho(z, gz) \sinh \tfrac{1}{2}\rho(z, fz) = \tfrac{1}{4} + (3/4y^2)$$

and, letting $y$ tend to $+\infty$, we see that the lower bounds of $\frac{1}{4}$ are best possible. ☐

**Example 11.6.3.** Let $g(z) = z + 1$ and let $h$ be the reflection in $|z + t| = t$ followed by reflection in $x = 0$ where $0 < t < \frac{1}{4}$. Thus $h$ is parabolic and fixes the origin: in fact,

$$h(z) = \frac{z}{(z/t) + 1}.$$

Using (11.6.1), we see that when $z = iy$,

$$\sinh \tfrac{1}{2}\rho(z, gz) \sinh \tfrac{1}{2}\rho(z, hz) = 1/4t.$$

Clearly, $\langle g, h \rangle$ is a non-elementary Fuchsian group of the second kind. Letting $t$ tend to $\frac{1}{4}$ we find that the lower bound of one in Theorem 11.6.1(1) is best possible.                                                                       $\square$

**Example 11.6.4.** Let $g(z) = z + 1$ and let $h$ be an elliptic element of order two fixing the point $iv$ where $0 < v < \frac{1}{2}$. Then $\langle g, h \rangle$ is discrete and non-elementary: for example,

$$\{z \in H^2 : |\operatorname{Re}[z]| < \tfrac{1}{2}, |z| > v\}$$

is a fundamental domain for $\langle g, h \rangle$. Now write $f = gh$: then $f$ is hyperbolic and is a reflection in $|z| = v$ followed by the reflection in $x = \frac{1}{2}$. It follows that

$$f(z) = \frac{(z/v) - v}{(z/v)}$$
$$= 1 - (v^2/z)$$

so

$$\sinh \tfrac{1}{2}\rho(z, gz) \sinh \tfrac{1}{2}\rho(z, fz) = \frac{|z - fz| \cdot |z/v|}{4y^2}$$
$$= \frac{|z^2 - z + v^2|}{4vy^2}.$$

Letting $y$ tend to $+\infty$ with, say, $x = 0$, this expression tends to $1/4v$. As $v$ can be arbitrarily close to $\frac{1}{2}$, and as $\langle g, h \rangle = \langle g, f \rangle$ we see that the lower bound of $\frac{1}{2}$ in Theorem 11.6.1(2) is best possible.                         $\square$

Next, we consider one elliptic and one parabolic generator: in this case we must use $M(g, h)$.

**Theorem 11.6.5.** *Let $g$ be parabolic, let $h$ be elliptic of order $q$ and suppose that $\langle g, h \rangle$ is discrete and non-elementary.*

(1) *If $q \geq 3$ then*

$$M(g, h) \geq \frac{\cos(\pi/q)}{[1 + 2\cos(\pi/q) - \cos^2(\pi/q)]^{1/2}}$$
$$\geq 1/\sqrt{7}.$$

(2) *If $q = 2$, then $M(g, h) \geq 1/\sqrt{8}$.*
(3) *If, in addition, $\langle g, h \rangle$ is not a Triangle group, then for $q \geq 2$ we have*

$$M(g, h) \geq \left[ \frac{1 + \cos(\pi/q)}{3 - \cos(\pi/q)} \right]^{1/2}$$

$$\geq 1/\sqrt{3}.$$

*All of these bounds are best possible.*

PROOF. Let

$$m(z) = \max\{\sinh \tfrac{1}{2}\rho(z, gz), \sinh \tfrac{1}{2}\rho(z, hz)\}.$$

We may assume that $g$ and $h$ act on $H^2$, that $g(z) = z + 1$ and that $h$ is a rotation of angle $2\theta$ (where $0 < 2\theta < \pi$) about a point $w$ of the form $iv$.

For any $z_1$, let $z_2$ be the point where the horizontal line (a horocycle at $\infty$) through $z_1$ meets the geodesic $L$ from $\infty$ through $w$. Now let $z_3$ be the point on the half-ray $[w, \infty)$ such that $z_2$ and $z_3$ are equidistant from $w$ (if $\operatorname{Im}[z_1] \geq \operatorname{Im}[w]$ then $z_2 = z_3$ but not otherwise). Then

$$\operatorname{Im}[z_1] = \operatorname{Im}[z_2] \leq \operatorname{Im}[z_3],$$
$$\rho(z_1, w) \geq \rho(z_2, w) = \rho(z_3, w)$$

and so (see Section 7.35),

$$m(z_1) \geq m(z_3).$$

As

$$M(g, h) = \inf_z m(z),$$

this means that we can confine out attention to $m(z)$ for those $z$ of the form $iy$ where $y \geq v$. As $y$ increases from $v$ to $+\infty$, so $\rho(z, gz)$ decreases to zero and $\rho(z, hz)$ increases from zero to $+\infty$: hence there is a unique $z$, say $z = it$, where

$$\sinh \tfrac{1}{2}\rho(z, gz) = \sinh \tfrac{1}{2}\rho(z, hz)$$

and where this common value is $M(g, h)$.

Now observe (from Section 7.35) that when $z = it$,

$$\sinh \tfrac{1}{2}\rho(z, gz) = 1/2t$$

and

$$\sinh \tfrac{1}{2}\rho(z, hz) = |\sin \theta| \sinh \rho(it, iv)$$

$$= \tfrac{1}{2}|\sin \theta| \left( \frac{t}{v} - \frac{v}{t} \right).$$

Thus

$$t^2 = v^2 + \frac{v}{|\sin \theta|}.$$

As $h$ is of order $q$ (and $h \neq I$), we must have $|\sin \theta| \geq \sin(\pi/q)$. By Corollary 11.2.3, if $q \geq 3$, then

$$v \leq \tfrac{1}{2} \tan(\pi/q)$$

and so

$$4t^2 \leq \tan^2(\pi/q) + \frac{2}{\cos(\pi/q)}.$$

As

$$M(g, h) = 1/2t,$$

the lower bound involving $q$ in (1) follows. By elementary calculus, this is an increasing function of $\cos(\pi/q)$ and the lower bound of $1/\sqrt{7}$ is the case $q = 3$. It is clear that this lower bound is best possible for each value of $q$: indeed, equality holds throughout this argument for the Hecke groups discussed in Section 11.3.

This argument fails if $q = 2$. However, in this case, $v \leq 1$ (the fixed point lies on the isometric circle and, by Jørgensen's inequality, $|c| \geq 1$) and $\theta = \pi/2$ and so $t^2 \leq 2$: this proves (2). This is also best possible: for example, take $g(z) = z + 1$, $h(z) = -1/z$ and $z = i\sqrt{2}$.

Now suppose that $\langle g, h \rangle$ is not a Triangle group. As $h$ is of order $q$, some power of $h$, say $h^n$, is a rotation of angle $2\pi/q$ about $iv$ and $\langle g, h^n \rangle$ $(=\langle g, h \rangle)$ is not a Triangle group. Exactly as above, we have

$$t^2 \leq v^2 + \frac{v}{\sin(\pi/q)}. \tag{11.6.2}$$

Now consider the quadrilateral (possibly with two free sides on the real axis) with sides lying on the lines $x = \tfrac{1}{2}$, $x = -\tfrac{1}{2}$ and the isometric circles of $h^n$ and $h^{-n}$. This quadrilateral is not bounded (Theorem 10.6.6), thus

$$\frac{v}{\sin(\pi/q)}[1 + \cos(\pi/q)] < \tfrac{1}{2}: \tag{11.6.3}$$

see Figure 11.6.1.

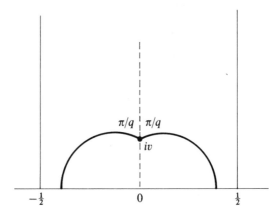

Figure 11.6.1

Using (11.6.2) and (11.6.3) we obtain

$$4t^2 \leq \frac{3 - \cos(\pi/q)}{1 + \cos(\pi/q)}$$

$$\leq 3$$

and the lower bounds in (3) hold. These lower bounds are best possible for we can construct groups from a quadrilateral as suggested by the proof: we omit the details.                                                                    □

Next, we consider two elliptic generators.

**Theorem 11.6.6.** *Let $g$ and $h$ be elliptic elements of orders $p$ and $q$ respectively and suppose that $\langle g, h \rangle$ is discrete and non-elementary. Then*

$$M(g, h) \geq \left[ \frac{4 \cos^2(\pi/7) - 3}{8 \cos(\pi/7) + 7} \right]^{1/2} = 0 \cdot 1318 \ldots.$$

*If, in addition, $\langle g, h \rangle$ is not a Triangle group, then*

$$M(g, h) \geq \left( \frac{[\cos(\pi/p) + \cos(\pi/q)]^2}{4 - [\cos(\pi/p) - \cos(\pi/q)]^2} \right)^{1/2} \geq \frac{1}{\sqrt{15}}.$$

*Both bounds are best possible.*

We shall need the following geometric result.

**Theorem 11.6.7.** *Let $g$ be elliptic of order $p$ with fixed point $u$, let $h$ be elliptic of order $q$ with fixed point $v$ and suppose that $\langle g, h \rangle$ is discrete, non-elementary but not a Triangle group. Then*

$$\cosh \rho(u, v) > \frac{1 + \cos(\pi/p) \cos(\pi/q)}{\sin(\pi/p) \sin(\pi/q)}.$$

PROOF OF THEOREM 11.6.7. Some $g_1$ in $\langle g \rangle$ has angle of rotation $2\pi/p$, some $h_1$ in $\langle h \rangle$ has angle of rotation $2\pi/q$ and $\langle g, h \rangle = \langle g_1, h_1 \rangle$. Thus we may assume that $g$ and $h$ have angles of rotation $2\pi/p, 2\pi/q$. Without loss of generality, $g$ and $h$ act on $\Delta$, $u = 0$ and $v > 0$. Now construct the isometric circles of $h$ and $h^{-1}$ and the segments from the origin making an angle $\pi/p$ with $(0, 1)$: see Figure 11.6.2. The rays $L$ and $L'$ are paired by $g$ and the rays $L_1$ and $L_1'$ are paired by $h$. If $L$ and $L_1$ meet, then $\langle g, h \rangle$ is a Triangle group (Theorem 10.6.6). If this is not so, then (Theorem 7.10.1) $\cosh \rho(u, v)$ is bounded below by the given bound.                                                   □

PROOF OF THEOREM 11.6.6. Write

$$m(z) = \max\{\sinh \tfrac{1}{2}\rho(z, gz), \sinh \tfrac{1}{2}\rho(z, hz)\}.$$

Clearly if $g$ or $h$ is replaced by a rotation about the same point but with a

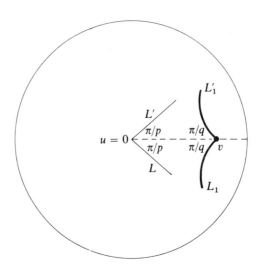

Figure 11.6.2

smaller angle of rotation, then the corresponding $m(z)$ decreases: thus we may assume that $g$ and $h$ have angles of rotation $2\pi/p$ and $2\pi/q$ respectively. We may assume that these act on $\Delta$, that $g$ fixes the origin and that $h$ fixes the point $v$ where $v > 0$. Exactly as in the proof of Theorem 11.6.5, the minimum value of $m(z)$ is attained at some point $x$ of the real segment $[0, v]$, where

$$\sinh \tfrac{1}{2}\rho(x, gx) = \sinh \tfrac{1}{2}\rho(x, hx)$$

and where this common value is $M(g, h)$.

Now write

$$\rho(0, x) = t, \qquad \rho(0, v) = d$$

so $\rho(x, v) = d - t$. Also, write

$$s_p = \sin(\pi/p), \qquad c_p = \cos(\pi/p)$$

and, similarly, for $s_q, c_q$. Then

$$s_p \sinh t = s_q \sinh(d - t)$$

(both sides are $M(g, h)$) so

$$\tanh t = \frac{s_q \sinh d}{s_p + s_q \cosh d}.$$

However,

$$
\begin{aligned}
M(g, h)^2 &= (s_p)^2 \sinh^2 t \\
&= \frac{(s_p)^2 \tanh^2 t}{1 - \tanh^2 t} \\
&= \frac{(s_p s_q)^2 [\cosh^2 d - 1]}{(s_p)^2 + (s_q)^2 + 2 s_p s_q \cosh d}.
\end{aligned}
\tag{11.6.4}
$$

By elementary calculus, this is an increasing function of $\cosh d$: thus if $\langle g, h \rangle$ is not a Triangle group, then Theorem 11.6.7 is applicable and

$$s_p s_q \cosh d \geq 1 + c_p c_q.$$

Substitution in (11.6.4) yields

$$M(g, h)^2 \geq \frac{(c_p + c_q)^2}{4 - (c_p - c_q)^2},$$

which is the lower bound stated in the theorem. This lower bound is an increasing symmetric function of $c_p$ and $c_q$ in the permitted ranges so taking say, $p = 3$ and $q = 2$ (if $p = q = 2$ then $\langle g, h \rangle$ is elementary) we obtain a lower bound of $M(g, h)^2$ equal to $\frac{1}{15}$. It is clear from this proof that one can construct groups to show that these bounds are best possible.

It remains only to establish the first (and smaller) lower bound in Theorem 11.6.6 in the case when $G$ is a Triangle group. Let $G$ be a Triangle group with signature $(0: m, n, r)$. Suppose that $g$ and $h$ are associated with the cyclic subgroups of orders $m$ and $n$ respectively (but they need not be of orders $m$ or $n$). The estimation of $M(g, h)$ must allow for, and cannot be smaller than, the estimation under the assumption that $g$ and $h$ have angles of rotation $2\pi/m$ and $2\pi/n$; thus we may assume that $m = p$ and $n = q$. In this case the fixed points $u$ and $v$ of $g$ and $h$ respectively must be separated by at least a distance along the side of a triangle with angles $\pi/p, \pi/q$ and $\pi/r$ (otherwise we could construct a fundamental domain with area less than the known value): thus by the Cosine Rule:

$$\cosh \rho(u, v) \geq \frac{c_p c_q + c_r}{s_p s_q}.$$

The identity (11.6.4) remains valid so

$$M(g, h)^2 \geq \frac{(c_p c_q + c_r)^2 - (s_p s_q)^2}{(s_p)^2 + (s_q)^2 + 2[c_p c_q + c_r]}$$

$$= \frac{c_p^2 + c_q^2 + c_r^2 + 2c_p c_q c_r - 1}{2 + 2c_r - (c_p - c_q)^2}.$$

We need to obtain the infimum of this expression over all $p, q$ and $r$ satisfying

$$\frac{1}{p} + \frac{1}{q} + \frac{1}{r} < 1.$$

In fact the infimum occurs when $r = 7$, $p = 2$ and $q = 3$ (or when $p = 3$, $q = 2$): in this case the lower bound is

$$\frac{4 \cos^2(\pi/7) - 3}{8 \cos(\pi/7) + 7} = 0 \cdot 0173 \ldots.$$

In general, we have

$$2 + 2c_r - (c_p - c_q)^2 \leq 4$$

so

$$M(g, h)^2 \geq \tfrac{1}{4} [c_p^2 + c_q^2 + c_r^2 + 2c_p c_q c_r - 1].$$

Assume for the moment that one of $p, q, r$, is at least 8: another is at least 3 and then

$$M(g, h)^2 \geq \tfrac{1}{4}[\cos^2(\pi/8) + \cos^2(\pi/3) - 1]$$
$$= 0.025\ldots.$$

Thus in our search for a lower bound on $M(g, h)$, we can assume that each of $p, q$ and $r$ is at most 7: this reduces the problem to a finite number of computations, however even most of these can be avoided.

If none of $p, q$ and $r$ are 2, then two are at least 3, the other being at least 4: then

$$M(g, h)^2 \geq \tfrac{1}{4}[2\cos^2(\pi/3) + \cos^2(\pi/4) + 2\cos^2(\pi/3)\cos(\pi/4) - 1]$$
$$= 0.088\ldots.$$

Thus we may assume that one of $p, q$ and $r$ is 2. If none are 3, then the others are at least 4 and 5 and then

$$M(g, h)^2 \geq [\cos^2(\pi/4) + \cos^2(\pi/5) - 1]/4$$
$$> 0.038.$$

We deduce that one of $p, q$ and $r$ is 2, another is 3 and the third is at most and (for positive area) at least 7. The lower bound is symmetric in $p$ and $q$ and the numerator is symmetric in $p, q$ and $r$. Thus we need only minimize

$$2 + 2c_r - (c_p - c_q)^2$$

over the possibilities

$$(p, q, r) = (2, 3, 7), (2, 7, 3), (3, 7, 2):$$

the details are omitted.                                                              □

We turn our attention now to hyperbolic elements. First, we establish geometric constraints which must be satisfied by any two hyperbolic elements in a discrete group. The motivation for the next two results is the distinction between simple and non-simple hyperbolic elements (Definition 8.1.5): however, the results are more generally applicable than this, indeed, they are concerned with whether or not the projection of the two axes cross on the quotient surface.

**Theorem 11.6.8.** *Let $g$ and $h$ be hyperbolic elements with axes and translation lengths $A_g$, $A_h$, $T_g$ and $T_h$ respectively. Suppose that $\langle g, h \rangle$ is discrete and non-elementary and that $A_g$ and $A_h$ cross at an angle $\theta$. Then*

(1)                   $\sinh(\tfrac{1}{2}T_g)\sinh(\tfrac{1}{2}T_h)\sin\theta \geq \cos(3\pi/7) = 0.2225\ldots.$

*In fact*

(2) $$\sinh(\tfrac{1}{2}T_g)\sinh(\tfrac{1}{2}T_h)\sin\theta \geq \tfrac{1}{2},$$

*except possibly when* $\langle g, h\rangle$ *has one of the signatures* $(0: 2, 3, q)$, $(0: 2, 4, q)$ *or* $(0: 3, 3, 4)$ *and*

(3) $$\sinh(\tfrac{1}{2}T_g)\sinh(\tfrac{1}{2}T_h)\sin\theta \geq 1$$

*if* $\langle g, h\rangle$ *has no elliptic elements or has an unbounded fundamental domain.*

*In particular, if* $g$ *is a non-simple hyperbolic element in* $\langle g, h\rangle$ *then* $\sinh(\tfrac{1}{2}T_g) \geq [\cos(3\pi/7)]^{1/2}(= 0{\cdot}47\ldots)$.

PROOF. Let $u$ be the point where $A_g$ and $A_h$ cross and construct points $v$ and $w$ on $A_g$ and $A_h$ respectively such that $\rho(u, v) = \tfrac{1}{2}T_g$, $\rho(u, w) = \tfrac{1}{2}T_h$ and such that the triangle with vertices $u, v, w$ has angle $\theta$ at $u$. Let $f_u, f_v$ and $f_w$ be elliptic elements of order two fixing $u, v$ and $w$ respectively. Replacing $g$ and(or) $h$ by their inverses as necessary, we may assume that

$$g = f_v f_u, \qquad h = f_w f_u, \qquad gh^{-1} = f_v f_w.$$

We deduce that every product of an even number of $f_u, f_v$ and $f_w$ is in $\langle g, h\rangle$: thus $\langle g, h\rangle$ is of index one or two in $\langle f_u, f_v, f_w\rangle$ and so this latter group is discrete.

Recalling the results of Section 11.5, we may write

$$\sinh(\tfrac{1}{2}T_g)\sinh(\tfrac{1}{2}T_h)\sin\theta = \lambda$$
$$= \tfrac{1}{2}|\operatorname{trace}(f_u f_v f_w)|$$

and Theorem 11.6.8 follows essentially from Theorem 11.5.2 and its proof. First, (1) is Corollary 11.5.3. If $\langle g, h\rangle$ has no elliptic elements, then (3) follows from Theorem 7.39.4: if $\langle g, h\rangle$ has an unbounded fundamental domain, then (3) follows from cases (1) and (2) of Theorem 11.5.2.

It remains to verify (2). According to Theorem 11.5.2(3), we see that the lower bound of $\tfrac{1}{2}$ in (2) holds except possibly in the cases when $\lambda$ is of the form $\cos(2\pi/q)$ or $\cos(3\pi/q)$. It is now necessary to examine the proof of Theorem 11.5.2 to see when this can arise. For brevity, we denote $\langle f_u, f_v, f_w\rangle$ by $G^*$ and $\langle g, h\rangle$ by $G$.

Referring to the proof of Theorem 11.5.2, we need only consider the cases $p = 2$ and $p = 3$. However, $G^*$ contains a product of three elliptic elements of order two which is a rotation of $2\pi p/q$. Thus if $p = 2$, there is a rotation $r$ of angle $2\pi/q$ such that $r^2$ is a product of three rotations of order two. As $r \in G^*$ we have $r \in G$: hence $G$ contains a rotation of order two. In this case, $G = G^*$ so $G$ has one of the signatures $(0: 2, 3, q)$ or $(0: 2, 4, q)$.

The remaining case is $p = 3$: here $G^*$ has one of the signatures $(0: 2, 3, q)$ where (see the proof of Theorem 11.5.2) $q = n = 7$ or 8. A tedious arithmetic exercise on areas shows that if $G$ has index two in $G^*$, then the only possible signature for $G$ is $(0: 3, 3, 4)$.

The last assertion concerning non-simple hyperbolic elements is an application of (1) in which $h$ is taken to be a conjugate of $g$. $\qquad\square$

**Theorem 11.6.9.** *Let $g$ and $h$ be hyperbolic with axes and translation lengths $A_g$, $A_h$, $T_g$ and $T_h$ respectively. Suppose that $\langle g, h \rangle$ is discrete and non-elementary and that no images of $A_g$ and $A_h$ cross. Then*

$$\sinh(\tfrac{1}{2}T_g) \sinh(\tfrac{1}{2}T_h) \cosh \rho(A_g, A_h) \geq \cosh(\tfrac{1}{2}T_g) \cosh(\tfrac{1}{2}T_h) - \tfrac{1}{2}.$$

*If $\langle g, h \rangle$ has no elliptic elements, we can replace $-\tfrac{1}{2}$ by $+1$ (and the lower bound by 2).*

If $g$ is a simple hyperbolic element in $\langle g, h \rangle$ this result can be applied with $h$ being any conjugate, say $fgf^{-1}$, of $g$. Thus (by elementary manipulation) we obtain the next inequality.

**Corollary 11.6.10.** *If $g$ and $h$ are hyperbolic elements generating a discrete non-elementary group and if $g$ is a simple hyperbolic element in this group, then for all $f$ in $\langle g, h \rangle$, either $f(A_g) = A_g$ or*

$$\sinh(\tfrac{1}{2}T_g) \sinh \tfrac{1}{2}\rho(A_g, fA_g) \geq \tfrac{1}{2}.$$

*This bound is best possible.*

The next example shows that the lower bound of $\tfrac{1}{2}$ is best possible.

**Example 11.6.11.** Construct the polygon $D$ as in Figure 11.6.3 where $f$ (elliptic of order two) and $g$ (hyperbolic) pair the sides of $D$. By Poincaré's Theorem, $D$ is a fundamental polygon for $\langle f, g \rangle$ and as $g$ pairs the sides of $D$, $g$ must be a simple hyperbolic element. Finally,

$$\sinh(\tfrac{1}{2}T_g) \sinh \tfrac{1}{2}\rho(A_g, fA_g) = \sinh \tfrac{1}{2}\rho(L, L') \sinh \rho(0, A_g)$$
$$= \cos(\pi/3). \qquad \square$$

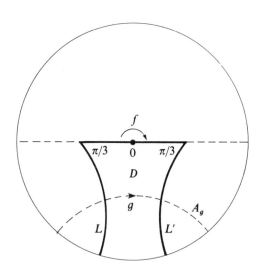

Figure 11.6.3

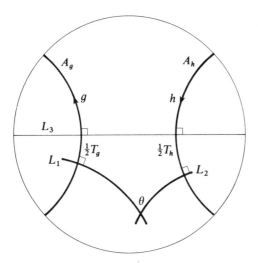

Figure 11.6.4

**Proof of Theorem 11.6.9.** Consider Figure 11.6.4. As $g$ (or $g^{-1}$) is $\sigma_3\sigma_1$ and $h$ (or $h^{-1}$) is $\sigma_2\sigma_3$ we see that $\sigma_2\sigma_1$ is in $G$. If $G$ has no elliptic elements, then $L_1$ and $L_2$ cannot intersect (this case is not illustrated) and from Theorem 7.19.2 we obtain

$$\sinh(\tfrac{1}{2}T_g)\sinh(\tfrac{1}{2}T_h)\cosh\rho(A_g, A_h) = \cosh(\tfrac{1}{2}T_g)\cosh(\tfrac{1}{2}T_h) + \cosh\rho(L_1, L_2).$$

This yields the second inequality.

If $L_1$ and $L_2$ intersect, say at an angle $\theta$, then $\theta = 2\pi p/q$ for some coprime integers $p$ and $q$. If $\theta > 2\pi/q$ we can rotate $A_h$ about the point of intersection to an image of itself which is closer to (but, by assumption, not intersecting) $A_g$. Thus if, in the argument above, we replace $h$ by a conjugate $fhf^{-1}$ of $h$ with the property that its axis $f(A_h)$ is as close as possible to (but distinct from) $A_g$, we find that

$$\rho(A_g, A_h) \geq \rho(A_g, fA_h)$$

and the corresponding $\theta$ satisfies $\theta \leq 2\pi/q \leq 2\pi/3$ as obviously $\theta < \pi$. Thus from Theorem 7.18.1 we obtain the first inequality, namely

$$\sinh(\tfrac{1}{2}T_g)\sinh(\tfrac{1}{2}T_h)\cosh\rho(A_g, A_h) \geq \cosh(\tfrac{1}{2}T_g)\cosh(\tfrac{1}{2}T_h) + \cos(2\pi/3). \quad \square$$

Theorems 11.6.8 and 11.6.9 yield the following bound on $P(g, h)$.

**Theorem 11.6.12.** *Let $g$ and $h$ be hyperbolic elements which generate a discrete non-elementary group. Then $P(g, h) \geq \cos(3\pi/7)$.*

PROOF. If the axes of $g$ and $h$ cross at $w$, say, then obviously (using the notation of Theorems 11.6.8 and 11.6.9)

$$P(g, h) = \sinh \tfrac{1}{2}\rho(w, gw) \sinh \tfrac{1}{2}\rho(w, hw)$$
$$= \sinh(\tfrac{1}{2}T_g) \sinh(\tfrac{1}{2}T_h)$$
$$\geq \cos(3\pi/7).$$

The same inequality holds if any images of $A_g$ and $A_h$ cross. If not, then Theorem 11.6.9 is applicable and we obtain

$$\sinh \tfrac{1}{2}\rho(z, gz) \sinh \tfrac{1}{2}\rho(z, hz) = \sinh(\tfrac{1}{2}T_g) \sinh(\tfrac{1}{2}T_h) \cosh \rho(z, A_g) \cosh \rho(z, A_h)$$
$$\geq \tfrac{1}{2}\sinh(\tfrac{1}{2}T_g) \sinh(\tfrac{1}{2}T_h) \cosh[\rho(z, A_g) + \rho(z, A_h)]$$
$$\geq \tfrac{1}{2}\sinh(\tfrac{1}{2}T_g) \sinh(\tfrac{1}{2}T_h) \cosh \rho(A_g, A_h)$$
$$\geq \tfrac{1}{2}[\cosh(\tfrac{1}{2}T_g) \cosh(\tfrac{1}{2}T_h) - \tfrac{1}{2}]$$
$$\geq \tfrac{1}{4}$$
$$> \cos(3\pi/7). \qquad \square$$

Finally, we consider $M(g, h)$ for one elliptic and one hyperbolic element.

**Theorem 11.6.13.** *Let $g$ be hyperbolic and let $h$ be elliptic of order $q$ ($q \geq 2$). If $\langle g, h \rangle$ is discrete and non-elementary, then $M(g, h) \geq 1/\sqrt{8}$.*

PROOF. If $g$ is a non-simple hyperbolic element of $\langle g, h \rangle$ then (from Theorem 11.6.8)

$$M(g, h) \geq \sinh(\tfrac{1}{2}T_g)$$
$$\geq [\cos(3\pi/7)]^{1/2}$$
$$> 1/\sqrt{8}.$$

We may now assume that $g$ is a simple hyperbolic element. In this case, the fixed point $v$ of the elliptic $h$ cannot lie on $A_g$ and a rotation of $A_g$ of an angle $2\pi/q$ about $v$ must map $A_g$ onto a disjoint image which we may assume is $h(A_g)$: see Figure 11.6.5.

From Section 7.17 we have

$$\cosh \rho(v, A_g) \sin(\pi/q) = \cosh \tfrac{1}{2}\rho(A_g, hA_g)$$
$$\geq \sinh \tfrac{1}{2}\rho(A_g, hA_g)$$

and, from Corollary 11.6.10 (applied to $\langle g, hgh^{-1} \rangle$),

$$\sinh(\tfrac{1}{2}T_g) \sinh \tfrac{1}{2}\rho(A_g, hA_g) \geq \tfrac{1}{2}.$$

Thus

$$\cosh \rho(v, A_g) \sin(\pi/q) \sinh(\tfrac{1}{2}T_g) \geq \tfrac{1}{2}.$$

This expresses a geometric constraint between the parameters $T_g$, $2\pi/q$ and the separation of $g$ and $h$ as measured by $\rho(v, A_g)$. Writing

$$m = \max\{\sinh \tfrac{1}{2}\rho(z, gz), \sinh \tfrac{1}{2}\rho(z, hz)\}$$

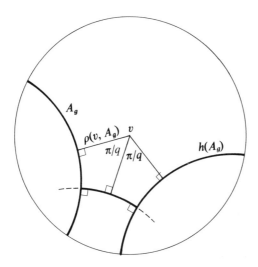

Figure 11.6.5

we have

$$\tfrac{1}{2} \le \sin(\pi/q) \sinh(\tfrac{1}{2}T_g) \cosh[\rho(v, z) + \rho(z, A_g)]$$
$$= \sin(\pi/q) \sinh(\tfrac{1}{2}T_g)[\cosh \rho(v, z) \cosh \rho(z, A_g) + \sinh \rho(v, z) \sinh \rho(z, A_g)]$$
$$\le m \sin(\pi/q)[1 + \sinh^2 \rho(v, z)]^{1/2} + m^2$$
$$\le m[\sin^2(\pi/q) + m^2]^{1/2} + m^2$$
$$\le m(1 + m^2)^{1/2} + m^2,$$

which certainly implies that $m \ge 1/\sqrt{8}$.    □

Collecting together all the results in this section we obtain a universal lower bound on $M(g, h)$.

**Theorem 11.6.14.** *If $g$ and $h$ generate a non-elementary discrete group, then $M(g, h) \ge 0{\cdot}1318\ldots$ and this lower bound is attained by two elliptic generators of the $(0:2, 3, 7)$-Triangle group.*

We end this section by completing an earlier proof.

PROOF OF THEOREM 11.2.4(2). We consider an accidental cycle of four vertices, say

$$v_1, \quad f(v_1) = v_2, \quad g(v_1) = v_3, \quad h(v_1) = v_4$$

on the boundary of a Dirichlet polygon: thus the $v_j$ lie on a circle with, say, centre $w$ and radius $r$. If $\langle f, g \rangle$ is non-elementary, then, as we have just seen,

$$M(f, g) \ge 0{\cdot}1318\ldots$$

and so for some $j$ ($=2$ or $3$),

$$0 \cdot 1318 \ldots \leq \sinh \tfrac{1}{2}\rho(v_1, v_j)$$
$$\leq \sinh \tfrac{1}{2}[\rho(v_1, w) + \rho(w, v_j)]$$
$$\leq \sinh r.$$

The same is true of $\langle g, h \rangle$ or $\langle h, f \rangle$ is non-elementary: thus it is sufficient to consider the case when all three groups $\langle g, h \rangle$, $\langle h, f \rangle$ and $\langle f, g \rangle$ are elementary.

We assume that these three two-generator groups are elementary. As $v_1, v_2$ and $v_3$ are concyclic, either $\langle f, g \rangle$ is cyclic with an elliptic generator or it is generated by two elliptic elements of order two. The first case cannot arise (else the elliptic generator fixes $w$): in the second case, one of $f$ and $g$, say, $g$, must be elliptic of order two. A similar argument holds for the other two groups so without loss of generality, we may assume that both $g$ and $h$ are elliptic and of order two.

If $f$ is hyperbolic, then as $\langle f, g \rangle$ and $\langle f, h \rangle$ are elementary, the axis of $f$ contains the fixed points of $g$ and $h$ and $\langle f, g, h \rangle$ is elementary. If $f$ is elliptic of order two, either the three fixed points $w_f$, $w_g$ and $w_h$ of $f, g$ and $h$ are collinear, and again $\langle f, g, h \rangle$ is elementary, or $w_f$, $w_g$ and $w_h$ are non-collinear in which case $\langle f, g, h \rangle$ is non-elementary.

If $\langle f, g, h \rangle$ is non-elementary, then from Section 11.5 we have

$$\sinh \rho(w_f, w_g) \sinh \rho(w_f, w_h) \geq \lambda$$
$$\geq \cos(3\pi/7).$$

However,

$$\rho(w_f, w_g) \leq \rho(w_f, v_1) + \rho(v_1, w_g)$$
$$= \tfrac{1}{2}\rho(v_1, v_2) + \tfrac{1}{2}\rho(v_1, v_3)$$
$$\leq \tfrac{1}{2}[\rho(v_1, w) + \rho(w, v_2) + \rho(v_1, w) + \rho(w, v_3)]$$
$$= 2r$$

so in this case,

$$\sinh^2 (2r) \geq \cos(3\pi/7).$$

There remain the cases in which $\langle f, g, h \rangle$ is elementary and we shall show that these cannot happen. We may suppose that the group acts on $H^2$ and that $\langle f, g, h \rangle$ leaves the positive imaginary axis invariant. The orbit of any point (not on the axis) is, say,

$$\{\ldots, z_{-1}, z_0, z_1, \ldots\} \cup \{\ldots, w_{-1}, w_0, w_1, \ldots\},$$

where this is illustrated in Figure 11.6.6 and where for each $j$,

$$\rho(z_j, z_{j+1}) = \rho(w_j, w_{j+1}) = t,$$

say.

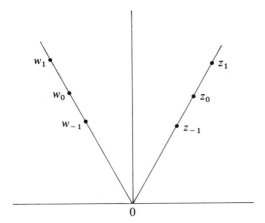

Figure 11.6.6

Now recall that in order that four points $v_1$, $v_2$, $v_3$, $v_4$ chosen from this orbit lie on the boundary of the Dirichlet polygon with centre $w$, it is necessary that these four points are the points in the orbit which are closest to (and equidistant from) $w$. Elementary metric and geometric considerations show that this can only happen when the centre $w$ lies on the positive imaginary axis and $|z_0| = |w_0|$ (after relabelling) with, say

$$\{v_1, v_2, v_3, v_4\} = \{z_0, z_1, w_0, w_1\}.$$

(consider the bisectors of the $[v_i, v_j]$: these must meet at $w$). Suppose that $v_1 = z_0$ and $v_2 = w_1$ (a similar argument holds for the other possibilities). Then $w$ is the mid-point of $[v_1, v_2]$ and $f$ (which maps $v_1$ to $v_2$) must be elliptic of order two: it follows that $f$ must fix $w$, a contradiction.  □

## Exercise 11.6

1. In the case of Theorem 11.6.1(1) we have $M(g, h) \geq \frac{1}{2}$. Use Example 11.6.2 to show that this is best possible.

2. Suppose that $\langle f, g \rangle$ is elementary. Prove that if $v, fv, gv$ are distinct points on a circle with centre $w$ then either

   (i) $f$ and $g$ are elliptic fixing $w$ or
   (ii) one of $f$ and $g$ is elliptic of order two (they cannot both be hyperbolic).

3. Consider Figure 11.6.3. Using reflections in $L$ and in the real and imaginary diameters of $\Delta$, show that $f^{-1}g$ is an elliptic element of order three fixing one vertex of $D$.

4. Let $G$ be a $(p, q, r)$-Triangle group. Suppose that $G$ contains $g$ of order $p$ fixing $u$ and $f$ of order $q$ fixing $v$. Prove that

$$\cosh\rho(u, v) \geq \frac{\cos(\pi/p)\cos(\pi/q) + \cos(\pi/r)}{\sin(\pi/p)\sin(\pi/q)}$$

   (this is used in the proof of Theorem 11.6.6). Hint: construct a quadilateral with angles $2\pi/p$ (at $u$), $2\pi/q$ (at $v$), $\theta$, $\theta$ which contains a fundamental domain for $G$.

5. Let $G$ be the Modular group and let $g$ in $G$ be hyperbolic with axis $A$ and translation length $T_g$. Let $N_g$ be the number of images of $A$ which intersect a fixed segment of length $T_g$ on $A$. Show that the average gap between images, namely $N_g/T_g$, can be arbitrarily small: more precisely, prove that

$$\inf_g N_g/T_g = 0.$$

6. Let $g$ be a non-simple hyperbolic element in a Fuchsian group without elliptic elements. Show that if $g$ has translation length $T$ then $\sinh(\frac{1}{2}T) \geq 1$.

## §11.7. Canonical Regions and Quotient Surfaces

The reader is invited to recall the geometric definition of a canonical region $\Sigma_g$ of an isometry $g$ (see Section 7.37): analytically,

$$\Sigma_g = \{z : \sinh \tfrac{1}{2}\rho(z, gz) < \tfrac{1}{2}|\text{trace}(g)|\}.$$

If $g$ is parabolic, then

$$\Sigma_g = \{z : \sinh \tfrac{1}{2}\rho(z, gz) < 1\}, \tag{11.7.1}$$

while if $g$ is hyperbolic with axis $A$ and translation length $T$, then

$$\Sigma_g = \{z : \sinh \rho(z, A) \sinh(\tfrac{1}{2}T) < 1\}, \tag{11.7.2}$$

because in this case $\Sigma_g$ is given by

$$\sinh \tfrac{1}{2}\rho(z, gz) = \sinh(\tfrac{1}{2}T) \cosh \rho(z, A)$$
$$< \cosh(\tfrac{1}{2}T). \tag{11.7.3}$$

Almost any Riemann surface $R$ is conformally equivalent to $\Delta/G$ for some Fuchsian group $G$ without elliptic elements. The hyperbolic metric on $\Delta$ projects to $\Delta/G$ and so transfers to $R$. With this in mind, the following result gives quantitative information on the metric structure of $R$.

**Theorem 11.7.1.** *Let $G$ be a Fuchsian group without elliptic elements, and suppose that $g$ and $h$ are in $G$.*

(1) *If $g$ and $h$ are parabolic elements with district fixed points, then $\Sigma_g$ and $\Sigma_h$ are disjoint.*
(2) *If $g$ is parabolic and $h$ is a simple hyperbolic element of $G$, then $\Sigma_g$ and $\Sigma_h$ are disjoint.*
(3) *If $g$ and $h$ are simple hyperbolic elements of $G$ whose axes do not cross, then $\Sigma_g$ and $\Sigma_h$ are disjoint.*

Essentially, this means that each puncture on $R$ lies in an open disc and each simple closed geodesic loop on $R$ lies in an open "collar": the discs do not intersect each other or the collars; two collars are disjoint if the corresponding loops are disjoint. Further, we know the sizes of the discs and collars (by computing the size of a canonical region) and each is the quotient of a horocyclic or hypercyclic region by a cyclic subgroup of $G$. Observe that Theorem 11.7.1 applies to boundary hyperbolic elements.

PROOF. For a Fuchsian group without elliptic elements, we have (Theorem 8.3.1)

$$\sinh \tfrac{1}{2}\rho(z, gz) \sinh \tfrac{1}{2}\rho(z, hz) \geq 1,$$

whenever $\langle g, h \rangle$ is non-elementary. In view of (11.7.1), this proves (1). For a geometric proof of (1), we may assume that

$$g(z) = z + 1, \qquad h(z) = \frac{z}{cz + 1}.$$

The isometric circles of $h$ and $h^{-1}$ must lie in the strip $|x| < \tfrac{1}{2}$ (else $G$ contains elliptic elements) and this implies that $\Sigma_g$ and $\Sigma_h$ (constructed geometrically) are disjoint.

We shall give a geometric proof of (2): an analytic proof is tricky and requires the inequality

$$\sinh(\tfrac{1}{4}T_h) \sinh \tfrac{1}{2}\rho(A_h, gA_h) \geq 1 :$$

see the proof of Theorem 8.2.1. We invite the reader to supply the details.

For the geometric proof, suppose that $g(z) = z + 1$ and construct the axis $A$ of $h$ and geodesics $L_1, L_2, L_3$ and $L_4$ as in Figure 11.7.1.

Clearly $\sigma_1\sigma_4$ and $\sigma_2\sigma_4$ are each $h$ or $h^{-1}$. Now $L_1$ cannot meet the line $x = x_0 + \tfrac{1}{2}$ and $L_2$ cannot meet the line $x = x_0 - \tfrac{1}{2}$ else $G$ would contain elliptic elements.

Moreover, $A_h$ cannot meet the lines $x = x_0 - \tfrac{1}{2}, x = x_0 + \tfrac{1}{2}$ as otherwise, $A_h$ has Euclidean radius greater than $\tfrac{1}{2}$ and then $A_h$ meets $g(A_h)$ (contradicting the fact that $h$ is simple). Thus the real interval $[w_1, w_2]$ lies strictly within the real interval $[x_0 - \tfrac{1}{2}, x_0 + \tfrac{1}{2}]$. The canonical region for

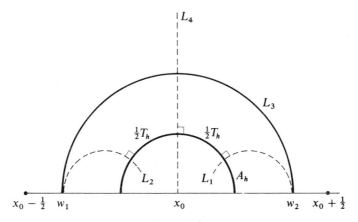

Figure 11.7.1

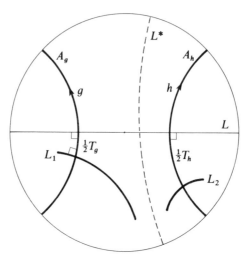

Figure 11.7.2

$h$ is bounded by the hypercycle which is tangent to $L_3$ and which ends at the end-points of $A_h$ (because $h(L_2) = L_1$): the canonical region for $g$ is above the geodesic with end-points $x_0 - \frac{1}{2}, x_0 + \frac{1}{2}$ so $\Sigma_g \cap \Sigma_h = \emptyset$. This proves (2).

To prove (3), consider Figure 11.7.2 with the geodesics $L, L_1, L_2$ as illustrated. Observe that

$$g^{-1}(A_h) = \sigma_1\sigma(A_h)$$
$$= \sigma_1(A_h).$$

As $h$ is a simple hyperbolic element, we see that $L_1$ cannot meet $A_h$ (else $\sigma_1(A_h)$ is an image of $A_h$ which meets $A_h$). Similarly, $L_2$ does not meet $A_g$. We know also that $L_1$ and $L_2$ do not meet (as $\sigma_2\sigma^2\sigma_1 \in G$). It follows that there is a geodesic $L^*$ with $L_1$ and $g(L_1)$ one side of $L^*$ and with $L_2$ and $h(L_2)$ on the other side of $L^*$. It is now immediate from geometric considerations that $\Sigma_g \cap \Sigma_h = \emptyset$.

For an analytic proof of (3) observe that as $L_1$ and $L_2$ do not meet, we have (Theorem 7.19.2),

$$\cosh \rho(A_g A_h) \sinh(\tfrac{1}{2}T_g) \sinh(\tfrac{1}{2}T_h) \geq 1 + \cosh(\tfrac{1}{2}T_g) \cosh(\tfrac{1}{2}T_h).$$

If $\Sigma_g \cap \Sigma_h \neq \emptyset$ then for some $z$ in the intersection, (11.7.2) and (11.7.3) hold (with $h$ as well as $g$) so

$$\sinh(\tfrac{1}{2}T_g) \sinh(\tfrac{1}{2}T_h) \cosh \rho(A_g, A_h)$$
$$< \sinh(\tfrac{1}{2}T_g) \sinh(\tfrac{1}{2}T_h) \cosh[\rho(z, A_g) + \rho(z, A_h)]$$
$$= \sinh(\tfrac{1}{2}T_g) \sinh(\tfrac{1}{2}T_h)[\cosh \rho(z, A_g) \cosh \rho(z, A_h)$$
$$+ \sinh \rho(z, A_g) \sinh \rho(z, A_h)]$$
$$< \cosh(\tfrac{1}{2}T_g) \cosh(\tfrac{1}{2}T_h) + 1$$

contradicting the application of Theorem 7.19.2.                                    □

It is possible to establish certain results for canonical regions even for Fuchsian groups with elliptic elements. For example, we have the following result.

**Theorem 11.7.2.** *Let G be a non-elementary, non-Triangle Fuchsian group. If g and h are elliptic or parabolic elements in G, then either $\langle g, h \rangle$ is cyclic or the canonical regions $\Sigma_g$ and $\Sigma_h$ are disjoint.*

PROOF. We may assume that $g$ and $h$ are primitive (this can only increase the size of $\Sigma_g$ and $\Sigma_h$). Construct the geodesic $L$ through (or ending at) the fixed point $u$ of $g$ and the fixed point $v$ of $h$. Construct geodesics $L_1$ and $L_2$ through $u$ which are symmetrically placed with respect to $L$ such that $g(L_1) = L_2$: repeat this construction using $L_3$ and $L_4$ through $v$ in the obvious way. Assume the $L_j$ are labelled so that $L_1$ and $L_3$ lie on the same side of $L$. If $L_1$ meets $L_3$, then $\langle g, h \rangle$ is a Triangle group and hence so is $G$ (Theorem 10.6.6). This is not so, thus $L_1$ and $L_3$ are disjoint. The geometrical construction of canonical regions now shows that $\Sigma_g$ and $\Sigma_h$ are disjoint.                                                                          □

EXERCISE 11.7

1. (i) Let $g$ be parabolic with canonical region $\Sigma_g$: show that h-area$(\Sigma_g/\langle g \rangle) = 2$.
   (ii) Let $g$ be hyperbolic with translation length $T$: show that $\Sigma_g/\langle g \rangle$ has area $2T/\sinh(\tfrac{1}{2}T)$.
   (iii) Let $g$ be elliptic with angle of rotation $2\pi/q$: show that $\Sigma_g/\langle g \rangle$ has area

$$\frac{2\pi}{q} \left[ \frac{1}{\sin(\pi/q)} - 1 \right]$$

   and this tends to 2 as $q \to +\infty$.

2. Let $G$ be a non-elementary Fuchsian group. At each fixed point $w$ of a parabolic element in $G$, let

$$H_w = \{z : \sinh \tfrac{1}{2}\rho(z, gz) < \tfrac{1}{2}\}$$

where $g$ generates the stabilizer of $w$. Show that for all parabolic fixed points $u$ and $v$,

$$H_u = H_v \quad \text{or} \quad H_u \cap H_v = \varnothing.$$

Prove also that for all $f$ in $G$,

$$f(H_u) = H_{fu}.$$

# §11.8. Notes

Some of the results in Section 11.6 occur in [59], [113]; for a completely algebraic approach, see [78], [79], [96]. For Section 11.7, see [12], [37], [43], [64], [87]: for a selection of geometric results on Fuchsian groups, consult [10], [75], [80], [81], [82], [84] and [93].

# References

[1]    Abikoff, W., The bounded model for hyperbolic 3-space and a uniformization theorem (preprint, 1981).

[2]    Abikoff, W., Appel, K. and Schupp, P., Lifting surface groups to SL(2, $C$); *Proceedings of Oaxtepec Conference, 1981.*

[3]    Ahlfors, L. V., Hyperbolic motions, *Nagoya Math. J.*, **29** (1967), 163–166.

[4]    Ahlfors, L. V., *Conformal Invariants*, McGraw-Hill, New York, 1973.

[5]    Ahlfors, L. V., Möbius transformations in several dimensions, *University of Minnesota Lecture Notes*, Minnesota, 1981.

[6]    Ahlfors, L. V. and Sario, L., *Riemann Surfaces*, Princeton University Press, Princeton, 1960.

[7]    Alexander, H. W., Vectorial inversive and non-Euclidean geometry, *Amer. Math. Monthly*, **74** (1967), 128–140.

[8]    Beardon, A. F. and Maskit, B., Limit points of Kleinian groups and finite-sided fundamental polyhedra, *Acta Math.*, **132** (1974), 1–12.

[9]    Beardon, A. F. and Jørgensen, T., Fundamental domains for finitely generated Kleinian groups, *Math. Scand.*, **36** (1975), 21–26.

[10]   Beardon, A. F., Hyperbolic polygons and Fuchsian groups, *J. London Math. Soc.*, **20** (1979), 247–254.

[11]   Beardon, A. F. and Waterman, P., Strongly discrete subgroups of SL(2, $C$), *J. London Math. Soc.*, **24** (1981), 325–328.

[12]   Beardon, A. F., Lie products, closed geodesics and Fuchsian groups, *Proc. Amer. Math. Soc.*, **85** (1982), 87–90.

[13]   Best, L. A., On torsion free discrete subgroups of PSL(2, $C$) with compact orbit space, *Can. J. Math.*, **23** (1971), 451–460.

[14]   Brooks, R. and Matelski, J. P., The dynamics of 2-generator subgroups of PSL(2, $C$); *Annals of Math. Studies 97*, Princeton University Press, Princeton, 1980.

[15]   Brooks, R. and Matelski, J. P., Collars in Kleinian groups, *Duke Math. J.* (to appear).

[16]   Bungaard, S. and Nielsen, J., On normal subgroups with finite index in $F$-groups, *Matematisk Tid. B*, 1951, 56–58.

[17]   Cassels, J. W. S., An embedding theorem for fields, *Bull. Aust. Math. Soc.*, **14** (1976), 193–198.

[18]  Cassels, J. W. S., An embedding theorem for fields: Addendum, *Bull. Aust. Math. Soc.*, **14** (1976), 479–480.

[19]  Chen, S. S., Greenberg, L., Hyperbolic spaces; *Contributions to Analysis* (edited by L. V. Ahlfors, I. Kra, B. Maskit and L. Nirenberg), Academic Press, New York, 1974.

[20]  Chevalley, C., *Theory of Lie Groups*, Princeton University Press, Princeton, 1946.

[21]  Coxeter, H. S. M., The inversive plane and hyperbolic space, *Abh. Math. Sem. Univ. Hamburg*, **29** (1966), 217–242.

[22]  Coxeter, H. S. M., Inversive distance, *Ann. Mat. Pura Appl.*, **71** (1966), 73–83.

[23]  Curtis, M. L., Matrix groups, *Universitext*, Springer-Verlag, New York, 1979.

[24]  De Rham, G., Sur les polygones générateurs des groupes Fuchsiens, *l'Enseigne-ment Math.*, **17** (1971), 49–62.

[25]  Dold, A. and Eckmann, B. (editors), A crash course in Kleinian groups, *Lecture Notes in Mathematics 400*, Springer-Verlag, New York, 1974.

[26]  Du Val, P., *Homographies, Quaternions and Rotations*, Clarendon Press, Oxford, 1964.

[27]  Edmonds, A. L., Ewing, J. H. and Kulkarni, R. S., Torsion free subgroups of Fuchsian groups and tesselations of surfaces, *Invent. Math.* (to appear).

[28]  Farkas, H. M. and Kra, I., *Riemann Surfaces*, Graduate Texts in Mathematics 71, Springer-Verlag, New York, 1980.

[29]  Fenchel, W. and Nielsen, J., On discontinuous groups of isometric transforma-tions of the non-Euclidean plane; *Studies and Essays Presented to R. Courant*, Interscience, New York, 1948.

[30]  Ford, L. R., *Automorphic Functions* (Second Edition), Chelsea, New York, 1951.

[31]  Fox, R. H., On Fenchel's conjecture about *F*-groups, *Matematisk Tid. B*, 1952, 61–65.

[32]  Gans, D., *An Introduction to Non-Euclidean geometry*, Academic Press, New York, 1973.

[33]  Greenberg, L., Discrete subgroups of the Lorentz group, *Math. Scand.* **10** (1962), 85–107.

[34]  Greenberg, L., Fundamental polygons for Fuchsian groups, *J. d'Analyse Math.*, **18** (1967), 99–105.

[35]  Greenberg, L., Finiteness theorems for Fuchsian and Kleinian groups; *Discrete Groups and Automorphic Functions* (edited by W. J. Harvey), Academic Press, London, 1977.

[36]  Gruenberg, K. W. and Weir, A. J., *Linear Geometry*, Van Nostrand, Princeton, 1967.

[37]  Halpern, N., A proof of the Collar Lemma, *Bull. London Math. Soc.*, **13** (1981), 141–144.

[38]  Heins, M., Fundamental polygons of Fuchsian and Fuchsoid groups, *Ann. Acad. Sci. Fenn.*, 1964, 1–30.

[39]  Higgins, P. J., An introduction to topological groups; *London Mathematical Society Lecture Note Series 15*, Cambridge University Press, Cambridge, 1974.

[40]  Jørgensen, T. and Kiikka, M., Some extreme discrete groups, *Ann. Acad. Sci. Fenn.*, *Ser. A*, **1** (1975), 245–248.

[41]  Jørgensen, T., On discrete groups of Möbius transformations, *Amer. J. Math.*, **98** (1976), 739–749.

[42]  Jørgensen, T., A note on subgroups of SL(2, *C*), *Quart. J. Math. Oxford Ser. II*, **28** (1977), 209–212.

[43]  Jørgensen, T., Closed geodesics on Riemann surfaces, *Proc. Amer. Math. Soc.*, **72** (1978), 140–142.

[44]  Jørgensen, T., Comments on a discreteness condition for subgroups of SL(2, *C*), *Can. J. Math.*, **31** (1979), 87–92.

[45] Jørgensen, T., Commutators in SL(2, *C*); Riemann surfaces and related topics, *Annals of Math. Studies 97*, Princeton University Press, Princeton, 1980.

[46] Keen, L., Canonical polygons for finitely generated Fuchsian groups, *Acta Math.*, **115** (1966), 1–16.

[47] Keen, L., On infinitely generated Fuchsian groups, *J. Indian Math. Soc.*, **35** (1971), 67–85.

[48] Knapp, A. W., Doubly generated Fuchsian groups, *Michigan Math. J.*, **15** (1968), 289–304.

[49] Knopp, M. I. and Newman, M., Congruence subgroups of positive genus of the Modular group, *Ill. J. Math.*, **9** (1965), 577–583.

[50] Kra, I., *Automorphic Forms and Kleinian Groups*, Benjamin, Reading, Mass., 1972.

[51] Lehner, J., Discontinuous groups and automorphic functions; *Mathematical Surveys, Number VIII*, American Math. Soc., Providence, 1964.

[52] Lehner, J., *A Short Course in Automorphic Functions*, Holt, Rinehart and Winston, New York, 1966.

[53] Lyndon, R. C. and Ullman, J. L., Groups of elliptic linear fractional transformations, *Proc. Amer. Math. Soc.*, **18** (1967), 1119–1124.

[54] Macbeath, A. M., Packings, free products and residually finite groups, *Proc. Cambridge Phil. Soc.*, **59** (1963), 555–558.

[55] Macbeath, A. M., The classification of non-Euclidean plane crystallographic groups, *Can. J. Math.*, **19** (1967), 1192–1205.

[56] Macbeath, A. M. and Hoare, A. H. M., Groups of hyperbolic crystallography, *Math. Proc. Cambridge Phil. Soc.*, **79** (1976), 235–249.

[57] Magnus, J. W., *Non-Euclidean Tesselations and Their Groups*, Academic Press, New York, 1974.

[58] Marden, A., On finitely generated Fuchsian groups, *Comment. Math. Helvitici*, **42** (1967), 81–85.

[59] Marden, A., Universal properties of Fuchsian groups in the Poincaré metric; Discontinuous groups and Riemann surfaces, *Annals of Math. Studies 79*, Princeton University Press, Princeton, 1974.

[60] Maskit, B., On Klein's combination theorem, *Trans. Amer. Math. Soc.*, **120** (1965), 499–509.

[61] Maskit, B., On Klein's combination theorem II, *Trans. Amer. Math. Soc.*, **131** (1968), 32–39.

[62] Maskit, B., On Poincaré's theorem for fundamental polygons, *Adv. in Math.*, **7** (1971), 219–230.

[63] Massey, W. S., *Algebraic Topology: An Introduction*; Graduate Texts in Mathematics 56, Springer-Verlag, New York, 1967.

[64] Matelski, J. P., A compactness theorem for Fuchsian groups of the second kind, *Duke Math. J.*, **43** (1976), 829–840.

[65] Matelski, J. P., The classification of discrete 2-generator subgroups of PSL(2, *C*), *Israel J. Math.* (to appear).

[66] Meschkowski, H., *Noneuclidean Geometry*, Academic Press, New York, 1964.

[67] Millman, R. S., Kleinian transformation geometry, *Amer. Math. Monthly*, **84** (1977), 338–349.

[68] Milnor, J., Hyperbolic geometry: the first 150 years, *Bull. Amer. Math. Soc.*, **6** (1982), 9–24.

[69] Montgomery, D. and Zippin, L., *Topological Transformation Groups*, Interscience, New York, 1955.

[70] Natanzon, S. M., Invariant lines of Fuchsian groups, *Russian Math. Surveys*, **27** (No. 4, 1972), 161–177.

[71] Nicholls, P. J. and Zarrow, R., Convex fundamental regions for Fuchsian groups, *Math. Proc. Cambridge Phil. Soc.*, **84** (1978), 507–518.

[72]   Nicholls, P. J. and Zarrow, R., Convex fundamental regions for Fuchsian groups II, *Math. Proc. Cambridge Phil. Soc.*, **86** (1979) 295–300.

[73]   Nicholls, P. J., Garnett points for Fuchsian groups, *Bull. London Math. Soc.*, **12** (1980), 216–218.

[74]   Patterson, S. J., On the cohomology of Fuchsian groups, *Glasgow Math. J.*, **16** (1975), 123–140.

[75]   Patterson, S. J., Diophantine approximation in Fuchsian groups, *Phil. Trans. Roy. Soc. London*, **282** (1976), 527–563.

[76]   Peczynski, N., Rosenberger, G. and Zieschang, H., Uber Erzeugende ebener diskontinuierlicher Gruppen, *Invent. Math.*, **29** (1975), 161–180.

[77]   Poincaré, H., Theorie des groupes Fuchsiens, *Acta Math.*, **1** (1882), 1–62.

[78]   Pommerenke, Ch. and Purzitsky, N., On the geometry of Fuchsian groups (preprint 1981).

[79]   Pommerenke, Ch. and Purzitsky, N., On some universal bounds for Fuchsian groups (preprint, 1981).

[80]   Purzitsky, N., Two generator Fuchsian groups of genus one, *Math. Zeit.*, **128** (1972), 245–251.

[81]   Purzitsky, N., Two generator discrete free products, *Math. Zeit.*, **126** (1972), 209–223.

[82]   Purzitsky, N., Correction to: two generator Fuchsian groups of genus one, *Math. Zeit.*, **132** (1973), 261–262.

[83]   Purzitsky, N., Canonical generators of Fuchsian groups, *Ill. J. Math.*, **18** (1974), 484–490.

[84]   Purzitsky, N., All two generator Fuchsian groups, *Math. Zeit.*, **147** (1976), 87–92.

[85]   Purzitsky, N., A cutting and pasting of non-compact polygons with applications to Fuchsian groups, *Acta Math.*, **143** (1979), 233–250.

[86]   Purzitsky, N., Quasi Fricke polygons and the Nielsen convex region, *Math. Zeit.*, **172** (1980), 239–244.

[87]   Randol, B., Cylinders in Riemann surfaces, *Comment. Math. Helvitici*, **54** (1979), 1–5.

[88]   Reade, M. O., On certain conformal maps in space, *Michigan Math. J.*, **4** (1957), 65–66.

[89]   Rosenberger, G., Eine Bemerkung zu einer Arbeit von T. Jørgensen, *Math. Zeit.*, **165** (1979), 261–265.

[90]   Rudin, W., *Real and Complex Analysis*, McGraw-Hill, New York, 1966.

[91]   Schwarzenberger, R. L. E., N-dimensional crystallography, *Research Notes in Mathematics 41*, Pitman, London, 1980.

[92]   Selberg, A., On discontinuous groups in higher-dimensional spaces; *Contributions to Function Theory*, Tata Institute, Bombay, 1960.

[93]   Shinnar, M. and Sturm, J., The maximal inscribed ball of a Fuchsian group; Discontinuous groups and Riemann surfaces, *Annals of Math. Studies 79*, Princeton University Press, Princeton, 1974.

[94]   Siegel, C. L., Discontinuous groups, *Ann. of Math.*, **44** (1943), 674–689.

[95]   Siegel, C. L., Bemerkung zu einem Satze von Jakob Nielsen, *Mat. Tidsskrift B*, 1950, 66–70.

[96]   Siegel, C. L., Uber einige Ungleichungen bei Bewegungsgruppen in der nichteuklidischen Ebene, *Math. Ann.*, **133** (1957), 127–138.

[97]   Singerman, D., Subgroups of Fuchsian groups and finite permutation groups, *Bull. London Math. Soc.*, **2** (1970), 319–323.

[98]   Singerman, D., On the structure of non-Euclidean crystallographic groups, *Proc. Cambridge Phil. Soc.*, **76** (1974), 233–240.

[99]   Sorvali, T., On discontinuity of Möbius groups without elliptic elements, *Univ. Joensuu, Ser. B*, **9** (1974), 1–4.

[100] Springer, G., *Introduction to Riemann Surfaces*, Addison-Wesley, Reading, Mass., 1957.

[101] Thurston, W., The geometry and topology of 3-manifolds, *Lecture notes*, Princeton, 1980.

[102] Tietze, H., Über Konvexheit in kleinen und im grossen und über gewisse den Punkten einer Menge zugeordnete Dimensionzahlen, *Math. Zeit.*, **28** (1928), 697–707.

[103] Tsuji, M., *Potential Theory in Modern Function Theory*, Maruzen, Tokyo, 1959.

[104] Tukia, P., On torsionless subgroups of infinitely generated Fuchsian groups, *Ann. Acad. Sci. Fenn., Ser. A*, **4** (1978–79), 203–205.

[105] Väisälä, J., Lectures on $n$-dimensional quasiconformal mappings; *Lecture Notes in Mathematics 229*, Springer-Verlag, New York, 1971.

[106] Wang, H., Discrete nilpotent subgroups of Lie groups, *J. Diff. Geometry*, **3** (1969), 481–492.

[107] Weyl, H., *Symmetry*, Princeton University Press, Princeton, 1952.

[108] Wielenberg, N. J., Discrete Möbius groups, fundamental polyhedra and convergence, *Amer. J. Math.*, **99** (1977), 861–877.

[109] Wielenberg, N. J., On the limit set of discrete Möbius groups with finite sided polyhedra (preprint, 1976).

[110] Wilker, J. B., Inversive geometry; *The Geometric Vein* (edited by C. Davis, B. Grünbaum and F. A. Scherk), Springer-Verlag, New York, 1981.

[111] Wolf, J., *Spaces of Constant Curvature*, McGraw-Hill, New York, 1967.

[112] Wylie, C. R., *Foundations of Geometry*, McGraw-Hill, New York, 1964.

[113] Yamada, A., On Marden's universal constant of Fuchsian groups, *Kodai Math. J.*, **4** (1981), 266–277.

[114] Zieschang, H., Vogt, E. and Coldewey, H., Surfaces and planar discontinuous groups; *Lecture Notes in Mathematics 835*, Springer-Verlag, New York, 1980.

# Index